HEALTH AND HUMAN DEVELOPMENT

HUMAN DEVELOPMENT

BIOLOGY FROM A HOLISTIC POINT OF VIEW

HEALTH AND HUMAN DEVELOPMENT

JOAV MERRICK - SERIES EDITOR –

NATIONAL INSTITUTE OF CHILD HEALTH
AND HUMAN DEVELOPMENT,
MINISTRY OF SOCIAL AFFAIRS, JERUSALEM

Adolescent Behavior Research: International Perspectives
Joav Merrick and Hatim A. Omar (Editors)
2007. ISBN: 1-60021-649-8

Complementary Medicine Systems: Comparison and Integration
Karl W. Kratky
2008. ISBN: 978-1-60456-475-4 (Hardcover)
2008. ISBN: 978-1-61122-433-7 (E-book)

Pain in Children and Youth
Patricia Schofield and Joav Merrick (Editors)
2008. ISBN: 978-1-60456-951-3 (Hardcover)
2008. ISBN: 978-1-61470-496-6 (E-book)

Challenges in Adolescent Health: An Australian Perspective
David Bennett, Susan Towns, Elizabeth Elliott and Joav Merrick (Editors)
2009. ISBN: 978-1-60741-616-6 (Hardcover)
2009. ISBN: 978-1-61668-240-8 (E-book)

Behavioral Pediatrics, 3rd Edition
Donald E. Greydanus, Dilip R. Patel, Helen D. Pratt and Joseph L. Calles, Jr. (Editors)
2009. ISBN: 978-1-60692-702-1 (Hardcover)
2009. ISBN: 978-1-60876-630-7 (E-book)

Health and Happiness from Meaningful Work: Research in Quality of Working Life
Søren Ventegodt and Joav Merrick (Editors)
2009. ISBN: 978-1-60692-820-2 (Hardcover)
2009. ISBN: 978-1-61324-981-9 (E-book)

Obesity and Adolescence: A Public Health Concern
Hatim A. Omar, Donald E. Greydanus, Dilip R. Patel and Joav Merrick (Editors)
2009. ISBN: 978-1-60692-821-9 (Hardcover)
2009. ISBN: 978-1-61470-465-2 (E-book)

Poverty and Children: A Public Health Concern
Alexis Lieberman and Joav Merrick (Editors)
2009. ISBN: 978-1-60741-140-6 (Hardcover)
2009. ISBN: 978-1-61470-601-4 (E-book)

Living on the Edge: The Mythical, Spiritual, and Philosophical Roots of Social Marginality
Joseph Goodbread
2009. ISBN: 978-1-60741-162-8 (Hardcover)
2011. ISBN: 978-1-61122-986-8 (Softcover)
2011. ISBN: 978-1-61470-192-7 (E-book)

Alcohol-Related Cognitive Disorders: Research and Clinical Perspectives
Leo Sher, Isack Kandel and Joav Merrick (Editors)
2009. ISBN: 978-1-60741-730-9 (Hardcover)
2009. ISBN: 978-1-60876-623-9 (E-book)

Rural Child Health: International Aspects
Erica Bell and Joav Merrick (Editors)
2010. ISBN: 978-1-60876-357-3 (Hardcover)
2010. ISBN: 978-1-61324-005-2 (E-book)

Children and Pain
Patricia Schofield and Joav Merrick (Editors)
2009. ISBN: 978-1-60876-020-6 (Hardcover)
2009. ISBN: 978-1-61728-183-9 (E-book)

Conceptualizing Behavior in Health and Social Research: A Practical Guide to Data Analysis
Said Shahtahmasebi and Damon Berridge
2010. ISBN: 978-1-60876-383-2

Chance Action and Therapy: The Playful Way of Changing
Uri Wernik
2010. ISBN: 978-1-60876-393-1 (Hardcover)
2011. ISBN: 978-1-61122-987-5 (Softcover)
2011. ISBN: 978-1-61209-874-6 (E-book)

Adolescence and Chronic Illness. A Public Health Concern
Hatim Omar, Donald E. Greydanus, Dilip R. Patel and Joav Merrick (Editors)
2010. ISBN: 978-1-60876-628-4 (Hardcover)
2010. ISBN: 978-1-61761-482-8 (E-book)

Adolescence and Sports
Dilip R. Patel, Donald E. Greydanus, Hatim Omar and Joav Merrick (Editors)
2010. ISBN: 978-1-60876-702-1 (Hardcover)
2010. ISBN: 978-1-61761-483-5 (E-book)

International Aspects of Child Abuse and Neglect
Howard Dubowitz and Joav Merrick (Editors)
2010. ISBN: 978-1-60876-703-8 (Hardcover)
2010. ISBN: 978-1-61122-049-0 (Softcover)
2010. ISBN: 978-1-61122-403-0 (E-book)

Pediatric and Adolescent Sexuality and Gynecology: Principles for the Primary Care Clinician
Hatim A. Omar, Donald E. Greydanus, Artemis K. Tsitsika, Dilip R. Patel and Joav Merrick (Editors)
2010. ISBN: 978-1-60876-735-9

Bone and Brain Metastases: Advances in Research and Treatment
Arjun Sahgal, Edward Chow and Joav Merrick (Editors)
2010. ISBN: 978-1-61668-365-8 (Hardcover)
2010. ISBN: 978-1-61728-085-6 (E-book)

Environment, Mood Disorders and Suicide
Teodor T. Postolache and Joav Merrick (Editors)
2010. ISBN: 978-1-61668-505-8

Positive Youth Development: Evaluation and Future Directions in a Chinese Context
Daniel T.L. Shek, Hing Keung Ma and Joav Merrick (Editors)
2011. ISBN: 978-1-60876-830-1 (Hardcover)
2011. ISBN: 978-1-62100-175-1 (Softcover)
2010. ISBN: 978-1-61209-091-7 (E-book)

Understanding Eating Disorders: Integrating Culture, Psychology and Biology
Yael Latzer, Joav Merrick and Daniel Stein (Editors)
2011. ISBN: 978-1-61728-298-0 (Hardcover)
2011. ISBN: 978-1-61470-976-3 (Softcover)
2011. ISBN: 978-1-61942-054-0 (E-book)

Advanced Cancer Pain and Quality of Life
Edward Chow and Joav Merrick (Editors)
2011. ISBN: 978-1-61668-207-1 (Hardcover)
2010. ISBN: 978-1-61668-400-6 (E-book)

Positive Youth Development: Implementation of a Youth Program in a Chinese Context
Daniel T.L Shek, Hing Keung Ma and Joav Merrick (Editors)
2011. ISBN: 978-1-61668-230-9 (Hardcover)

Social and Cultural Psychiatry Experience from the Caribbean Region
Hari D. Maharajh and Joav Merrick (Editors)
2011. ISBN: 978-1-61668-506-5 (Hardcover)
2010. ISBN: 978-1-61728-088-7 (E-book)

Narratives and Meanings of Migration
Julia Mirsky
2011. ISBN: 978-1-61761-103-2 (Hardcover)
2010. ISBN: 978-1-61761-519-1 (E-book)

Self-Management and the Health Care Consumer
Peter William Harvey
2011. ISBN: 978-1-61761-796-6 (Hardcover)
2011. ISBN: 978-1-61122-214-2 (E-book)

Sexology from a Holistic Point of View
Soren Ventegodt and Joav Merrick
2011. ISBN: 978-1-61761-859-8 (Hardcover)
2011. ISBN: 978-1-61122-262-3 (E-book)

Principles of Holistic Psychiatry: A Textbook on Holistic Medicine for Mental Disorders
Soren Ventegodt and Joav Merrick
2011. ISBN: 978-1-61761-940-3 (Hardcover)
2011. ISBN: 978-1-61122-263-0 (E-book)

Clinical Aspects of Psychopharmacology in Childhood and Adolescence
Donald E. Greydanus, Joseph L. Calles, Jr., Dilip P. Patel, Ahsan Nazeer and Joav Merrick (Editors)
2011. ISBN: 978-1-61122-135-0 (Hardcover)
2011. ISBN: 978-1-61122-715-4 (E-book)

Climate Change and Rural Child Health
Erica Bell, Bastian M. Seidel and Joav Merrick (Editors)
2011. ISBN: 978-1-61122-640-9 (Hardcover)
2011. ISBN: 978-1-61209-014-6 (E-book)

Rural Medical Education: Practical Strategies
Erica Bell, Craig Zimitat and Joav Merrick (Editors)
2011. ISBN: 978-1-61122-649-2 (Hardcover)
2011. ISBN: 978-1-61209-476-2 (E-book)

Advances in Environmental Health Effects of Toxigenic Mold and Mycotoxins
Ebere Cyril Anyanwu
2011. ISBN: 978-1-60741-953-2

Child and Adolescent Health Yearbook 2009
Joav Merrick (Editor)
2011. ISBN: 978-1-61668-913-1

Public Health Yearbook 2009
Joav Merrick (Editor)
2011. ISBN: 978-1-61668-911-7

Child Health and Human Development Yearbook 2009
Joav Merrick
2011. ISBN: 978-1-61668-912-4

Alternative Medicine Yearbook 2009
Joav Merrick (Editor)
2011. ISBN: 978-1-61668-910-0

The Dance of Sleeping and Eating among Adolescents: Normal and Pathological Perspectives
Yael Latzer and Orna Tzischinsky (Editors)
2011. ISBN: 978-1-61209-710-7

Child and Adolescent Health Yearbook 2010
Joav Merrick (Editor)
2011. ISBN: 978-1-61209-788-6

Child Health and Human Development Yearbook 2010
Joav Merrick (Editor)
2011. ISBN: 978-1-61209-789-3

Public Health Yearbook 2010
Joav Merrick (Editor)
2011. ISBN: 978-1-61209-971-2

Alternative Medicine Yearbook 2010
Joav Merrick (Editor)
2011. ISBN: 978-1-62100-132-4 (Hardcover)
2011. ISBN: 978-1-62100-210-9 (E-book)

The Astonishing Brain and Holistic Conciousness: Neuroscience and Vedanta Perspectives
Vinod D. Deshmukh
2011. ISBN: 978-1-61324-295-7

Randomized Clinical Trials and Placebo: Can You Trust the Drugs are Working and Safe?
Søren Ventegodt and Joav Merrick (Editors)
2012. ISBN: 978-1-61470-067-8 (Hardcover)

Translational Research for Primary Healthcare
Erica Bell and Joav Merrick (Editors)
2011. ISBN: 978-1-61324-647-4

Drug Abuse in Hong Kong: Development and Evaluation of a Prevention Program
Daniel T.L. Shek, Rachel C.F. Sun and Joav Merrick (Editors)
2011. ISBN: 978-1-61324-491-3

Human Development: Biology from a Holistic Point of View
Søren Ventegodt, Tyge Dahl Hermansen and Joav Merrick
2011. ISBN: 978-1-61470-441-6 (Hardcover)
2011. ISBN: 978-1-61470-541-3 (E-book)

Our Search for Meaning in Life
Søren Ventegodt and Joav Merrick
2011. ISBN: 978-1-61470-494-2 (Hardcover)
2011. ISBN: 978-1-61470-519-2 (E-book)

Building Community Capacity: Minority and Immigrant Populations
Rosemary M. Caron and Joav Merrick
2012. ISBN: 978-1-62081-032-3 (Hardcover)
2012. ISBN: 978-1-62081-032-3 (E-book)

HEALTH AND HUMAN DEVELOPMENT

HUMAN DEVELOPMENT

BIOLOGY FROM A HOLISTIC POINT OF VIEW

SØREN VENTEGODT
TYGE DAHL HERMANSEN
AND
JOAV MERRICK

NOVA BIOMEDICAL
New York

For permission to use material from this book please contact us:
Telephone 631-231-7269; Fax 631-231-8175
Web Site: http://www.novapublishers.com

Library of Congress Cataloging-in-Publication Data

Human development : biology from a holistic point of view / editors, Sxren Ventegodt, Tyge Dahl Hermansen, Joav Merrick.
p. cm.
Includes bibliographical references and index.
ISBN 978-1-61470-441-6 (hardcover : alk. paper) 1. Human biology. I. Ventegodt, Sxren. II. Hermansen, Tyge Dahl. III. Merrick, Joav, 1950-
QP34.5.H846 2011
612.6--dc23

2011023224

Published by Nova Science Publishers, Inc. † New York

Contents

Preface

This book is about human life and how to make the most of it. It is about the miracle of consciousness and how body and mind are connected. It is about our inner powers for healing, love and understanding. It is about our body shape and inner structure, how it is made and how it was developed during evolution. It is also about the universe we live in and the natural forces and laws that have made not only us, but all that exists. In the end of our analysis we come to a new understanding of Antonovsky's concept of salutogenesis – existential healing - and meaning of life, as it is experienced as *sense of coherence*. We understand how consciousness can interfere with body shape and mind, as it can with our immunological defences. We wish you a good journey though this strange and new-thinking book about human development.

Foreword

How can we understand and explain human behavior such as substance abuse? This is a basic question that is asked by students taking courses in psychology, medicine and health studies as well as colleagues in human services. Generally speaking, human behavior is understood in terms of three broad perspectives – biology, psychology and sociology.

Under the biological perspective, human behavior is seen as a product of biological structures and mechanisms. For example, substance abuse is seen as a result of problems in neurotransmitters and genetic makeup (i.e., biological determinism). Under the psychological perspective, human behavior is seen as a result of mental processes, which may or may not have biological bases. For example, cognitive psychologists uphold the belief that "men are not influenced by the outer world, but by their interpretations of it", which clearly suggests the importance of transcendence of "mind" over "matter". For substance abuse, it is a matter of "beliefs" about the meaning of taking drugs. Finally, sociologists believe that external social structures and processes determine human behavior. External environmental factors such as poverty, income inequality and alienation contribute to substance abuse.

There are two questions we should ask ourselves when we use the above perspectives to look at human behavior. First, what are the strengths and limitations of using the above perspectives to understand human behavior? With specific reference to the biological perspective, while it is a scientific and objective approach, it has been criticized as reductionist and it fails to explain many human phenomena. For example, the use of drugs alone cannot treat substance abuse and prevent relapse. Second, it is theoretically and practically important to ask how different domains (e.g., biological domain and psychological domain) interact. For example, in times of adversity, how one's psychology (e.g. courage to fight and perseverance) may strengthen bodily strength and endurance is an important question to explore.

Against this background, I am very pleased to note that Ventegodt, Hermansen and Merrick have written this book titled "Human development. Biology from a holistic point of view". There are several unique features of this book. First, it outlines the fundamental problems of biology, medicine and neuropsychology. Besides, the nature of matter, life and consciousness as well as the brain-mind and body-mind issues are highlighted. Second, it gives a new perspective on life and biologic information, including living cell information system, relationship between biochemistry and morphogenesis, and the origin of cellular order. Third, discussion on the relationship between consciousness and human brain functions, including holistic human development, consciousness, and metamorphosis is included. Fourth, there is stimulating discussion on resilience, pathogenesis and salutogenesis. Finally, the theory for accelerated, spontaneous existential healing and related applied implications of holistic health for medical disciplines are thought-provoking.

This is an excellent book that gives the readers a new perspective on the nature of consciousness and how the body is connected to the mind. It also provides a new understanding of Antonovsky's concept of salutogenesis in terms of existential healing, meaning of life and sense of coherence. This book is truly humanistic in nature which sees great potentials in human beings. It is a book that cannot be missed by colleagues in the fields of health, medicine and holistic care.

Daniel TL Shek, PhD, FHKPS, BBS, JP
Chair Professor of Applied Social Sciences, The Hong Kong Polytechnic University, Hunghom, Hong Kong, PRC
Associate Director, Public Policy Research Institute, The Hong Kong Polytechnic University, Hunghom, Hong Kong, PRC
Advisory Professor, East China Normal University, Shanghai, PRC
Honorary Professor, Kiang Wu Nursing College of Macau, Macau, PRC
Adjunct Professor, University of Kentucky College of Medicine, Lexington, Kentucky, United States of America
E-mail: ssdaniel@inet.polyu.edu.hk

Section I: Research problems

Chapter I

Fundamental problems of biology, medicine and neuro-psychology

In this book we address a number of unsolved problems in biology today. First of all, the unsolved enigma concerning how the differentiation from a single zygote to an adult individual happens has been object for severe research through decades. By uncovering a new holistic biological paradigm that introduces an energetic-informational interpretation of reality as a new way to experience biology, these papers try to solve the problems connected with the events of biological ontogenesis from a single cell involvement in the fractal hierarchy, to the function of the human brain. The problems discussed are interpreted within the frames of a universe of roomy fractal structures containing energetic strings that are able to deliver the energy necessary for biological information. We think biological organization is guided by energetic changes on the level of quantum mechanics, interacting with the intension that again guides the energetic conformation of the fractal structures to gain disorders or healthiness. Furthermore we introduce two new concepts, "Metamorphous Top down" evolution and "Adult Human Metamorphosis". The first is a new evolutionary theory involving metamorphosis as a main concept of evolution. The last are tightly linked to the evolutionary principle and explains how human self-recovery is governed. Other subjects of special interest we shall look deeper into are: the immunological self-nonself discrimination, the structure and function of the human brain, the etiology and salutogenesis of mental and somatic diseases, and the structure of the consciousness of a human being. We shall criticize Szentagothai's model for the modulated structure of the human cerebral cortex, and Niels K. Jerne's theory of the immunological regulatory anti-idiotypic network.

Introduction

Several well-known phenomena in the living world have not yet found an adequate explanation. What is human consciousness and how is it connected to the brain and the wholeness of the organism? How is the differentiation of the cells brought about in the developing organism during ontogenesis? Basically, how does a fertilized egg develop into an adult organism? What are the forces behind the evolution of the living world? How is biological information stored and distributed in the organism – is everything really in the DNA? How is information managed in a living organism? How does the phenomenon of self-nonself discrimination, the basis of immune system regulation, arise? How does an organism maintain and re-establish biological order and thus health and physical well being during its lifetime? What are the structural relations of the living world with the physical energy of the world's particles, atoms, and molecules? And what happens when the cellular order is disturbed and suddenly reappears in spontaneous healing? We shall address these persistent questions in the next paper devoted to the unsolved topics, and discuss their relevance for clinical holistic medicine.

These questions and many more have been discussed by the research team of the Quality of Life Research Center in Copenhagen over the past many years and we have put the thoughts, questions and discussions into paper form and relate it to clinical holistic medicine. We have published our thoughts in scientific papers already published:

1. Hermansen TD, Ventegodt S, Rald E, Clausen B, Nielsen ML, Merrick J. Human development I. Twenty fundamental problems of biology, medicine and neuropsychology related to biological information. ScientificWorldJournal 2006;6:747-59.
2. Ventegodt S, Hermansen TD, Nielsen ML, Clausen B, Merrick J. Human development II. We need an integrated theory for matter, life and consciousness to understand life and healing. ScientificWorldJournal 2006;6:760-6.
3. Ventegodt S, Hermansen TD, Rald E, Flensborg-Madsen T, Nielsen ML, Clausen B, Merrick J. Human development III. Bridging brain-mind and body-mind. Itroduction to "deep" (fractal, poly-ray) cosmology. ScientificWorldJournal 2006;6:767-76.
4. Ventegodt S, Hermansen TD, Rald E, Flensborg-Madsen T, Nielsen ML, Clausen B, Merrick J. Human development IV. The living cell has information-directed self-organization. ScientificWorldJournal 2006;6:1132-8.
5. Ventegodt S, Hermansen TD, Flensborg-Madsen T, Nielsen ML, Clausen B, Merrick J. Human development V: Biochemistry unable to explain the emergence of biological form (morphogenesis) and therefore a new principle as source of biological information is needed. ScientificWorldJournal 2006;6:1359-67.
6. Ventegodt S, Hermansen TD, Flensborg-Madsen T, Nielsen M, Merrick J. Human development VI: Supracellular morphogenesis. The origin of biological and cellular order. ScientificWorldJournal 2006;6:1424-33.
7. Ventegodt S, Hermansen TD, Flensborg-Madsen T, Rald E, Nielsen ML, Merrick J. Human development VII: A spiral fractal model of fine structure of physical energy could explain central aspects of biological information, biological organization and biological creativity. ScientificWorldJournal 2006;6:1434-40.
8. Ventegodt S, Hermansen TD, Flensborg-Madsen T, Nielsen ML, Merrick J. Human development VIII: A theory of "deep" quantum chemistry and cell consciousness: Quantum chemistry controls genes and biochemistry to give cells and higher organisms consciousness and complex behavior. ScientificWorldJournal 2006;6:1441-53.
9. Ventegodt S, Hermansen TD, Flensborg-Madsen T, Rald E, Nielsen ML, Merrick J. Human development IX: A model of the wholeness of man, his consciousness and collective consciousness. ScientificWorldJournal 2006;6:1454-9.

10. Hermansen TD, Ventegodt S, Merrick J. Human development X: Explanation of macroevolution — top-down evolution materializes consciousness. The origin of metamorphosis. ScientificWorldJournal 2006;6:1656-66.
11. Hermansen TD, Ventegodt S, Kandel I. Human development XI: The structure of the cerebral cortex. Are there really modules in the brain? ScientificWorldJournal 2007;7:1922-9.
12. Ventegodt S, Hermansen TD, Kandel I, Merrick J. Human development XII: a theory for the structure and function of the human brain. ScientificWorldJournal 2008;8:621-42.
13. Ventegodt S, Hermansen TD, Kandel I, Merrick J. Human development XIII: The connection between the structure of the overtone system and the tone language of music. Some implications for our understanding of the human brain. ScientificWorldJournal 2008;8:643-57.
14. Ventegodt S, Hermansen TD, Kandel I, Merrick J. Human development XIV: Degeneration and regeneration of existence. Metamorphosis changes the purpose of life, the level of existential responsibility, and the depth of consciousness (the ray-number of the cosmology). J Altern Med Res 2009;1(4):409-18.
15. Ventegodt S, Hermansen TD, Kandel I, Merrick J. Human development XV: The biochemical hypothesis for the etiology of the mental diseases is not substantiated. J Altern Med Res 2009;1(4):419-32.
16. Ventegodt S, Hermansen TD, Kandel I, Merrick J. Human development XVI: General etiology of mental diseases and the role of adult metamorphosis in spontaneous recovery. J Altern Med Res 2009;1(4):433-8.
17. Ventegodt S, Hermansen TD, Kandel I, Merrick J. Human development XVII: Jerne's anti-idiotypic network theory cannot explain self-nonself discrimination. J Altern Med Res 2009;1(4):439-44.
18. Hermansen TD, Ventegodt S, Kandel I, Merrick J. Human development XVIII: A theory for immune system regulation (the "self-nonself discrimination"). J Altern Med Res 2009;1(4):445-52.
19. Ventegodt S, Hermansen TD, Kandel I, Merrick J. Human development XIX: The aetiology of somatic diseases. The energetic healing crisis ("kidney-crisis") of patients in Antonovsky-salutogenesis with clinical holistic medicine and holistic sexology. J Altern Med Res 2009;1(4):453-64.
20. Ventegodt S, Hermansen TD, Kandel I, Merrick J. Human development XX: A theory for accelerated, spontaneous existential healing (salutogenesis): Human adult metamorphosis. J Altern Med Res 2009;1(4):465-74.
21. Hermansen TD, Ventegodt S, Kandel I, Merrick J. Human development XXI: A theory of top-down macro-evolution and metamorphosis: How the fly got its wings and how an organism's conscious intent can materialize biological order and body form. J Altern Med Res 2009;1(4):475-80.
22. Hermansen TD, Ventegodt S, Kandel I, Merrick J. Human development XXII: Holistic therapy induces Antonovsky salutogenesis and spontanous healing and recovery. Human genes for metamorphosis might control salutogenesis. If the expression of such genes are controlled by our (sub)consciousness this explains the placebo effect and possibly also spontanous remission from cancer. J Altern Med Res 2009;1(4):481-90.

And this book is a revision and update discussing a theory for life, biological information, healing and adult human metamorphosis for use in clinical holistic medicine.

Biology, medicine and neural sciences have fundamental problems that are related to the fundamental problem of biological information. This problem in science can be worded simply (1): What is life? This is how basic it is. Everything about life is mysterious: its joy, its wisdom and tacit knowledge and competence, its ability to assume form and to be creative and intelligent, and of cause its fundamental ability to be conscious. The mystery of the two sides of life, the objective material and structural side, and the subjective experiential problem might be the hardest of all the problems. This problem is often referred to as *the hard problem* (2). Penrose, Hameroff and other people have addressed the hard problem using the concepts, formalisms and ideas from quantum mechanics. The problem using concepts from physics is that physics has got as many fundamental problems as does biology, and many physical theories are in conflict these days, like the theory of relativity and the theory of quantum mechanics. Niels Henrik David Bohr (1885-1962) stated, that you should not make pictures of the quantum levels energy, because quantum mechanics does not explain about reality but only something about how we describe reality, which has made it difficult to proceed.

We believe that physics should provide us with pictures and understanding of the energy of the universe, but this has not happened yet. The super symmetrical strings (super string theory), wormholes, and the like have provided us with some pictures and ideas of the energy of the world, but still not usable for biology. In 1996 in discussions between Holger Bech Nielsen (professor at the Niels Bohr Institute in Copenhagen), Søren Ventegodt and the Dalai Lama had a discussion on “science and spirituality”, where the Dalai Lama explained that when his friend meditated he could make knots on hard needles. Holger Beck Nielsen concluded that the whole physics had to be changed if that was true and he has often said (as member of the study group “Matter, life and consciousness”) that he doubts if consciousness really exists, because there are no laws in physics that make this kind of integrative and non-local phenomenon possible.

Brain research has for many years been unable to explain the most simple things like memory and perception, actually the analysis into feature detectors in the brain have made the problem on qualia much more obvious. Qualia is the basic quality of perception – like the colour red - which the brain really cannot provide, as the neurons that make the brain uses signals that are simple electrical impulses without any quality attached to it, according to contemporary physical and biochemical laws! Biology is full of fundamental problems, like the sad fact that chemistry cannot provide the spatial information to build an embryo according to the known laws of chemistry and diffusion, as we are going to demonstrate in the papers in this series.

When this is said nobody can be surprised that complicated things like cellular order of the body, morphogenesis, evolution and phylogenesis, perception and psychosis, disease and healing are not understood.

In this introduction we want to list some of the problems that we are going to address and give our solution to. Hopefully after reading this book you will actually have an idea of biological information, consciousness, health, disease and healing. If we succeed in our mission and make at least some physicians develop an understanding of phenomena like spontaneous healing and human adult metamorphosis, we can hope to have strengthened the general belief in the patient’s inner healing powers. This understanding is important in order to mobilise in the patient the all-important fight against unhappiness, disease and poor ability.

In this book we would like to address some of the following fundamental issues and problems of holistic biology:

- There are lots of fundamental problems in biology, medicine and neuro-psychology; most of these problems are related to biological information
- The problem of science is that the disciplines use different concepts and methods, making it very difficult to establish the needed interdisciplinary and integrated theory for matter, life and consciousness. Without such a multidisciplinary approach there is little hope ever to get to an understanding of complex phenomena like healing and human adult metamorphosis
- We have a body-mind and brain-mind, each carrying its own perspective of life and reality. How can we unite these almost opposite and contradicting interpretations of the world into one wise synthesis?
- The living cell cannot be understood though the concept of self-organisation, because more information is needed than the molecules apparently can provide. And consciousness remains a riddle.
- Biochemistry cannot provide the spatial information needed to explain morphogenesis.
- Supra-cellular morphogenesis is mysterious. Nobody seems to understand the origin of biological and cellular order.
- Physical energy is deeply mysterious. How can we proceed when the physicist will not give us the explanatory tools we need as biologists?
- Quantum chemistry could be the key to cell consciousness. Can a deeply structured quantum field give cells memory, coherence, intent, consciousness, and choice?
- What is the wholeness of man? How can 10.000 billion cells say they are me?
- Evolution has breed a fly with eyes and humans with a large wonderful brain. But how did the organisms materialise these things? Is creative metamorphosis a product of the conscious wishes of the plants, animals, and the human being?
- How can we understand the structure of the cerebral cortex, the most complex structure in the known universe? Can we reduce it to something simple like a number of modules in the brain?
- How does the brain think? Can it think without the body? Can it think without the wholeness (including the soul)? How does it remember? How does it organise a model of reality?
- Can a fractal music theory based on holistic brain theory explain the foundation of minor major?
- We survive by degenerating our existence. Can adult human metamorphosis upgrade our purpose of life, our level of existential responsibility, and the depth of our consciousness if we are low lifes?
- Why do we get mentally ill?
- How can the aetiology of mental diseases be explained by the mental and spiritual crises of metamorphosis?
- How is the immune system regulated?
- How does the immune system function? How does the immune system make self-nonself discrimination?

- Why do we get physically ill?
- What is healing and how can we get well aging when we are ill? What is spontaneous healing? Is human adult metamorphosis really a possibility?

Acknowledgements

These studies and the background for this book were supported by grants from IMK Almene Fond. Our research in quality of life has been approved by the Copenhagen Scientific Ethical Committee under number (KF)V.100.2123/91.

References

[1] Schrodinger E. What is Life? Cambridge: Cambridge University Press, 1992.
[2] Hameroff S, Penrose R. Conscious events as orchestrated space–time selections. JCS 1996;3(1):36-53.

Chapter II

Matter, life and consciousness

For almost a decade we have experimented with supporting the philosophical development of severely ill patients to induce recovery and spontaneous healing. Recently we have observed a new pattern of extremely rapid spontaneous healing that apparently can facilitate even the spontaneous remission of cancer and the spontaneous recovery of mental diseases like schizophrenia and borderline. Our working hypothesis is that the accelerated healing is a function of the patient's brain-mind and body-mind coming closer together due to the development of what we call "deep" cosmology.

To understand and describe what happens at a biological level, we have suggested naming the process adult human metamorphosis, a possibility that is opened by the human genome showing full generic equipment for metamorphosis. To understand the mechanistic details in the complicated interaction between consciousness and biology we need an adequate theory for biological information. We suggest that a relatively simple model based on interacting wholenesses instead of isolated parts can shed a new light on a number of difficult issues we need to explain and understand in biology and medicine in order to understand and use metamorphosis in the holistic medical clinic.

We aim to give a holistic theoretical interpretation of biological phenomena at large, morphogenesis, evolution, immune system regulation (self-nonself discrimination), brain function, consciousness, and health in particular.

We start at the most fundamental problem: what is biological information at the sub-cellular, cellular and supra-cellular levels if we presume that it is the same phenomenon on all levels (using Occam's razor), and how can this be described scientifically?

The problems we address are all connected to the information flow in the functioning, living organism: function of the brain and consciousness, the regulations of the immune system and cell growth, the dynamics of health and disease.

We suggest that life utilises an unseen fine structure of the physical energy of the universe at a sub particular or quantum level to give information directed self-organisation; we give a first sketch of a possible fractal structure of the energy able to both contain and communicate biological information and carry individual and collective consciousness. Finally thorough our analysis we put up a model for adult human metamorphosis.

Introduction

Recently we have observed what seems to be a completely new pattern of highly accelerated spontaneous healing in the holistic medical clinic, where we help people to improve their quality of life, health, and ability by developing their consciousness and the value they create by using more of their inborn talents. The new pattern of healing resembles the process of metamorphosis of the butterfly, when the larvae turns into pupae and in only ten days melts totally down to reappear as the adult butterfly. We have noticed that the human genome contains full equipment for metamorphosis and actually very much the same genes for metamorphosis as the insects and the lower vertebrates, like the amphibian. The frog goes trough a radical metamorphosis from tadpole to frog, as we were frogs not long ago on the evolutionary scale. The genes for human metamorphosis are known to be active during the ontogenesis, but we have observed adult human metamorphosis (1). We have seen the change in the whole character of the person, from purpose of life, to cognitive patterns of the interpretation of the world, to the actual physical form of the body. The patients that enter metamorphosis seem to have two things in common: 1) they have been arrested in their psychosexual development early in life, often due to severe sexual or violent abuse or neglect, and 2) they have come to a sudden deeper understanding of life. This deeper understanding must be of the form we would call "a deepening of their cosmology".

We have observed that our understanding of the world can go though evolution and revolution. The evolution is the gradual growth of insight and understanding and the revolution is a crisis, where everything in a way looses its normal meaning and significance and in this void a completely new fundamental pattern of understanding of the world appears. What seem to happen is that the structure of the conceptual tree in the brain-mind changes; in our analysis this structure is fractal, and what happens is that the bifurcation number changes (2). When it goes down, we run away from our existential responsibility and this event seem often to happen in our early childhood, where responsibility becomes to painful (3). When the person as an adult searches for deeper understanding this reduced cosmology relates (4), so survival is re-exchanged with a deeper, more complex cosmology, looking very much like the original "natural" and deep, emotionally rich cosmology of the small child (4). Interestingly, the brain-mind and the body-mind – the informational system of the organism - seems to be separated by the cosmology going down in number, separating also mind and body, and making emotions less intense (4). As an adult, the opposite happens: when a new and deeper cosmology appears, a new emotional and experiential life being, with more intense living, more focus on the body, on enjoying, on sexuality, and on existence (4). So it seems that deepening the cosmology is a hallmark of existential healing. It seems that this healing is bringing the brain-mind and the body-mind together, making the person more whole and more emotional. When body and mind merges the totality of the being surfaces, and the self is perceived more transcendently. We say that the person recovers the existential coherence, because life inside is reconnected with the world outside and quality of life (QOL), health both mental and physical, and all the abilities of the person is often radically improved.

Adult human metamorphosis is strange in that this process happens in weeks, not during several years as normal therapy. The patient undergoing metamorphosis becomes often completely introverted and enters a highly visionary state of existence, known from Native American rituals like the Mitote (using the hallucinogenic cactus peyote) (5) and the Vision quest (using no drugs but focused intent on transformation)(6). More peculiar is the very ill patient's tendency to seemingly suddenly recover from even cancer (7-9) and schizophrenia (10,11), when QOL improved during and after transformation.

What we have found is highly unexpected and it seems to call on radical new thinking; it simply needs a new more holistic theory of biology and the role of consciousness both in ontogenesis and phylogenesis. As we see it, it calls for a deep understanding of biological information, for the informational flows and the appearance of the hidden person almost in one magical stroke is really what needs explanation. Of cause, we already know metamorphosis from the animal world, so it cannot be that much of a shock for the observant reader, but to see human beings melting down, enter the state of the human pupae, and reappearing as a completely upgraded version of themselves after a few weeks, even recovering from severe mental or somatic disease is really unexpected, even for the most open-minded physician.

Our clinical experience and observations

It might be that we are completely mistaken or mistaken in our interpretations of what we observe. On the other hand, during the last years we have been studying first human development, then the phenomenon of holistic healing, after that human transformation, and now the accelerated process of human metamorphosis, which seem to explain also the slower processes of spontaneous healing in biological terms.

It is quite clear from studying the patients entering the transitional state of metamorphosis that the patients seem to open up to what Carl Gustav Jung used to call "the collective consciousness" (12): a layer of immense information about being human, that all human beings can dive into for inspiration and renewal. We have learned just recently that many pre-modern cultures like the Native American cosmology of the Delicate Lodge - in our analysis carrying the bifurcation-number 8 (an "8-ray cosmology") - in its ontology have both the collective consciousness of plants (called "The White Buffalo Woman"), animals (called "Sacred Sweet Medicine") and man. So we know that different people and cultures have appreciated these deep information-carrying levels of human existence. Western medicine is just now opening up to their existence in research (13,14).

Professor Daniel Shek from Hong Kong (personally communication) has talked about "the three T's of holistic medicine: transformation, transcendence, and totality". This is very accurately the focus of our holistic medical endeavour of healing our patients, which are not only the traditional ill patients, but everybody that needs an improvement of QOL. The accelerated human growth of metamorphosis seem to be very promising for helping also the most troubled and harmed of the people.

Holistic understanding of life

Our interest in theoretical biology started in the 1980s, but our work seemed for two decades to be to idiosyncratic to be published. Only now after first testing the models and noticing their positive consequences for holistic medicine have we have gained sufficient confidence in these models to publish them. The induction of spontaneous healing from unsolvable health problems, like cancer and severe mental diseases at our Research Clinic for Holistic Medicine (7-10) for almost a decade has also given us the belief that it is time to publish our unconventional perspective on biology and human development now. We have recently presented the first ideas of holistic biology (15,16).

The idiosyncratic feeling stems from the radical belief that we need to go all the way down to the fundamental energy of the universe to understand biology. Instead of using established mathematical structures as the basis of our model we looked in 1985 for a more intuitive correct presentation of biological informational systems of the living organisms, from single cells to large ecosystems. Unfortunately, the structures that followed from this quest were not compatible with any known mathematical structures at that time. As you might have guessed we discovered, as did Benoit B Mandelbrot and many other at that time, the fractal nature of the living world. We made our first fractal model for the organism in 1986 (2). After 20 years everybody seems to have accepted the fractals, but biology has not been better explained after all. So we are still missing the point. What we try to bring forward now is what we believe to be the missing parts of the puzzle.

Our major point is that the well-known order of the universe described by contemporary science in a way is illusory, and our certainty of many things might make us overlook a hidden order in the universe, which guides not only how the sub-atomic energy and particles are organized into atoms and molecules (17), but also how matter is organized into the living cell, with its dynamic organelles and highly advanced supra-molecular structures (18). It might be the same principles that guide the cells into a complex organism (19) and the organisms into extremely complex and "intelligently" created biospheres including planet Earth at large (20). And it might even be the same principles that count for "white" and dark matter organising the dust, planets and stars into galaxies and clusters of galaxies.

Our suggestions

We suggest that there exists one hidden but highly active, fundamental, organising principle, which we call the fine structure of the physical energy, and we believe that a first approximation to a description of this energy is a spiral fractal (2). We suggest a simple geometry in which all levels interact with each other and with every part existing on each level. Unfortunately, this geometry has been difficult to formalise, as with most other fractals it cannot be build on logic and set theory. Interestingly the description we need is fundamentally different from the principles of logic and the sets of the set theory as both logic and the set theory is based on the idea of a well-defined element distinctly separated from its surroundings, in that it has qualities that is unique to this element and not dependent on the state of the space or of other co-existing elements. Our normal logic only has meaning if such elements exist. The "dependent" elements of the real world have, as you might know from quantum mechanics, complex and mysterious qualities that have been extremely difficult to define.

Our fundamental postulate that started our thinking presented is that independent elements do not really exist. In the world of non-living, "dead" objects we know that independent elements do not exist on the level of the particle, but large dead elements seems to function almost independent of each other (one is red and square, another blue and round), and for many years researchers have believed that large living organisms could be understood in parallel to large dead objects in this respect.

Our position is that living beings has been specialised in using the deep coherence and connectedness of everything, and that consciousness has risen from this connectedness. The coherence of living organisms might be important for evolution. It could help the animal of a species to develop simultaneously (when one fly gets its wings and becomes successful this way, the other flies are encouraged through this connectedness also to turn winged) (20).

We also believe the single cell to be conscious (as do other thinkers in the field: Ben Jacob personally communication) (21) and able to make decisions on its own, changing its behaviours or other characteristics of the whole cell. We also believe that the consciousness of the whole organism rises from the multiple consciousnesses of its cells (12). These cells have the same ability to cohere as all living beings, and they use this coherence to get information to make valuable decisions and thus change their own character and also the character of the whole organism. The cells are thus efficiently using the fundamental coherent nature of the world, which is so deeply mysterious that we do not have a theory for it yet. This is the theory we want to make.

We thus believe that science needs a completely new language to describe and analyse what is happening in the living world, and the present work is a first try to provide such a language. We hope that the solution we have found will shed light on many of the unsolved problems of biology. Most interestingly many of the human diseases and the patient's spontaneous healing seem to be understandable on top of this new understanding of biology, the body, immunology, the human brain, and consciousness.

May this work be of value to all living

What we want to model is the internal and mechanistic structure of this coherence of the elements and beings of the living world on the many levels of the biological universe, from molecule over the organelle and cell to the whole organism, its family, its species and the eco-system it inhabits. The phenomenon we have identified seems to be a principle almost only used by living organisms. We have called it "information-directed self-organization". We have analysed a number of biological systems and situations and we have found that the concept of "information-directed self-organisation" can explain yet unexplained biological phenomena.

Most interestingly the new symbolic language seem to be able to explain what wholeness is. For instance: man is a wholeness, living in a wholeness of the world. We all feel our autonomous freedom to be, do and think. But are we free to make choices? Are we free to perceive? Are we determined of our history and our world, or are we really causal to our world? Can we create our own life, healthy and happy, independently of the surrounding world and our personal past, gene pool etc.? Can we change our own destiny and get well again even from metastatic cancer? Can we go through metamorphosis and both change mind and body, purpose of life and our fundamental experience of the world – our worldview or cosmology? We believe that we can. But where is then the theoretical limit for our human abilities? It is all such questions we want to answer and this is the quest that has motivated the present analysis of life and man.

References

[1] Ventegodt S, Hermansen TD, Kandel I, Merrick J. Human development XX: A theory for accelerated, spontaneous existential healing (salutogenesis): Human adult metamorphosis. J Altern Med Res 2009;1(4):465-74.

[2] Ventegodt S, Hermansen TD, Flensborg-Madsen T, Rald E, Nielsen ML, Merrick J. Human development VII: A spiral fractal model of fine structure of physical energy could explain central aspects of biological information, biological organization and biological creativity. ScientificWorldJournal 2006;6:1434-40.

[3] Ventegodt S, Hermansen TD, Kandel I, Merrick J. Human development XIV: Degeneration and regeneration of existence. Metamorphosis changes the purpose of life, the level of existential responsibility, and the depth of consciousness (the ray-number of the cosmology). J Altern Med Res 2009;1(4):409-18.

[4] Ventegodt S, Kromann M, Andersen NJ, Merrick J. The life mission theory VI: A theory for the human character: Healing with holistic medicine through recovery of character and purpose of life. ScientificWorldJournal 2004;4:859-80.

[5] Anderson EF. Peyote: The divine cactus. Tucson, AZ: University Arizona Press, 1996.

[6] Moondance W, Starhawk S. Vision quest: Native American magical healing. New York: Sterling, 2004.

[7] Spiegel D, Bloom JR, Kraemer HC, Gottheil E. Effect of psychosocial treatment on survival of patients with metastatic breast cancer. Lancet 1989;2(8668):888-91.

[8] Ventegodt S, Morad M, Merrick J. Clinical holistic medicine: Induction of spontaneous remission of cancer by recovery of the human character and the purpose of life (the life mission). ScientificWorldJournal 2004;4:362-7.

[9] Ventegodt S, Solheim E, Saunte ME, Morad M, Kandel I, Merrick J. Clinical holistic medicine: Metastatic cancer. ScientificWorldJournal 2004;4:913-35.

[10] Ventegodt S, Andersen NJ, Neikrug S, Kandel I, Merrick J. Clinical holistic medicine: Mental disorders in a holistic perspective. ScientificWorldJournal 2005;5:313-23.

[11] Ventegodt S, Andersen NJ, Neikrug S, Kandel I, Merrick J. Clinical holistic medicine: Holistic treatment of mental disorders. ScientificWorldJournal 2005;5:427-45.

[12] Ventegodt S, Hermansen TD, Flensborg-Madsen T, Nielsen ML, Merrick J. Human development VIII: A theory of "deep" quantum chemistry and cell consciousness: Quantum chemistry controls genes and biochemistry to give cells and higher organisms consciousness and complex behavior. ScientificWorldJournal 2006;6:1441-53.

[13] Grof S. Psychology of the future: Lessons from modern consciousness research. Suny Series in Transpersonal and Humanistic Psychology. New York: State University of New York Press, 2000.

[14] Wilber K. Integral psychology: Consciousness, spirit, psychology, therapy. Boston, MA: Shambhala, 2000.

[15] Ventegodt S, Merrick J, Andersen NJ. QOL philosophy III: Towards a new biology. ScientificWorldJournal 2003;3:1186-98.

[16] Ventegodt S, Andersen NJ, Merrick J. QOL philosophy IV: The brain and consciousness. ScientificWorldJournal 2003;3:1199-1209.

[17] Ventegodt S, Hermansen TD, Rald E, Flensborg-Madsen T, Nielsen ML, Clausen B, Merrick J. Human development III. Bridging brain-mind and body-mind. Introduction to "deep" (fractal, poly-ray) cosmology. ScientificWorldJournal 2006;6:767-76.

[18] Ventegodt S, Hermansen TD, Rald E, Flensborg-Madsen T, Nielsen ML, Clausen B, Merrick J. Human development IV. The living cell has information-directed self-organization. ScientificWorldJournal 2006;6:1132-8.

[19] Ventegodt S, Hermansen TD, Flensborg-Madsen T, Nielsen M, Merrick J. Human development VI: Supracellular morphogenesis. The origin of biological and cellular order. ScientificWorldJournal 2006;6:1424-33.
[20] Hermansen TD, Ventegodt S, Merrick J. Human development X: Explanation of macroevolution — top-down evolution materializes consciousness. The origin of metamorphosis. ScientificWorldJournal 2006;6:1656-66.
[21] Penrose R, Gardner M. The Emperor's new mind: Concerning computers, minds, and the laws of physics. Oxford: Oxford University Press, 2002.

Chapter III

Brain-mind and body-mind

Reality can be interpreted in many ways, but two distinctly different ways are the mental and the emotional interpretation. The traditional way of thinking in science today is the first: an often simple and mechanical interpretation of reality that empowers us to handle the outer physical world with great, often brutal efficiency. The development of a mind that enables us to handle the outer physical world and survive makes a lot of sense from an evolutionary perspective; the problem is that the mental reason and linear logic reduces all phenomena to well-defined interacting objects, which might not exist from a deeper perspective of reality. A more intuitive way to interpret the world makes much more sense, when it comes to our human relations. So to function as a human being, we need both these two ways of seeing the world, and two different modi operandi. In many patients, we find an internalized conflict between logical and mental reasoning on one hand, and emotional and sexual approach to reality and human needs on the other. We speculate that this conflict causes the deep emotional problems that really are the basis of most human diseases. Only by merging brain-mind and body-mind will we be whole and free and truly ourselves. We need to develop our mental understanding, deepen our cosmology, and develop our sexuality and body-mind in order to make them meet and merge. To facilitate this existential healing, we propose a third integrative way of looking at our human nature, which we call "the energetic-informational interpretation of reality". What it does is allows us to look at both brain-mind and body-mind as a highly structured field of "energy and information". Energy and information are actually the same from a scientific point of view; when the world is seen through the body-mind, it looks more like energy; when seen though the brain-mind, it looks more like information.

Introduction

Science is about describing and interpreting reality. An interpretation of reality is a brain representation of reality, which carries a specific structure. Reality can be "broken down" and interpreted in several possible ways. Usually, before we develop our understanding of life and consciousness, we are able to interpret reality and its phenomena in at least one of two very different ways: a rational/mental or an emotional/intuitive way. The rational/mental way is often a mechanistic way of looking at nature, as chemistry and logical patterns of behavior, while the emotional/intuitive way is a more naive and immediate interpretation of reality with offspring in our bodily experiences. Often, we are able to use the first understanding of the world when we work, and the second understanding when we are most private, like taking care of our children or while we are making love to our partner.

The rational interpretation of the world is seen as more "clever" and more discriminative and more analytically intelligent than the intuitive interpretation, which on its own is more "synthesizing" and emotionally intelligent (1), making us see meaning in things and feel love and wholeness. When the rational and mental interpretation of reality has been accepted by the biological and the medical sciences, as it has happened with molecular biology and biomedicine, this will lead to many anomalies and paradoxes and leave many important phenomena incomprehensible.

The human consciousness is maybe the most difficult thing to explain with a mechanical approach, as the "light" of consciousness has a coherent, unifying, and global quality, which seemingly cannot be matched by any known natural, physical law in contemporary natural science, even when tried with the most brilliant mental intelligence.

When we look at the world through our feelings, we see relations and connections with no place in the rational, mental, and formal view of the world. It seems that the emotional interpretation of reality in many ways is closer to human biology than the mental interpretation (1). In this book on human development, we propose a third way of looking at nature that integrates these two perspectives. We call this third perspective the "energetic-informational interpretation of reality".

The search for an integrated perspective is really the search for a philosophy of life that integrates the mental and the emotional-physical side of man. This interesting search is a door to a more general understanding of the many ways the world can be understood: the varieties of cosmologies that exist now. They have been developed through the times by the many human civilizations on many continents, from the African Sangomas to the Australian aboriginals, Asian Indian sages, and American Maya Indians. Most interestingly, medicine seems to be one of the primary purposes for developing the cosmologies; the most complex of cosmologies, the cosmologies with the greatest concepts and most profound mystical depth, are often held by the shamans and the medicine men for the purpose of healing the people. This is not a coincidence.

Bridging the brain-mind and the body-mind is to make man whole; it is a way to heal both mental and physical diseases. The integration is not easy because it takes a deep philosophical understanding of life and self to mobilize all the hidden resources in man (2). The new integrative interpretation of the world is what allows us to heal and even sometimes to transform completely and go into the process of radical reorganization of the whole existence — the very intense and rapid process of spontaneous self-healing we call "adult human metamorphosis". The understanding of the background of this pattern of healing is what motivates us for the quest and, interestingly, the quest to understand life better was what motivated and gave one of the authors (SV) the faith to follow the first patients through the dramatic, intense, and often very painful process of human metamorphosis 7 years ago, when it looked like a common psychosis, just with more intensity from the deep layer of existence that Carl Gustav Jung (1876–1961) called the "collective consciousness" (3). To make the reader understand and believe what we have seen and found though the last 20 years of research in the connection between health and human consciousness, we will introduce the reader to the different aspects of a new holistic view of the world and its major phenomena:

matter, life, and consciousness. This interdisciplinary understanding of human development seems to be what is needed to explain healing.

Holistic biological paradigm

When we approach the world though the mind, we interpret it. We cannot interpret the world without a personal purpose in life to source our values and intentions. In order to make sense out of chaos, we need to be a part of the world. This makes all science highly biased, which also is the point of many philosophers of science (Popper, Kuhn, Chalmers, Fireaben, Gøtzche). In a way, it is also nice because it empowers every one of us to look for our own truth.

We have been experimenting with different interpretations of reality, as they rose when we looked at the world though different "glasses": through the brain, the world looks one way, and through emotions and intuition, the world looks quite different. A mental perspective makes it possible to analyze the world into different pieces. An emotional perspective empowers us to see the relations between people and things. To train the ability to see the world in different ways is an exciting experiment that gives us respect for the process of interpretation. Every person has his own way of looking at and understanding the world, and is simultaneously able to relate to a set of different ways to look at the world because our human organism embodies a set of faculties of intelligences that make this possible.

This modern view makes science complicated. For what can be absolutely truth, if we are so subjective and our worldview so easily manipulated? We believe that when we have understood mind and emotions well, a new set of eyes opens, allowing us to see the world in a deeper and more objective way. This might be as illusory as everything else of course, but we invite you to follow us on this journey, and we believe that in doing so, you will share our experience of a deeper understanding of life. We have called this deeper perspective the energetic-informational interpretation of reality or simply the holistic paradigm, as it brings us closer to the wholeness of the world and its phenomena than do both the mental and the emotional interpretations of the world. When we look at biology in this way, it seems possible to obtain a deeper understanding of life and human development, in the to papers that will follow.

Reflections on Science

Scientists often suppose that they reach their conclusions by use of mental reasoning, and scientists will normally try to avoid bias from emotions, intentions, and beliefs. Mental reasoning is the ability to analyze and group phenomena, built on logic and set theory of mathematics.

Classically, science is performed by dividing reality into its parts and then describing the characteristics of these parts and their interactions. The parts can normally be subdivided again and again. By and large, the different scientific disciplines are defined by the different sizes of their study objects (see Figure 1) and, to some extent after, if the phenomenon is subjective or objective (compare: Wilber's four quadrants (4)).

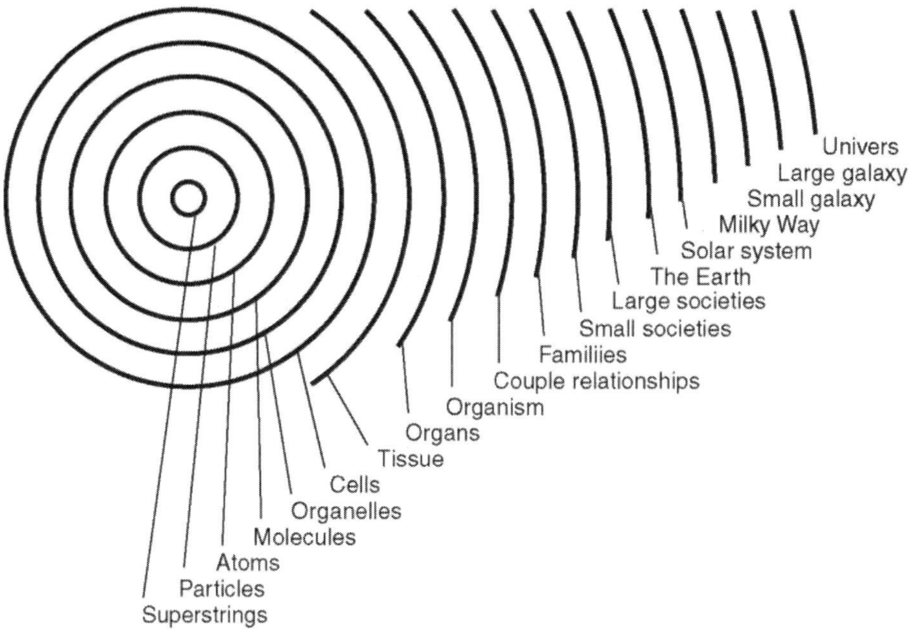

Figure 1. How reason has divided reality into scientific disciplines according to the size of the objects of investigation.

Between the disciplines, strong demarcation lines exist, new regularities between interactions of units emerging at each level, respectively; this is based on the idea that reality consists of levels and each level has its own sphere with its own class of phenomena and associated laws.

Mental reason reduces all phenomena to well-defined interacting objects. We become terribly frustrated when a description (for example, of the electron as a well-defined, isolated independent particle) is not valid. Then, we try to comfort ourselves with the fact that the description of nature as particles at least persists regarding bigger things than particles, even when science shows that the quantum mechanical laws also rules for big molecules (5). But most unfortunately, we cannot understand biological organisms by reducing them to "things". To understand why this is so, we must accept that an interpretation of reality always happens to serve a specific purpose.

A simple and mechanical (gr. mechane = tool) interpretation of reality empowers us as human beings to handle cleverly the outer physical world through the use of tools and technologies. The ability of our reason to reduce reality to interacting elements is reflected deep down in the roots of our language, demonstrating that the brain and language have been busy with this kind of survival for many generations. The development of a mind matching the outer world can be seen from an evolutionary perspective. The complicated use of tools and language makes us as human beings fit for survival, which is exactly what sets us apart from the apes. When we became human and developed logical reasoning, in addition to the feelings and emotional reasoning we already had as apes, we were forced into a conflict between emotions and logical reason. As apes, we had conflicts between reality and simple biological needs. The modern man developed a brain that allowed him to a far higher degree to comply with the physical world; however, the conflict between the biological needs and the outer world persists, but today the conflict has partly moved inwards in the organism and has became a conflict between the brain and the rest of the biological wholeness of man. Man's excellent ability to adapt has become his heel of Achilles.

We assume that this internalized conflict between logical and mental reason and emotional and sexual needs is the basis of all illnesses in the human being, both concerning mental diseases like schizophrenia (only known in man) and the largest fraction of the somatic illnesses (also found much more frequently in man than in apes and other animals). Man is generally very vulnerable compared to animals; brain surgery may be carried out without problems in rats under nonsterile conditions; in man this can hardly be done.

The project of science is to explore reality by means of study and experiments. However, experiments always have to be constructed, and the results always have to be interpreted. The rationalists assume that the road to truth goes through the interpretation of experimental experiences, while the emotions constitute an "irrational element", diverting reason from truth. But reason betrays, too. Reason notices most easily phenomena that make good sense. This inevitably leads to confirmation of previously established ideas and expectations. Logic is able to provide consistency in a logical system, but every logical system rests upon axioms, which we get from our intuitive understanding of the world. Reason, therefore, is in its essence nurtured by the tacit, emotional dimensions of existence, not the other way around. When discriminative reason analyzes reality into parts, it requires a "synthetic" function of emotion to group them into a meaningful synthesis. Meaning is supplied by emotionality and intuition.

Emotions betray, because a great amount of conditioned learning (learning that serves adaptation to reality at a given point in time normally motivated by survival) lies between the individual and a direct experience of reality. We have a strong tendency to mistake past for present because we do not update our adaptation to our shifting reality. We feel what we have learned to feel instead of what we really feel. Reason betrays, because it has a great tendency to notice what gives immediate meaning, which is a phenomena in accordance with previously learned fundamentals and axioms. It is self-affirmative. We are slaves of our learning and "betrayed" by both our emotions and our reason.

In order to solve these problems and obtain true knowledge, we must handle the delusion of reason as well as the delusion of emotion. Reason must be expanded through a constant attempt to introduce more roomy axioms, and to handle these in a noncontradictory way. The reality of the emotions must be cleansed of conditioned learning in order to make us experience reality without the influence of pollution from past learning. The ideal is a complete unification of emotions and reason, corresponding to man's greatest potential of knowledge. The more we cultivate our mind, the better we represent a matter in our consciousness and mental understanding. We call this the double cultivation of emotions and reason, and the application of their synthesis in the interpretation of reality, we call the interpretation paradigm.

The Mental-rational interpretation of reality

When we want to handle the outer world, we use logic, set up theories, and interpret the phenomena of the world as things, e.g., bodies or particles. The particles move in space and when they move, time is created. Their interactions reveal forces that may cause the bodies to accelerate. The resistance to acceleration corresponds to their mass. However, the forces and the masses, as well as the particles themselves, are not explained by this description.

Modern physics, including the theories of relativity and quantum mechanics, has shown that a simple, mechanical interpretation of reality, such as Newton's, is not only insufficient, but also incorrect. We have to admit that our rational interpretation of reality is in fact a simplification of reality. Relativity theory shows that gravitational forces between particles equivalently can be described as curvature in a four-dimensional space-time (6). Space and time has to be broken down in another way. The notion of curved space-time does not supply a real understanding, but indicates that a completely new interpretation of reality is necessary for a deeper understanding of reality.

Quantum mechanics describes the electron as sometimes a wave (4,7) and sometimes a particle. This is the famous Copenhagen interpretation of quantum mechanics by Niels Henrik David Bohr (1885–1962). This interpretation violates our straightforward impression of a particle as something well defined in space. Both relativity theory and quantum mechanics points to the fact that a new interpretation of reality is needed to arrive at a proper understanding.

The Emotional-intuitive interpretation of reality

In our daily lives, we usually interpret reality with a starting point in ourselves, introspectively in our inner world, and in the outer world by looking at living organisms – above all people we love. Our attention is drawn to fluidly changing totalities. We think by association and in pictures. There are no fixed boundaries between the world and us, and there are no fixed boundaries between the parts of the world. There are in fact no parts, only totalities moving to create meaningful patterns. In the rational interpretation of reality, delimited elements are a necessary condition for interpretation, but in the emotional interpretation, we find significance and system states instead. Focus is at the patterns of rays extending throughout the levels (see Figure 2).

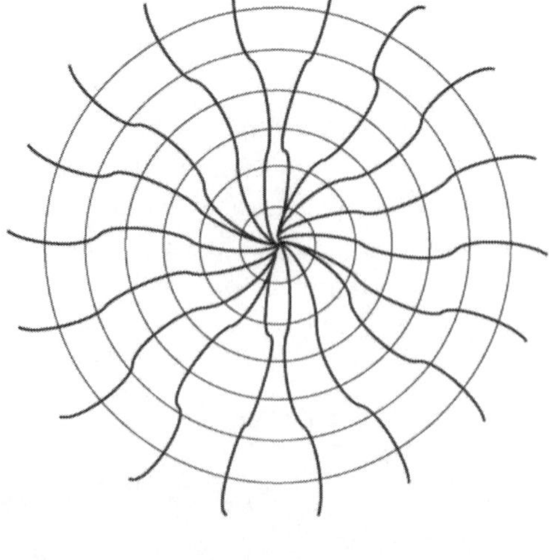

Figure 2. The aspects illustrated across the organizational levels.

The rational interpretation is fixed, local, and logical, and it thinks in terms of space, time, and mass. The emotional interpretation is fluid, nonlocal, nonlogical, and not involved with concepts of space, time, and mass. It is the reality of the unconscious, of feelings and urges, and of intention and love.

The Energetic-informational interpretation of reality

In human life, the rational and the emotional interpretations of reality often conflict or seem to mutually exclude each other. The conflict between power and love is a classical conflict, well known to many people in their own life. In philosophy, the incommensurability of the body and the soul is a manifestation of the same conflict. In order to obtain true knowledge, we have to handle these discrepancies and create a synthesis of the two interpretations. We have to give solidity to the fluid emotions, and we have to make firm reason more liquid. In other words, we have to exchange the axioms of reason with axioms that can incorporate our emotional reality. Hereby the formerly disparate rational and emotional interpretations become transformed to a deeper interpretation cleansed of all the conditionings that oppose each other in the two separate interpretations. Energy, organization, and information seem to be sufficiently elastic concepts of the simultaneous description of both the rational and the emotional interpretation of reality. The concepts of space, time, particle, and mass, therefore, have to be exchanged with the concepts of information, organization, and energy. The particles of physics lose their particle status and have to be described as patterns of structure and information containing energy. What we, according to the rational interpretation, would call a particle in space and time is, in the energetic-informational interpretation, a pattern of structure and information that can never be separated from the totality of which it is a part.

The theory of superstrings (8) has done admirable work by expanding the principle of equivalence to all basic forces of nature. But the multidimensional nature of the universe contradicts our common experience of space as three-dimensional. How can space-time have 10 or more dimensions? When mathematical formality requires extra dimensions, these are simply given status as reality. The superstrings are conceived as being very small and are able to form particles. The concept of particles belongs to the rational interpretation of reality and should be abandoned in the energetic-informational interpretation of reality. Even strings seem to be too mechanical to fit in. It is still a simplistic model of reality.

The energetic-informational interpretation of reality must employ a much more complex breakdown of reality. We must seek for the complex and coherent structure of the universe to really understand life and consciousness. This is the challenge we face in the present book on holistic biology.

Discussion

We believe that the internalized conflict between logical and mental reason and emotional and sexual needs is the basis of all human illnesses. The severe incongruence of the brain-mind and the body-mind destroys coherence (9-11). We conclude that in order to solve these problems and obtain true knowledge, we must handle the delusion of reason as well as the delusion of emotion. Reason can be expanded through a constant attempt to introduce more spacious axioms, and to handle these in a noncontradictory way. This is the evolution of reason. A more radical way is to allow chaos and confusion and let a completely new matter of understanding appear – this is the revolution of mind that lead to deeper cosmology (12). The more we cultivate our mind, the better we represent the world in our consciousness and mental understanding. We call this the double cultivation of emotions and reason, and the application of their synthesis in the interpretation of reality, the integrative, or "deep", cosmology.

In the concept of the rational interpretation of reality, the relativistic notion of curved space-time and the uncertainties and nonlocality of quantum mechanics indicate that a completely new interpretation of reality is necessary for a deeper understanding of reality. Both relativity theory and quantum mechanics point to the fact that what appears for the eye is not so; a new, more complex interpretation of reality is needed for a proper understanding of the world. Interestingly, both of them are irresponsible in a way; the overintellectual escapes the painful feelings and the overemotional escapes the narrowing bands from clear responsible thinking.

A true, trustworthy, mature, and "deep" cosmology will deal with all phenomena of the world in a orderly way, giving a graduated spectrum of phenomena from the dense material to the almost nonexisting spiritual (13). In human life, the rational and the emotional interpretations of reality often conflict or seem to mutually exclude each other. In order to obtain true knowledge as well as mental, physical, social, and spiritual health, we need to handle these discrepancies and create a synthesis of the two interpretations, leading to an interpretation of reality based on concepts of energy and information. In this, the physical particles lose their particle status and feelings lose their personal and historical significance. The world is described as patterns of more or less materialized information, and more or less free energy.

What we would call "a particle in space and time" in the rational interpretation of the world and what we would call "a subjective experience" in the emotional/intuitive interpretation will, in the deep cosmology (in the energetic-informational interpretation of the world), be seen as a pattern of structure and information that can never be separated from the totality of which it is a part.

Conclusion

We become terribly frustrated when a description (for example, of the electron as a well-defined, isolated particle) is not valid. Then, we try to comfort ourselves with the fact that the description of nature as particles persists, at least regarding bigger things than particles, even when science clearly demonstrates that also much larger things than small molecules, atoms and particles are ruled by the strange laws of quantum mechanics. Recently quantum interference experiments by Olaf Nairz, Markus Arndt, and Anton Zeilingerb with very large molecules have been documenting this quite convincingly (14). But most unfortunately, we cannot understand biological organisms by reducing them to "things". To understand why this is so, we must accept that an interpretation of reality always happens to serve a specific purpose.

We assume that the internalized conflict between logical and mental reason and emotional and sexual needs is the basis of all illnesses in the human being, both concerning mental illnesses like schizophrenia (only known in man) and the largest fraction of the somatic illnesses (also found much more frequently in man than in apes and other animals).

Emotions betray because a great amount of conditioned learning exists between the individual and a direct experience of reality. Also, reason betrays because it has a great tendency to notice what gives immediate meaning, which is a phenomena in accordance with previously learned fundamentals and axioms.

In order to solve these problems and obtain true knowledge, we must handle the delusion of reason as well as the delusion of emotion. Reason must be expanded through a constant attempt to introduce more spacious axioms, and to handle these in a noncontradictory way. The reality of the emotions must be cleansed of conditioned learning in order to make us experience reality without the influence of pollution from past learning. The ideal is a complete unification of emotions and reason, corresponding to man's greatest potential of knowledge. The more we cultivate our mind, the better we represent a matter in our consciousness and mental understanding. We call this the double cultivation of emotions and reason, and the application of their synthesis in the interpretation of reality, we call the interpretation paradigm.

The rational interpretation is fixed, local, and logical, and it thinks in terms of space, time, and mass. The emotional interpretation is fluid, nonlocal, nonlogical, and not involved with concepts of space, time, and mass. It is the reality of the unconscious, of feelings and urges, and of intention and love. The energetic-informational interpretation of reality (the holistic paradigm) must employ a much more complex breakdown of reality. We must seek for the complex and coherent structure of the universe to really understand life and consciousness.

References

[1] Goleman D. What makes a leader? Clin Lab Manage Rev 1999;13(3):123–31.
[2] Ventegodt S, Andersen NJ, Merrick J. Quality of life philosophy: when life sparkles or can we make wisdom a science? ScientificWorldJournal 2003;3:1160–3.
[3] Jung CG. Man and his symbols. New York: Doubleday, 1969.
[4] Wilber K. A theory of everything: An integral vision for business, politics, science and spirituality. Boston, MA: Shambhala, 2001.
[5] Arndt M, Nairz O, Voss-Andreae J, Keller C, van der Zouw G, Zeilinger A. Wave-particle duality of C60. Nature 1999;401:680–2.
[6] Einstein A. Relativity: The special and general theory. New York: Henry Holt, 1920.
[7] Stoney GJ. Of the "electrone", or atom of electricity. Philos Mag 1894;38:418–20.
[8] Green MB, Schwarz JH, Witten E. Superstring theory. Cambridge: Cambridge University Press, 1987.
[9] Ventegodt S, Andersen NJ, Merrick J. The life mission theory V. Theory of the anti-self (the shadow) or the evil side of man. ScientificWorldJournal 2003;3:1302–13.
[10] Antonovsky A. Health, stress and coping. London: Jossey-Bass, 1985.
[11] Antonovsky A. Unravelling the mystery of health. How people manage stress and stay well. San Franscisco: Jossey-Bass, 1987.
[12] Ventegodt S, Hermansen TD, Kandel I, Merrick J. Human development XV: The biochemical hypothesis for the etiology of the mental diseases is not substantiated. J Altern Med Res 2009;1(4):419-32.

[13] Ventegodt S, Flensborg-Madsen T, Andersen NJ, Merrick J. Life mission theory VII. Theory of existential (Antonovsky) coherence: a theory of quality of life, health, and ability for use in holistic medicine. ScientificWorldJournal 2005;5:377–89.
[14] Nairz O, Arndt M, Zeilingerb A. Quantum interference experiments with large molecules. Am J Phys 2003;71(4):319-25.

Section II: Understanding life and biologic information

Chapter IV

Living cell information system

In this chapter, restricted to describe the ontogenesis of the cell, we discuss the processing of DNA through RNA to proteins and argue that this process is not able to transfer the information necessary to organize the proteins in the cell, but only to transfer the information necessary to form the shape of the proteins. We shortly describe the structure of the information carrying field recruited by the cells that we think is responsible for building the organelles and other cellular structures.

We use the cells superior control of its cytoskeleton as an example of how the cell is using an informational field giving the positional information guiding all the local chemical processes behind the cell movement. We describe the information-directed self-organization in cells and argue that this can explain the ontogenesis of the cell. We also suggest the existence of an undiscovered phenomenon behind the information transmitting cell interactions.

We conclude that during evolution the cell has developed into an information-guided self-organizing structure. The mystery we want to solve is: what is the mechanical cause and nature of biological information?

Introduction

Caused by lack of understanding of the cell itself the contemporary biological theory of ontogenesis does not explain how the information for determination and differentiation of cells are recruited.

Using the existing theory of DNA as donor of all cellular information, it has not been possible to explain the powers controlling the cell ontogenesis. We believe that these powers are transmitted through positional information. For instance the, cell movement and axon- and dendrite-excrescence could be caused by a superior control of the cytoskeleton through positional information making it possible for the cell to induce information directed self-organization.

We believe that in general the shapes of organelles (as demonstrated in Tetrahymena) (1) are not formed by self-organization of proteins alone, although the self-organising property of proteins are well known.

We found from analysis of the overall morphological dynamics of i.e. the cytoskeleton that more information is needed than what the molecules themselves is able to provide. We propose ontogenesis and cell determination to be explained by the cells using information directed self-organization. e.g. the organelles and other cellular structures are build, because the cells can recruit and apply complex information though an information carrying field. In existing biological theory it is particularly the determination and differentiation of the cells that are not accounted for, probably due to the lack of understanding of the cell itself.

Generally speaking, the ontogenesis is a phenomenon that is poorly understood and makes a puzzle for the established science. Based on the existing knowledge, we conclude that modern theories are not able to explain the ontogenesis. We propose that a holistic biology using holistic organizing principles is needed to explain this astonishing enigma of the biology.

We propose a holistic biological paradigm based on distribution of information through roomy fractal structures, which seems to be in much better accordance with the real structure of biological organisms and ecological systems. In this book, we have chosen to split the description of the ontogenesis in two parts: "The ontogenesis of the cell" and "the supra cellular ontogenesis".

This chapter is restricted to only describe the ontogenesis of the cell and review much of the existing data on this matter.

What we think the cell is

If the information of biological systems has to be understood, it is obvious to study how the complexity of the isolated biological system has been developed. This is the case for the development of all levels of the biological systems as e.g. the organization of the cell structure including organelles, Golgi apparatus, endoplasmatic reticulum (ER) etc., as well as the supra cellular structures (described in the next paper of this series). This process, creation of structure and organization, is called the morphogenesis; or a more comprehensive phrase, ontogenesis – the formation of the being.

Figure 1A and B shows a generalized representation of the morphogenesis as complex bifurcation, where the different levels of organization are specified as M1, M2, etc. An example of such structure in nature is a green Romanesco cauliflower (see an example of this on: http://www.fourmilab.ch/images/Romanesco/)

We know very little about the mechanisms behind formation of shape, from molecular organization to cell-organelles, and further to the superior levels. Therefore, we assume that the same principle stands behind the formation of shape on all biological levels. This counts for organization of organelles in cells, and cell organization to tissue, and this to the superior shape of organisms.

The English philosopher, William of Occam (app. 1285-1349) wrote, "when a phenomenon – for example, biological formation of shape – has to be explained, you have to use an absolutely minimum of assumptions", (Occam's razor) (2,3). Therefore the cell is involved in organization of the ontogenesis of all levels. In this chapter however we specifically describes the cellular ontogenesis and give a theory for the cell.

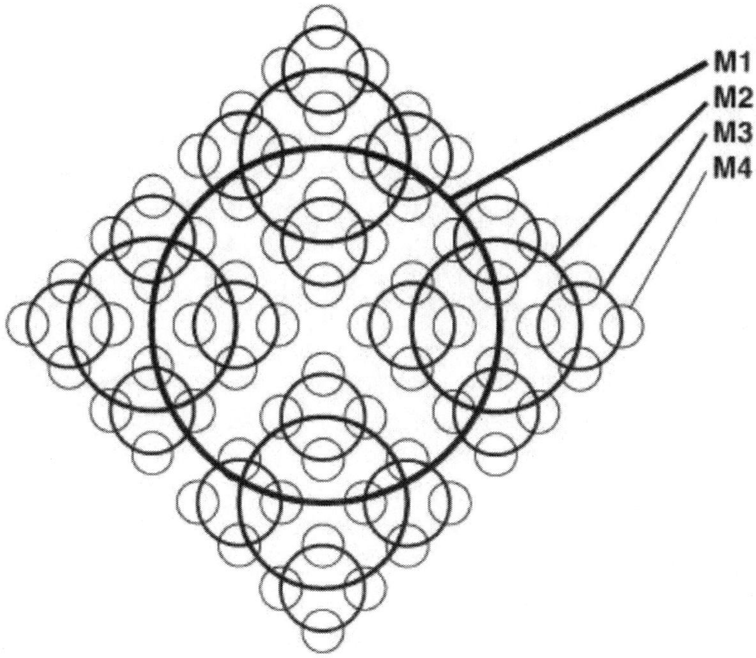

A

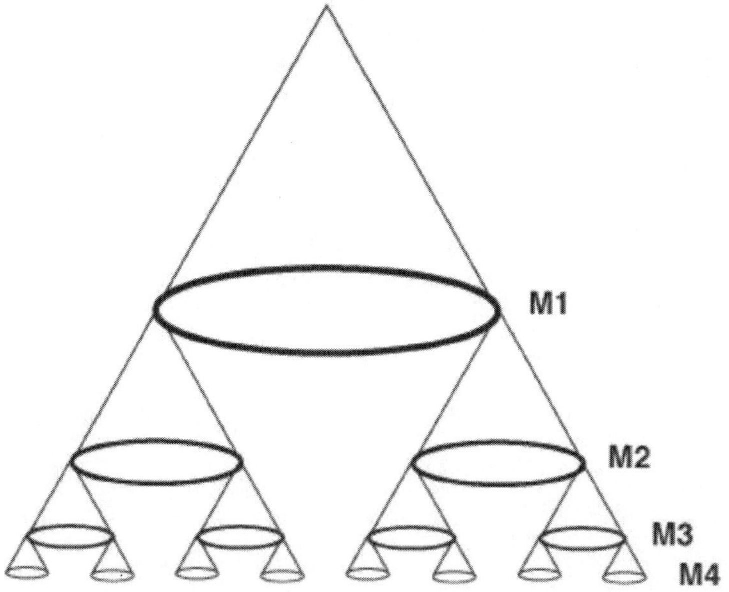

B

Figure 1. A,B. This figure gives a generalized representation of the morphogenesis as complex bifurcation: First, the zygote is created. This divides – eventually only by division of the core as with the insects – to form a heap of cells. This again divides (as fore instance with *Hydra*) and when the new group of cells are suitable enough (variegating between different species) it performs its first organization corresponding to the superior construction of the organism (establishment of the germinal layer). After this, the embryo begins to grow, and the germinal layers are divided into organ systems. These again, are divided into organs that again are divided into theirs structures on many levels. This goes on until all details of the organism are specified. All these levels of organization are specified in A and B as M1, M2, etc. http://www.fourmilab.ch/images/Romanesco/).

The ontogenesis of the cell

The mechanisms behind organization and morphogenesis are completely un-comprehended. The organization of cells takes place at different levels from molecules to organelles that again organizes to form the full cell. A good question is what kind of powers the cell is organizing on these sub-cellular levels. Based on the existing knowledge concerning DNA as storage of information, it is obvious to imagine that DNA-information is responsible for the organization of the cell. The transfer of information from DNA to proteins is carried out through a straight pathway, from the DNA through the RNA to proteins, where the RNA translates the DNA-code to amino acids that again are gathered to form the proteins of the cell-cytoplasm (1). Thus, the DNA is merely able to transfer the information for the shape of the proteins, not to transfer the information necessary to organize the proteins in the cell. Consequently it might be that the organizing mechanism comes from an unknown source. The mechanisms behind cell (as well supra-cellular) organization and morphogenesis thereby seem to be very difficult to describe in terms of a conventional (particular) molecular biological frame of reference. It is very interesting that while the details of biological mechanisms as the replication, transcription, and translation are very well described in molecular biological terms, the regulation of the eukaryotic gene expression – logically, a step in the morphogenesis – is completely un-comprehended even though it has been thoroughly investigated through decades. Therefore, it seems that principles not yet known has to be discovered (this was also the position of Schrødinger)(4).

The affinity of proteins is directed against themselves and each other, giving a kind of simple self-organization as for instance the virus-capsules (1). Here, it is tempting to suppose that simple self-organization can explain the whole cell and its behaviour. However, organelles that are 100-1,000 protein diameters in diameter do not show well-ordered, repetitive structures in the size of 10 protein diameters (common micrographs results). This is for example seen concerning the difference in mitochondria shape that depends on the type of cells (1), which not support the simple Self-organization idea. The viscous structure that would be caused by this kind of organization is conflicting with the great mobility that is characteristic for the organelles. Below, we will discuss a number of concrete phenomena, not explainable by conventional knowledge, but able to enlighten the principles of the cell-organization explained by our model.

The shape of organelles is not formed by self-organization of proteins

The formation of virus capsules shape takes place through self-organization of proteins (1). A similar process would be expected in our cells, but this is not the case. Even the polymerization of the simplest structures of the organelles, as actin and microtubules, seems to be subjects of superior and exact control from the cells (1). For example, cells are able to send out micro-spikes, a kind of feelers, and draw them back again, if needed.

Generally speaking, cell-organelles are shaped precisely after the needs of the specific type of cell. Examples of this are the endoplasmatic reticule, the Golgi apparatus, the vesicle – cell membrane systems, and mitochondrion, that takes their shape after the specific type of cell (1). This means that the shapes of organelles are often specified a lot more carefully then necessary in condition to the function it is mend to have. Seen in superior perspective, the mechanisms behind designing of organelle shape, mutual organelle movement and precisely mutually placement between the organelles (as for example vesicular transport to the right sack in the Golgi apparatus) are completely unknown.

At the Tetrahymenae, the cell structure is not decided by the proteins they consist of; this could be a general phenomenon in the world of the eukaryotes: "To a first approximation ... Tetrahyminae pyriformis cells and T. vorax microstomes maintain the same morphology with different proteins, whereas T. vorax microstomes and T. vorax macrostomes maintain different morphologies with the same proteins" (5). Also, the genetic variegating Tetrahyminae, often have the same shape. This means that proteins have not a superior meaning for the formation of shape in this. Therefore, it is hardly the proteins that give the protozoans their manifold and magnificent forms (6).

The cells of a body are almost alike concerning population of proteins

Regardless of which two of a bodies approximately 300 functionally different cells that are chosen, they will have about 98% of the most ordinary proteins in common in concentrations variegating less then a factor 5, and only 2% or less of the proteins will variegate more in concentrations, or, only be expressed in one of the cells (1). This indicates that the molecular differences between two randomly chosen cells are not qualitative, but quantitative. So, one single specific cell may be characterized on the expression of lots of genes. With this, we think it does not seem plausible that cellular formation of shape should be a function of the relations of the protein mix. Consequently, we have no molecular explanation of the cells morphology.

The foundation of cell movement and axon- and dendrite-excrescence

We assume that a superior control of the cytoskeleton through positional information causes cell movement and axon- and dendrite-excrescence. In spite of great steps forward in the research of the cytoskeleton, the mechanisms behind movements as phagocytosis, mitoses, saltatory movements of mitochondrion and vesicles, and migration of cells, are still not very well known (1) and cell movements is one of the most challenging problems for the modern molecular biology today (1).

Cell movement, as for example in connection with fibroblasts is extremely complex, because the formation and degradation of a great amount of different structures, happens at the same time. It is coordinated across a huge area of the cell, during the formation of a pseudopodium – a kind of foot drawing a cell forward. Likewise, the growth conus moves itself ahead, corresponding to the pseudopodium, drawing the axon instead of the cell after itself (7).

Following phenomenon indicates a connection between cell migration and axon excrescence. When a cell in vitro proliferates, the two daughter cells can often be seen as a mirror image of each other though longer time (1) (see the 1983 edition, from the introduction Figure b, page XXXVii). When the axon proliferates through growth in vitro, the same kind of symmetrically mirror image can be observed (7). This indicates that cell movement and axon excrescence could be influenced by superior inside powers of organization able directly to steer the cytoskeleton organization of the cell. This could also be the case for the dendrites, since different nerve cell dendrites of vertebrates in vitro are observed to outgrow on a way that is typically for this type of cells (7). Since we know nothing about the mechanisms behind the distribution of information to the cytoskeleton, we chose to define this phenomenon as being mediated through positional information on cell level.

Discussion

The same molecules generally build different cells and their different structures. The different ways of organization may be due to different information. Containing 20% of protein, the cytoplasm may be understood as a liquid viscous crystalline phase, and the cytoskeleton that takes care of all movement and transport of vesicles in the cell may be considered as a more solid part of this phase. Self-organization that is known from the most simple molecular organization – for instance virus capsules –, may be assumed to be a principle that also can be seen in the cell, but self-organization in a liquid crystal phase, has to be a lot more complicated then in simple, solid systems. Apparently it is obvious that supply of information is necessary to avoid that the superior level sinks into a phase of chaos; try to imagine the pattern behind cell movement. For example is the well organised walk of a fibroblast an inconceivable well-organised incident; the cytoskeleton in maybe thousand of different places have to react chemically in ways that are different from the close neighbouring areas. All this happens extremely coordinated. To get a picture of the capability of the cell in this respect, just think about the patterns behind the mitosis or behind the internal organization of the skeletal mussel cell.

The information necessary for this organization can be outlined as a pattern imprinted to the biological system, and as we, as mentioned above, do not know any physical laws that are able to generate such complex patterns in a reproducible way, we may assume that an undiscovered nature phenomenon exists behind the transmission of information.

Cells involvement in the morphogenesis

Information-transmitting interactions gives information directed self-organization in biological systems. When the information-transmitting interactions have given the information for the organization to the cells that thereby can organize themselves through complex behaviour – as fore example cell movement and advanced reorganization of the cytoskeleton to desmosomes – these cells can through new information be organised at a higher level. By this, the information directed self-organization that was able to explain the cell ontogenesis, also without problems is able to explain the organization of the higher levels (8). As it was the case for the molecular conditions compared to the organelles, it is the case for the cells that, despite their organization apparently quite often are disordered with a 5-10 cell diameters level, the organization always comes through to a higher level typically containing about 100 cell diameters (compare e.g. the liver lobulus)(9,10), when needed. However, the organization is never tighter then needed.

The mechanistic nature of the information system seems to be an undiscovered dimension of physical energy, a patterned aspect of it so to speak, that has not yet been discovered in physics as it only manifest itself clearly in biological systems. You could call it an information carrying global quantum field if you like. No doubt it must be continuous with the molecule orbitals of the proteins of the cells to regulate the activity of the proteins as we see it in cell walk as discussed. The idea of cytoplasm as a liquid crystal has been suggested before (11,12).

Conclusions

We have five concluding remarks:

1. The regulation of eukaryotic gene expression is not understood at all, which is why we think it is necessary to look for another explanation than the current. We do not think that the processing of DNA through RNA to proteins is able to transfer the information necessary to organize the proteins in the cell, but only to transfer the information necessary to form the shape of the proteins, which is why we think the organizing mechanism comes from a former unknown source, but is now able to be explained by our holistic way of thinking.
2. The mechanisms behind designing of organelle shape, mutual organelle movement and precisely mutually placement between the organelles are completely unknown. It has not been proofed that proteins have a superior function for these mechanisms. We think that an information-carrying field recruited by the cells is responsible for building the organelles and other cellular organizations.
3. When the axon proliferates through growth in vitro a symmetrically mirror image can be observed indicating that cell movement and axon excrescence could be influenced by superior inside powers of organization able to steer the cytoskeleton organization of the cell. This could also be the case for the dendrites. We think that a superior control of the cytoskeleton through positional information causes cell movement and axon- and dendrite-excrescence of cells. To explain this protein activity we chose to describe the cytoplasm as a liquid crystal phase.
4. In general, different cells and cell-structures are made by the same molecules. The different ways of organization may therefore be due to varying information. We think the information necessary for this organization can be outlined as a pattern imprinted to the biological system. Therefore, we assume that a former undiscovered nature phenomenon exists behind the transmission of information for the cell structure. We call our explanation the holistic paradigm.
5. Information-transmitting interactions give the information for the organization to the cells so these can organize themselves through complex behaviour. Through new information these cells can be organised at a higher level. We think the information directed self-organization in cells is able to explain the cell ontogenesis.

References

[1] Alberts B, Johnson A, Levis J, Raff M, Roberts K, Walker P. Molecular biology of the cell. London: Garland Science, Taylor Francis Group, 1983.

[2] Thorburn WM. Occam's razor. Mind 1915;24:287-8.

[3] Thorburn WM. The myth of Occam's razor. Mind 1918;27:345-53.

[4] Schrödinger E. What is Life? Cambridge: Cambridge University Press, 1958.

[5] Buhse HEJr, Williams NE. A comparison of cortical proteins in Tetrahymena vorax microstomes and macrostomes. J Protozool 1982;29(2):222-6.

[6] Lee JJ, Hutner SH, Bovee EC. Illustrated guide to the protozoa. Lawrence, KS: Allen Press, 1985.

[7] Purves D, Lichtman JW. Principles of neural development. Sunderland, MA: Sinauer, 1985.

[8] Ventegodt S, Hermansen TD, Flensborg-Madsen T, Nielsen M, Merrick J. Human development VI: Supracellular morphogenesis. The origin of biological and cellular order. ScientificWorldJournal 2006;6:1424-33.

[9] Cormack DH. Ham's histology. Philadelphia, PA: JB Lippincott, 1987.

[10] Hammersen F. Histology. Color atlas of microscopic anatomy. Baltimore, MD: Urban Schwarzenberg, 1985.

[11] Ventegodt S, Andersen NJ, Kromann M, Merrick J. Quality of life philosophy II: What is a human being? ScientificWorldJournal 2003;3:1176-5.

[12] Ventegodt S, Andersen NJ, Merrick J. Quality of life philosophy III. Towards a new biology: Understanding the biological connection between quality of life, disease, and healing. ScientificWorldJournal 2003;3:1186-98.

Chapter V

Can biochemistry explain morphogenesis?

Today's biomedicine builds on the conviction that biochemistry can explain the creation of the body, its anatomy and physiology. Unfortunately there are still deep mysteries strangely "fighting back" when we try to define and understand the organism and its creation in the ontogenesis as emerging from biochemistry. In analysing this from a theoretical perspective using a mathematical model focusing on the noise in complex chemical systems we argue that evolving biological structure cannot in principle be a product of chemistry. In this paper we go through the chemical gradient model and argue that this is not able to explain the ontogenesis. We discuss the used gradients as information carriers in chemical self-organizing systems and argue that by use of the "Turing structures" we are only able to model the most simple biological systems. The bio-chemical model is only able to model simple organization but not to explain the complexity of biological phenomena. We conclude that we seemingly have presented a formal proof (a NO-GO theorem) that the self-organizing chemical systems that are using chemical gradients are not able to explain complex biological matters as the ontogenesis. We need a fundamentally new, information-carrying principle to understand biological information and biological order.

Introduction

A way to understanding the cell is looking at what the cell definitely is not. One popular metaphor in understanding the nature of the living cell has been the cybernetic "controlling machine", which besides just simply producing proteins and when needed exposing these on the surface of the cell membrane i.e. to express its membership of a specific type of tissue, it also have the unique capability of organizing the information for the ontogenesis of the tissue into complete functional structures through genetic control.

In this chapter, we argue that the simple gradient model proposed for morphogenesis of protostome animal lines (lower animals that develop mouth before anus) does not find its parallel with the deuterostome lines (Animals that develop anus first); chemical gradients do not explain even the simplest biological functions in vertebrates. The theoretical reason for this is quite obvious: 1) chemical gradients do not work as information donors in self-organizing biological systems, as the law of diffusion causes a distinct lack of precision through this process, making the iteration needed for development of complex animals impossible. Hence, using the theoretically possible reaction-diffusion structures (modelling of signal molecules distribution in time-space network) as basis for morphogenesis does not work for the same reason; 2) using this model on embryos does not reveal the predicted disturbance of the development; 3) the developmental stages of, for example, the giant tadpole, seem to exclude the possibility of diffusion (chemical signalling) as informational backbone in developing systems. Therefore, chemical gradients and signalling molecules in general fail to explain complex biological functions. In this paper, we discuss the involvement of the cell in the ontogenetic process and the failure of the chemical gradient model to explain the ontogenesis.

Gradients fail to explain the ontogenesis

Gradients as information promoter of self-governmental systems: As apparent, hormones can be used by biological systems to determine sizes of head and body of the cnidarian *Hydra vulgaris* (1), and the adjustment of the amount of hormones determine the number of motor neurons through NGF (nerve growth factor) (2). But as implied by the problems described below, the information necessary for such organization hardly comes from a time-space network of signal molecules.

The information for the cell-determination in a developing embryo could be sourced from the genes that could distribute the information through a complex time-space network of signal molecules. Such systems are known from the cybernetics (3), but the reliability of these systems involves three main problems. 1) To avoid a chaotic outcome through time, all steps (specially the first steps) have to be extremely precise. But, unfortunately, the law of diffusion – that normally is used to describe the distribution of signalling molecules in space – causes a distinct lack of precision in such systems (4). 2) Using this model on embryos does not disturb the development. Not even after violent manipulations (4), as this model implies. Therefore, the dynamics of developing systems makes this model less reliable. 3). Tyroidea hormones generation of the metamorphosis in the giant tadpole (4), indicates that the morphogenetic hormones themselves, does not transform the information to form the shape. Instate, it seems like these, through positional information, are involved in a superior control of the manifestation of the genetic information. E.g., the final positional organization of a cell may be build by a sequence of positional informational items registered through time (4). But because of the huge distance in such biological systems (max 2 mm owing to the time) also such stages of development seem to exclude the possibility of diffusion.

The biological complexity arises from sequential recruitment of global organization called reaction-diffusion structures (or "Turing structures"). After the inclusion of Fick's diffusion-law the organization of the "Touring structures" became roomy and caused sensation (5). Therefore, it was proposed that such organization delivers potential information through morphogenesis. Since then, a lot of scientists have used variations of this equation to explain roomy organization in biological systems (For example: JD Murray, HG Othmer, GF Oster, A Hunding, B Goodwin, L Wolpert and H Meinhart). These researchers have also used enzyme reaction systems in modelling of biological systems. Such models could result in roomy organization. Also "Touring analogue" models have been developed based on active transport and interactions between molecules (6). Also the thermo dynamic equations have been used in such models, but without result (7). Through such modelling "Turing structures" have been developed to a subgroup of structures arising from self-organization of systems ahead from equilibrium. Very few of the last mentioned model types but none of the first mentioned have yet been discovered.

Understanding matter, life and consciousness – making No-Go theorems for life as chemical, self-organizing systems

That living organisms is organized in a fractal manner on a number of levels (7) makes it possible to test our assumptions concerning living organisms in a fractal manner. We will here, by use of No-Go theorem, try to show the formal impossibility of well-organized multi-cellular life based on the ordinary recognized molecular biological assumptions concerning living organisms as self organizing molecular systems. It seems that No-Go theorems can make it necessary to revise basic assumptions that normally are granted.

It is hard to understand the informational systems of the living, and it does not seam that the quantum mechanics can explain the conditions behind the existence of the fundamental substances in living systems. The disciplines of consciousness (cognitive, religious and philosophical disciplines) are not able to explain consciousness's ability to create wholeness, coherence and meaning in a chaotic detailed world.

Because our consciousness is developed from the living, it is obvious to consider, which kind of information that is able to create the superior functions of the consciousness. The conditions of information in biological systems are yet very poorly understood. Living multi celled organisms seem to exist by means of an information system interacting with individual cells and with the body as a whole, simultaneously. This information system seems to be fractal organized (8) with structure parallel to the fractal organization of living organisms. But until now, the traits that link the organization of organisms of all levels is not understood. This information system could very well be the foundation of consciousness. Classically this trait is described as the quality of the biological information system that makes it possible to establish superior connections as e.g. size relations, (despite the unstable or chaotic relations of the sub-levels). An example is the small tubules in the kidney, able to maintain the kidneys exact size despite a huge enlargement of the tissues necessary for formation of the kidney. This organization may be directed by "positional information" (see Figure 1) (9). Such organization most likely is controlled by roomy organized biological information. The classical expression for this has been "positional information" or "the biological (information) field" (4). It has been proposed that diffusion-dissipation-structures could deliver such roomy positions (7). But below we will show (even if it is theoretical accepted) that the roomy structure that could be organized by non-linear, chemical processes as the dynamic, fractal organised information system (including the superior regulatory functions of biological systems) could not in an appropriate way be described by this.

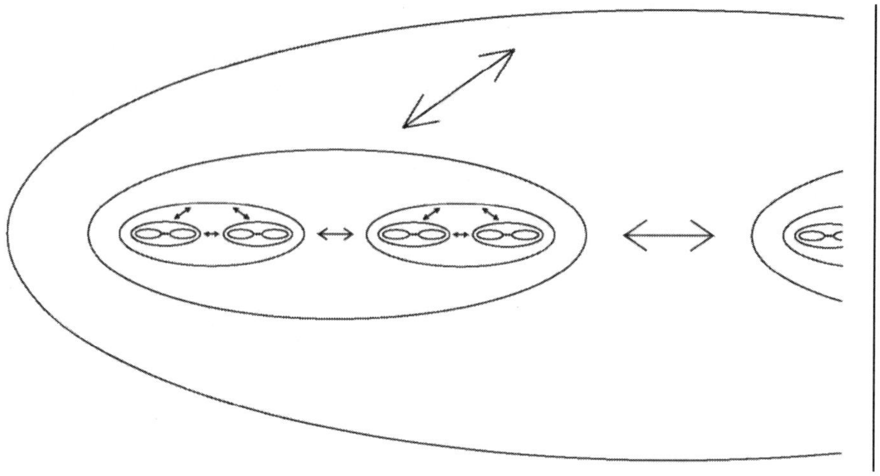

Figure 1. The fractal organization of the living with parts that interacts and by this forms a number of levels. The arrows illustrate information transmitting interactions and are generalized of the empiric found. This illustrates how biologic systems acts as a fractal kind of Chinese box containing the information transmitting interactions between the different parts of the same level.

A No-Go theorem for the living as self-organizing, molecular systems

Here, we will try to show that chemical self-organizing systems are not capable of giving the observed precision of structures of the superior biological levels. In the fractal organized system (see Figure 1) the parts interacts and thereby gives a wholeness on a number of levels, in total corresponding to a complete individual multi celled organism. We will try to show that a level never can be more precise then the under-laying levels. When the last mentioned levels fluctuate in time, we think it gives even more inaccuracy. However, No-Go theorem only counts for static situation. A No-Go theorem for the (chaotic-) dynamic balance is under construction. Because the information is coded by the DNA the signalling agents may carry the roomy organization. Such information is only able to escape the DNA through RNA transcription and translation into enzymes, and the enzyme activity, etc. Therefore, we are able to project the living into a multi dimensional room of chemical gradients as shown on Figure 2. The living is represented as a lane (trajectory) in this room. The lane may never cross itself, because this would involve a collapse of a representation necessary for creating and keeping up the organism.

The known physics and chemistry involves that all interactions happens locally. The multi-dimensional gradient room therefore, may have a metric that makes it possible to maintain global order, because local interactions alone would involve to big an uncertainty and huge distances on the superior levels (see Figure 3). Such metric however, will in itself block for the dynamics existing in a living organism, because a specific structure with a specific size (e.g. kidney-tubules or a human leg of a specific length), is necessary to make a construction based on the under-laying metric. The under-laying level therefore, may not fluctuate more than the superior level.

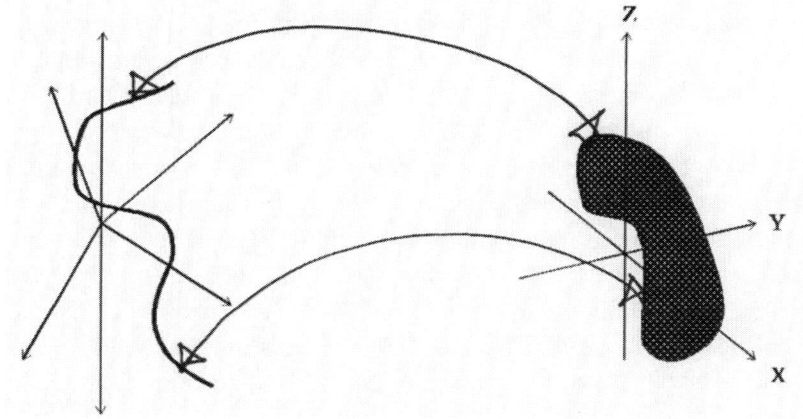

Figure 2. The living organism is projected to a multi-dimensional room of chemical gradients.

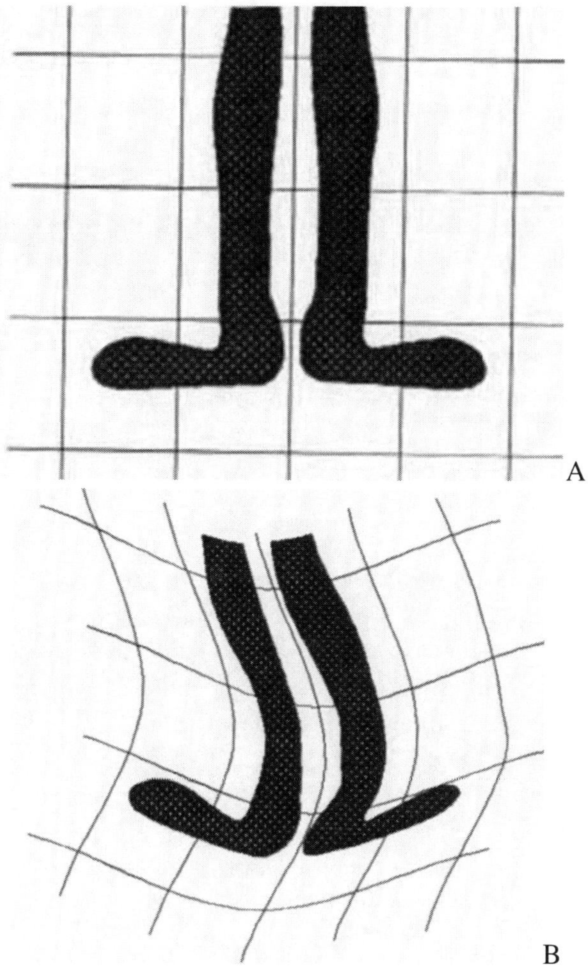

Figure 3. AB. Without firm metric (right) the organisms form and other high-level- qualities, will not be able to be established or maintained in a system only based on local interactions.

Conclusively, a matrix in the multi-dimensional room of chemical gradients having the necessary firmness will change the system dynamic in a degree struggling against the biological empery. A looser metric will allow the observed dynamic, but will not be able to function as "bag-bone" for exact organization on higher levels and over huge distances. An example of the consequence of this is that a humans leg length of about one meter, is very exact within a few per mille on both legs of the same person. This demands a metric that needs a development of legs of the same length with all people, and this is not the case. Thereby, it seems to be excluded (through static terms) that the known laws of the molecular biology can explain the biological informational conditions. This means that the biological information systems could be based on constant changes of until now unknown kind, or an unknown physical-chemical law exists, including organizing qualities, as the living has been able to use.

NO-GO theorem for phenomenological observed precision in biological systems at macroscopic scales

Examples are given of large-scale precision in biological systems. Following example is a transcription of Bennett et al (10): it is inexplicable given that the known fluctuations in the number of underlying fractal structures exceed the tolerances to which macroscopic precision is observed. As a concrete example, we have in mind the precision (5 o/oo) to which the two legs of an average human being have the same length.

The assumption that this macroscopic precision is inexplicable is based on a model for biological organization using the inverse of a mapping *f* from the 3-dimensional geometrical space R^3 in which the organism is embedded to a very high dimensional space *K* spanned by the concentrations of all the chemical components of the organism.

If the mapping $f: R^3 \rightarrow K$ leads to a "blueprint" (residing in the space *K*) from which a biological organism can be constructed from R^3, the mapping *f* must certainly be injective, else the mapping *f* would not be invertible. Not being injective would create an ambiguity in position of R^3 when attempting to use the "blueprint" to create the organism in R^3. A potentially interesting speculation is that such ambiguities could be an important factor in certain pathological conditions; for example, an ambiguity as where a biological structure starts and where it stops could lead to uncontrolled replication.

However, more than the injectivity of the map *f* is required. In order for *f* to have an inverse; there may be a way of providing a restriction in the high dimensional space *K* so it coincides with the image of the mapping *f*. A possible restriction that could be used comes from the reasonable assumption that the observed stability of biological organization necessarily must be supported by complicated entropic biochemical processes (necessary, for example, to avoid "demise by diffusion"). Such complicated biochemical processes may be expected to place severe constraints on relationships between concentrations of chemical components. Stable processes able to resist small violations of such constraints must be used in biological processes [for example, such stability is seen in "buffered" systems; maybe we really will be able to show the relevance of our multiple point principle in this biological context as sketched in our project descriptions; maybe a more fashionable way to talk about this is to say "buffering by commodity conservation as a way of enforcing the presence of two or more phases of first order, may define a manifold embedding of *K*. Such a manifold would correspond to an attractor dictated by stable biochemical processes necessary for the maintenance of biological organization.

Let *M* designate the manifold (assumed differentiable) embedded in *K* that corresponds to this attractor necessary to insure the invertibility of the map *f*. Because our biological "blueprint" is forced to reside on the manifold *M* (i.e., the limit cycle), we can adopt it as a metric on *M* that induces the tangent spaces of *M* by the directional derivatives along various chemical components directions. Thus endowed with a metric, the surface *M* becomes a geometric surface and one might then hope that the inverse mapping $f^{-1} : K|_M \rightarrow R^3$ could be used as a "blueprint" from which the biological system in R^3 could be constructed.

Consider now the trajectories $R(t)$and $L(t)$in R^3 along a right and left leg that transform into each other through symmetrical operations in R^3. It is further assumed that these trajectories start and end at macroscopically corresponding points on the two legs in accordance with having two legs of the same length. For example, we can think of trajectories $R(t)$and $L(t)$that transform under reflections as $\hat{A}R(t) = L(t)$ (t is a parameter; $\hat{A}$ is the lateral reflection operator in the mid plane of the human body). We would expect that the trajectories $R(t)$and $L(t)$would be mapped by *f* into the parallel trajectories $f(R(t))$ and $f(L(t))$ in *M*. This would insure that integration of the metric (induced in *M*) along the trajectories in *M* would be equal (in accord with the assumption of (symmetric) trajectories in R^3 along left and right legs of the same length). The reason for expecting this equality:

$$\int_{f(R(t))} \sqrt{g}\, ds = \int_{f(L(t))} \sqrt{g}\, ds \qquad (1)$$

is that the integrals along trajectories $f(R(t))$ and $f(L(t))$ that are "close together" in *M* "feel" the essentially same variation in the metric.

Now the following question must be addressed: given the magnitude of known fluctuations in the numbers of self-similar structures (e.g., cells) that make up a leg, we must ask if our candidate for a biological "blueprint" residing in *M* can accommodate these fluctuations and still function as a map in R^3. Such map can lead to a right and a left leg with same length within a tolerance known to be less than what would correspond to the fluctuations in the number of underlying cellular building blocks. The essence of the NO-GO theorem is that it is not possible without "fine-tuning" of additional assumptions.

The idea behind the "proof" is roughly as follows. The trajectories $f(R(t))$ and $f(L(t))$ in *M* (corresponding to the trajectories $R(t)$ and $L(t)$ in R^3 described above) would be expected to be spirals. The periodicity comes from the fractal-like (cellular) nature repeatedly encountered on a trajectory (in R^3) of biological material. The number of cells encountered along the trajectory may be topologically encoded in *M* as the winding number of the spiral in the trajectory. Now, the naturally occurring fluctuations in the number of cells encountered in the trajectory along the right and left leg would correspond to different winding numbers of the spirals for the two trajectories in *M*. However we know that the two trajectories $f(R(t))$ and $f(L(t))$ may be parallel. The essence of our NO-GO is the impossibility of having two parallel spirals (in *M*) with different winding numbers: the known relative fluctuation in the number of underlying fractal structures, making up a macroscopic organism, leads to different winding numbers for the trajectories $f(R(t))$and $f(L(t))$ thereby precluding that these can be parallel in M as required by (1). This mismatch in our biological "blueprint" prevents the construction in R^3 of a right and left leg that have the same length.

Discussion

Simple gradient models do not seem to be able to explain the ontogenesis (2). With the existing knowledge about DNA and genes, it is rational to think that information in biological systems is directed from the genomes, through a chemical arrangement. Transport of information in such systems has been explained through gradient models, trying to state how the information is conveyed from genes to shape – some examples is the axon excrescence through a gradient of NGF, AND morphogenesis through a gradient of sticky-molecules as N-CAM and NG-CAM (11). In the simplest versions – as the "French flag model" (12,13) – a gradient of a signal molecule ("morphogene") is formed, through a "source" and a "drain". Cells in between these could be imagined to have concentration sensitive receptors with thresholds that can only be reached in some areas of the system, but this is still hypothetical. The idea is that activation of the receptor should control the specific gene expression. But even the simplest forms of the gradient model (2) have shown to be un-functional without the assumption that the threshold variegates locally. By assuming this, the cells themselves could have the information to create the complex biological patterns. The gradient model therefore, is able to explain simple biological functions, but not capable of explaining complex matters as the morphogenesis by itself.

The existing knowledge concerning protein processing makes it logical to think that information in biological systems is distributed from the genes. We do not think this could explain the information for the ontogenesis alone. For example the information responsible for the organization of agents deciding the sizes of body and head of *H. vulgaris* or the decision of the amount of motor neurons through NGF, are hardly directed from a time-space network of signal molecules. Therefore it does not seem reasonable that simple gradient models should be able to explain the complete information pattern necessary to organize the ontogenesis. Even the simplest forms of the gradient model have shown to be un-functional. However, by assuming that the thresholds vary locally the cells could get the information to create the complex biological patterns by themselves. This, on the other hand, is only an assumption. Even when we use the "Turing structures", we are only able to make models of the simplest biological systems. Therefore, this model is not able to explain the complexity of biological phenomenons. This means that the chemical gradient model is able to explain simple biological functions. But by itself it is not able to explain complex matters as the morphogenesis.

The idea that sequential recruitment of global organizations is able to activate the morphogenesis is a good idea, but the question is if the reaction-diffusion systems really are able to create the necessary "pre-patterns". The "Touring structures" have since the middle 1970s been criticised, because of problems like: 1) Experimentally, no one have succeeded in proving these structures as a roomy organization of a three dimensional system, 2) Theory and praxis have shown that it is impossible to model complex, reproducible, reaction – diffusion structures (Axel Hunding, personally comm.). If a structure has to be defined as roomy, using a solution (length of structure/wavelength) of only 5 x 5 x 5, the reproducibility will vanish completely. The border for creation of roomy organization seems to be divided into about 10 compartments (Axel Hunding, personally comm.). This means, that we by use of the "Turing structures", only are able to modelling the most simple biological systems. Therefore, this model is only able to model simple organization as for instance, in the fruit fly *Drosophila*, egg-polarization, gastrulation, and compartmentalization of the eggs, but it is not able to explain the complexity of biological phenomenons.

The point of view of contemporary molecular biology is that information to the organization of living beings is tied to genetic activity (4). But the paradox of the C-value (C = complexity) (2,14), and the macroevolution (15) makes this less likely. According to the C-value paradox, the beings of high evolutionary levels are more complex, but, on the other hand, they have not much more DNA, than simple (including single celled) organisms (4). The macroevolution – the evolution of living beings shape – seems to be independent of the microevolution – the evolution of molecular structures. For example, the DNA of human beings and chimpanzees are 99% identical. This is the same difference as seen between twin-species of mammalians and fruit flies respectively (4). Based on this, the NO-GO theorem makes it possible that something else stands behind the biological organization.

The presented NO-GO theorem makes it likely that something else then biochemistry is behind the biological information capable of organizing the morphogenesis.

References

[1] Meinhardt H. Models of biological pattern formation. London: Academic Press, 1982.
[2] Purves D, Lichtman JW. Principles of neural development. Sunderland, MA: Sinauer, 1985.
[3] Ashby WR. An introduction to cybernetics. London: Chapman Hall, 1956.
[4] Alberts B, Johnson A, Levis J, Raff M, Roberts K, Walker P. Molecular biology of the cell. Fourth edition. New York: Garland, 2002.
[5] Turing AM. The chemical basis of morphogenesis. Phil Trans Roy Soc B 1952;237:37-72.
[6] Babloyantz A. Self-organization phenomena resulting from cell-cell-contact. J Theor Biol 1977;68: 551-61.
[7] Prigogine I, Stengers I. [Den nye pagt mellem mennesket og universet]. Copenhagen: Forlaget ASK, 1985. [Danish]
[8] Ventegodt S, Hermansen TD, Flensborg-Madsen T, Rald E, Nielsen ML, Merrick J. Human development VII: A spiral fractal model of fine structure of physical energy could explain central aspects of biological information, biological organization and biological creativity. ScientificWorldJournal 2006;6:1434-40.
[9] Frankhauser G. The role of nucleus and cytoplasm. In: Willier BH, Weiss PA, Hamburger V, eds. Analysis of development. Philidelphia, PA: Saunders, 1955:126-50.
[10] Bennett D, Ventegodt S, Nielsen HB. NO-GO theorem for phenomenologically observed precision in biological systems at manroscopic scale. Copenhagen: NBI preprint, 1994.
[11] Edelman G.M. Cell adhesion molecules in the regulation of animal form and tissue pattern. Ann Rev Cell Biol 1986;2:81-116.
[12] Wolpert L. The French flag problem: A contribution to the discussion on pattern formation and regulation. In: Waddington CH, ed. Towards a theoretical biology. Edinburgh: University Press, 1968:125-33.
[13] Wolpert L. Pattern formation in biological development. Sci Am 1978;239(4):154-64.

[14] Greilhuber J, Dolezel J, Lysak MA, Bennett MD. The origin, evolution and proposed stabilization of the terms 'genome size' and 'C-value' to describe nuclear DNA contents. Ann Bot (Lond) 2005;95(1): 255-60.
[15] Kutschera U, Niklas KJ. The modern theory of biological evolution: an expanded synthesis. Naturwissenschaften 2004;91(6):255-76.

Chapter VI

The origin of biological and cellular order

The uninterrupted morphogenesis shows the informational potentials of the biological organisms. The experimentally disturbed morphogenesis shows the compensational dynamics of the biological informational system, which is the rich informational redundancy. In this paper we use these data to describe the morphogenesis in terms of the development of supra-cellular levels of the organism and we define "complex epi-genesis" and supra-cellular differentiation.

We review the phenomena of regeneration and induction of *Hydra* and amphibians, and the higher animal's informational needs for developing their complex nervous systems. We argue – also building on the NO-GO theorem for ontogenesis as chemistry - that the traditional chemical explanations of high-level informational events in ontogenesis like transmutation, regeneration, and induction are insufficient. We analyze the informational dynamics of three embryonic compensatory reactions to different types of disturbances: 1) Transmutations of the imaginal discs of insects. 2) Regeneration after removal of embryonic tissue. 3) Embryonic induction, where two tissues that normally are separated experimentally are made to influence each other. We describe morphogenesis as a complex bifurcation, and the resulting many morphological levels of the organism as organized in a fractal manner and supported by positional information. We suggest that some kind of real non-chemical phenomena must be taking the form in living organisms as an information-carrying dynamic fractal field, causing morhogenesis and supporting the organism's morphology through time. We argue that only such a phenomena that provides information-directed self-organisation to the organism is able to explain the observed dynamic distribution of biological information through morphogenesis and the organism's ability to rejuvenation and healing.

Introduction

The existing morphogenetic theory does not explain how the cells recruit the information to organise the different structures of the embryo. Caused by missing understanding of the cell, the levels of the organism and the cell determination and differentiation have not been explained. Our papers about the ontogenesis try to change this fact by giving a new theory for the cell (1) and explain the supra-cellular ontogenesis and the order in biological systems (in this chapter). We have discussed the fact that it has not been possible to explain the powers of the cells by use of chemical gradient models or the existing theory of DNA as donor of all cellular information in order to make the cells able to organise the ontogenesis (2). The development from a single undifferentiated cell, the zygote, to all the different types of cells creating complex tissues and organs like nerve, brain, muscle, eye, bone or nail has not yet been understood. The ability of the cells to communicate and orient themselves in the embryo and organism in order to enable to form and maintain the organs and the wholeness of the body has long been a complete mystery. We give our explanation of these astonishing biological enigmas elsewhere (1,2). We chose to describe the mechanism behind the organization of the biological system at all levels as *information directed self-organization*. This way, the living being can be described as an information directed self organizing, complex, dynamic system (complex as combine (put together), dynamic as motion by itself), in this series of papers often referred to as, the complex dynamic. We propose that cell determination and ontogenesis can be explained, if cells of all developmental levels use *information directed self-organization*. For example, the building of organelles, tissues and organs is organised by cell recruitment of complex information that is supplied to the organelles, tissues and organs though an information carrying field. For example, this can be illustrated as a developing flower where the information carrying fields develops different structures in an illustrative way (e.g. (3) page 38, Figures 22-34).

We think the existence of information in biological systems is realised through the high degree of organization in these. From the beginning, the curiosity of the human being has aimed to understand this biological enigma, for instance through an expression of a specific élan vital – a movement founded by the French philosopher Henri Bergson (1859-1941). As the fact that today these kinds of unwieldy ideas have been abandoned due to the idea that nature is ruled by "ordinary" and simple principles, and in spite of the discovery of the DNA, it has not yet been possible for the science to explain the mechanism behind the creation of living being's morphogenesis. The ontogenesis is often described by use of non-particular ideas as "morphological fields" (4) or "positional information" (4) that means, information given as a consequence of position in a developing biological system (3). Of course, these relief concepts do not explain anything, but express that the formation of the patterns in biological systems apparently involves the whole spaciousness of the biological system, in a complex and incomprehensible way. In this chapter, we discuss the supra cellular terms of the ontogenesis in the view of our hypothesis. In another paper (1) we described the ontogenesis of cellular development, but in this chapter we will discuss the supra cellular conditions of the ontogenesis.

What we think the cell is not

First, the cell is not just chemical machinery. The point of view of contemporary molecular biology is that information to the organization of living beings, as well at cell level as superior levels, is committed to the genes (4). But the paradox of the C-value (C = complexity) (5), the macroevolution (6) and the self-organization (discussed below) (4), makes this less likely. According to the C-value paradox, the beings of high evolutionary levels are more complex, but on the other hand, they have not much more DNA, than simple (including single celled) organisms (4). Humans only contain three hundred times as much DNA as large bacteria, and only one third compared to the Lilly (4). The macroevolution – the evolution of living being's shape – seems to be independent of the microevolution – the evolution of molecular structures. For example, the DNA of human beings and chimpanzees are 98% identical. This is the same difference as seen between twin-species of mammalians and fruit flies respectively (4). On the other hand, all kinds of vertebrates have the same fundamental building plane. For example, they have identical (bones), muscles, organization of the brain, and so on (7), only the form is different. This means that it is highly questionable that the structure and organization should be directly related to the molecular functions, and with this, the existence of another explanation is more realistic. The problems concerning the understanding of information in biological systems has also without much luck been tried to explain by use of different mathematical interpretations (8,9). But most scientists feel that there has not been a theory yet to explain the distribution of information in biological systems. The existing models do simply not explain the organization of the extreme complexity, stability and reproduction related to interacting cells. This also concerns for the chemical gradient models that fail to explain the development in biological organisms. These act on the idea that information in biological systems is directed through a chemical arrangement from the genes. Such models calculate how the information is transferred from genes to shape. These models are subject of this chapter.

The ontogenesis of the supra cellular levels

Supra-cellular ontogenesis: Figure 1 in (1) shows a generalized representation of the morphogenesis as a complex bifurcation, where the different levels of organization are specified as M1, M2, etc. It is worth to notice, that the whole organism, in parallel with the gradual development of the complexity, is able to go through a radical reorganization of structure on all levels, as it is known from the metamorphosis of vertebrates (for instance amphibians). In tadpoles, thyroidea hormones trigger the metamorphosis. If the thyroidea gland is removed, the result is a giant tadpole. This also goes through a process of metamorphosis to form ordinary frogs after injection of hormone (3). In this way the systems, seemingly, are uninfluenced by the size. When the giant tadpoles are much longer then 2 mm, it is not likely that the morphological information is promoted through diffusion (for instance as a reaction-diffusion structure).

The phenomenon of compensation

Transmutations and positional information through the morphogenesis of insects

In a convincing way, transmutations (4) show that information can be found at tissue level, but not as expected at the cell level. The imaginable disk that normally develops to one of the fruit flies organs, for example a wing is able to spontaneously change its development into a leg or an antenna. Also known from the fruit fly is the special, homeotic or atavistic mutations that gives morphological disturbances resembling the transmutations. In the opinion of some scientists these mutations are interpreted as cell determination directed by the genes (4), but the mechanism behind this is not understood in molecular biological terms. What can be observed is that one complete structure replaces another. So it looks as if the whole package of positional information is exchanged (10). However, we believe that this interpretation is doubtful. The imaginal dishes can experimentally be shown to match with a morphogene field – an area of tissue that functions as a whole, concerning positional information (look below: the amphibian regeneration) (10).

Phenomenon of regeneration

Hydras regeneration – positional information in a simple 3-dimentional system: The little freshwater polyp that is the most simple of all cnidarians is assumed to be the first creature on earth with a nerve system (4). Its body wall contains three layers of tissue. The outer one is the epidermis with sense cells. The innermost layer is the gastrodermis with secretory and nutrient absorbing cells. In between these two layers is a layer (mesogloea) traversed by an uncomplicated network of simple nerve cells (4,9). This nerve tissue is able to send ring shaped impulses – in all directions along the nerve network – from the contact point (11). This makes Hydra capable of making complicated and un-understandable behavioral patterns. It is, for example, able to catch water flies, swim and make somersaults. If hydra is separated in two, both parts regenerate immediately (12). This means that *Hydra* can be compared to a morphogene field (see next section). A very good question is how a new-formed *Hydra* is able to control these patterns of movements? We think that this can only be explained by positional information.

The amphibian regulation of morphogenetic fields has been investigated by scientists (13,14). Parts of an amphibian embryo can be removed without any injury. If, for example an eye or another part is removed, the surrounding tissue compensates for this. This is called regulation. There is a limit for how large a part that can be removed, without injuring the embryo. Early in the scientific history a "morphogene field" was operationally defined as a part of the embryo inside which it is possible to make regulations.

Gradually as the embryo develops, it gets more and more fields that make up a minor and minor part of the whole. This is the case for all vertebrates, but with the mammalians, the amounts of embryological fields reach its maximum. In the foetus area, inside the "field of limbs", it can be removed, but in adults, it is only possible to remove tissue from the "field of the little finger tip". Gradually as the amounts of cells expand through the development, one area is divided into several (4). This corresponds to an ongoing delivery from the biological system, of information through positional information. In newts, limbs can grow out again through formation of a normal lower leg in a way that is directed by positional information. This is the case in spite of the fact that the upper leg is turned surgically before outgrowth (4). Regeneration and metamorphosis show that the morphological information in general is accessible, also after completion of the morphogenesis.

Phenomenon of induction: To induce means to influence with each other. A long row of experiments have been carried out, e.g. (15,16) and interpreted as if the tissues in developing systems interacts on an abstract way that can be described as an induction or redefinition of positional information. A very good example of this is a total lack of development of the nervous system that results in a blockage of the morphogenetic development of the embryo (16). Another classical example is the induction of the neural tissue of the eyes lens (16) that also gives a review over experiments that throw light on the information in biological systems. If the tissues are not positioned correctly compared to each other, the development stops, and the information can no longer be recruited from the biological system.

Early in this century it was shown that transplantation of a dorsal lip from one amphibian blastula to a ventral area of another one, could induce the formation of a smaller secondary embryo ventral in the last one. This was at first interpreted as if the information to the embryo was lying in this region. From this comes the expression, organizer region (14). But this interpretation had to be rejected, because it turned out that many stranger substances could have the same effects. Examples of this are other tissues – also boiled and fixed –, steroids, organic systems, and cotton with turpentine. Here the formation of sub-cellular patterns is induced (14), which can be described as a redefinition and duplication of the potential information of the system.

As described through all these examples, the distribution of information necessary for organizing the different levels of biological systems cannot be interpreted in conventional terms. In our model, the biological system can be interpreted as a Chinese box model of information transmitting interactions. Through cell morphogenesis the information of the biological system can manifest itself through positional information.

The morphogenesis of the nervous system

Morphogenesis of the brain

The morphogenesis of the nervous system is, because of the enormous amount of information this needs, without any doubt the most impressive achievement of biological nature. Obviously, it is difficult to measure the amount of information that has to be used for this performance. However, anyone that studies a map of the housefly's brain (17) may be excited and astonished by its complexity. This brain only takes up 1/3 mm^3 and weigh 0.4 mg, but contains 340,000 of the most complex and varying neurons of the animal kingdom connected in ways that are immensely precise and variegating, as "in all probability one cell body may give rise to several sets of arborisations, which functions as separate integrative units" (17). The details of the brain of a housefly, gives the impression of the existence of an extremely specific determined connectivity. And in accordance with this the housefly is able to do all neural functions from the beginning without learning. Today, the development of such a system can only be explained in conditions of transfer of information as positional information that is read by the cells and their axons and dendrites.

Contrary to insects and lower invertebrates, it is the case for lots of mammalians, especially humans, that the nervous system is developed further after birth – humans goes though a process called myelinisation at least until the child is two years old (18). The human has a considerable ability to learn, but we know from experiments with apes that much perception and behaviour are embedded in the biological information (15,19). It is not known if this information is manifested through the connectivity or if it is just transferred to a functionally level of the nervous system. But apparently the cortical reproduction of the body (in maps) is not "hard wired", but able to go through momentary re-organization after nerve-racking. This means that the representation exists as functionally patterns in the brain (15), in agreement with the last of the two alternatives.

Axon growth in vivo is directed through positional information

Excrescence of axons in vitro: As earlier mentioned, the axon excrescence in vitro is influenced by interior factors. However, frequently the axon excrescence in vitro is dominated by the influence from outside factors – as for example a sticky lane on the button of the Petri dish (4,14), or a gradient of NGF (4). This has lead to an intensive *in vivo* research, after "epi-genetic" factors – factors that are involved with the genetic regulation of developmental pathways, leading to the cell differentiation – through evolution.

Concerning excrescence of axons in vivo, special invertebrate neurons can be recognised between different individuals, and develop almost the same axons and dendrites (20,21). Also the exterior stimuli leading to the well-known grasshopper neurons, in which axons at first have an identical growth and since grow in different directions have been studied and concluded: "It is highly unlikely that the growth cones we have been studying are directed passively by mechanical guidance cues" (22).

In vertebrates, many examples are known concerning the precision of axon growth following complicated patterns not explainable by mechanical (or chemical) footprints. The retina of the gold fish is extremely precise projected to tectum (a part of the eye). This organization occurs through addition of rings of retina through the development, while the tectum develops through regional outgrowth. The impression of dynamics in this system is completely conformed by experiments. By these, a correct mapping of the retina on the tectum is re-created.

This was possible in spite of cutting of nerves, blockade of axon excrescence, and re-movement of tissue from retina and tectum etc (4). Through development of vertebrate limbs, the nerves can be observed to grow in complicated patterns that seem extremely determined. As for example, the pattern of nerves in wings is almost completely symmetrical (4).

It is also possible to show that if a pair of spinal cord segments is turned around in an early stage of the development of a chicken, then the motor neuron of a specific segment still finds the right muscle group to enervate (4).

If a spinal cord segment is turned during the development of an amphibian, axons that usually grow caudal, suddenly first grow in a rostra manner, but soon after it turns and continue its growth in the normal cordally way (14).

We conclude that it is not yet understand what mechanism – seemingly extremely dynamic – that leads axons through development. But generally it does not seem to be "epi-genetic factors". The excrescence of axons, seemingly, can only be described by positional information. This means that they get the information for the organization, dependent on their position in the developing organism.

Positional information as a source for development of tissue

Potential information of vertebrates' development of limbs was reviewed (4,14). Accordingly in humans the buds that develop to arms and legs are visible in the beginning of the 5th week. On these buds is a zone that polarizes the development of arms and legs, called ZAP (zone of activity polarization). Such zone also exists in amphibians where it is called an organizer zone. If this zone is transplanted from one wing bud of a chicken to another, too many boons of underarms, hands, and fingers will be developed. It has been shown that developmental buds of different tissues can have such effect. The results of these experiments can be described as re-definition of the positional information during the development of limbs.

In other experiments a block of cells known to develop a thigh from the leg bud, was transplanted to a wing bud, resulting in the development of a toe on the tip of the wing (4). The information for such development may be abstract because the transplanted cells already were polarized to a specific development of leg when they were transferred to the wing bud. Therefore, the cells may have polarized with the leg tissue before the transplantation, indicating that the information for the polarization of specific tissue is transferred to all cells included in a tissue simultaneously, a form of development called non-equivalent (4). By this, the information from the genes seems to be dependent on a cells location in the wholeness of a body. To understand all this it is necessary to emphasise that all cells before polarization seemingly, are identical, and if such (not yet polarized) cells are transplanted to another tissue, they will polarize with the new tissue and undergo a natural development with this tissue.

The information necessary for such development is following the model explained in this book understand as positional information, an expression also used by other scientists (14). Such development counts for all types of tissue including the nerve system. E.g. *Hydra's* nerve net does not have any superior structures, but anyway, a new-developed *Hydra* searches its food immediately. We think development of *Hydra's* nerve net may be sourced from the biological system through the same kind of positional information as described above, and as the nerve system has evolved through the evolution, it is reasonable to think that positional information plays a similar role in development of all nerve systems including complex nerve systems as the human brain.

Morphogenesis operates through "complex epi-genesis"

When the zygote has been formed, it proliferates to several un-polarized cells. Through polarization of these cells directed by positional information to all cells simultaneously (to the wholeness of these cells), the embryo is organized. As the embryo is small, the needed positional information is limited at this stage of development. Gradually as the embryo grows, more complex patterns are developed. This of cause increases the need for positional information, and as tissues are developed and new patterns form in the embryo, positional information is supplied to a more and more complex body structure with numerous levels of organization. Such development may function as a feed back mechanism where development of more complex tissue messages back to the source of positional information that more complex information is needed, after which the needed amount of information is supplied.

In investigations of the development of a multi celled organism, the timing of development of more and more complexity has been one of the biggest enigmas. Such development happens through a process that distributes the biological complexity in higher organisms to several levels, with the brain as the superior controlling organ and the whole body as the superior level. As the development of the embryo may be directed by feedback to the source of positional information when more complex information is needed (as described above), why then, does this development never pace backwards when all mathematical equations are reversible (except the thermo dynamic)? However, even though, we think development through the above described "self-organization" through a feed back system of positional information makes sense.

We call such self-organization directed by positional information of tissues, developed through a lover level (less complex) of self-organization also directed by positional information, for epi-genesis. The strength of "complex epi-genesis" in complex multi celled organisms is, that it allows development of more and more etc complex patterns formed through supply of more and more etc complex positional information through feed back control.

Interaction between positional information and biological matter may be sourced by power not yet noticed in the biological system (maybe comparable with e.g. the chaos theory) (23). The complexity of multi celled biological systems and their source of positional information, has been fine tuned through the evolution. We think that such development has made the evolution of the human being possible.

Information for biological complexity can be described as a system of Chinese boxes

Complex biological organization through positional information is very different from the molecular system description, as the information is supplied in a system of roomy patterns of several levels able to interact locally with all individual parts. In such system individual cells orient themselves in a position (needed for specific biological development) directed by the supplied positional information. If an embryo is disturbed in its development before polarization of the cells, this will not have any influence on the development, as the individual cell is not addicted to a specific type of tissue before the positional information for polarization is supplied. Positional information may be able to correct disturbances in the growth through supply of correct information to new (un-polarized) cells produced from stem cells. If failure in development is caused by abnormal growth after cell polarization, the positional information may be able to read and correct such development through information-transmitted interactions between the superior level (the whole body), and the inferior levels (tissues, organs, cells, etc).

Insects are simple organisms and positional information in those is only supplied at a single level, the wholeness. In *Hydra*, the regeneration indicates that the same positional information specifies the form and nerve system. Amphibians and more complex organisms are as described above organized in complex structures of several levels. Such organization can be compared to a system of Chinese boxes, but here with the wholeness as the largest box, and the next box is organs, and then cells, etc, and this organization is specified by the positional information. In our model, this corresponds to the existence of information-transmitting interactions between the different levels of the embryo. In our model of the embryo described above, this implies the existence of information-transmitting interactions between the different levels under development.

If we regard the development of complex organisms as "complex epi-genesis" using *information directed self-organisation*, it seems that we actually can understand it. We therefore conclude that the biological system can be described as a Chinese box system of interactions between various levels of positional information (1). Through the development of cells and the various cellular levels in a complex body, the potential information of a biological system is able to maintain itself through feedback from the developing body as also described in another paper (1).

Criticism of our ideas concerning ontogenesis

Professor Claus Emmeche (University of Copenhagen) commented our discussion of the embryonic development [Claus Emmeche 1989 Personal contact] and below we will discuss his criticism in the mirror of our hypothesis. He comments the idea that the same force coordinates all levels of organization in biological systems in an ordered fractal manner by describing a very busy day concerning Christmas shopping. He describes different levels of organization as: movements between people on the shopping street; movement of the arms holding out for different things; movement of the fingers holding out for the purse and the money in it. He thinks this hardly may involve that the coordinated ordered organization of a specific level should be dependent of the organization of the underlying level. We do not agree with this.

Firstly, the body, arms, fingers movements, as a riverbank where it is possible to move deeper and deeper into finer and finer structures, describes a kind of fractal levels. Seemingly, the patterns of the single levels could be chaotic compared to each other. But if we compare with the butterfly effect, where lots of levels are involved from the butterfly itself to huge hurricanes, seen in the light of Henry Feigenbaum's chaos theory, the result of the most disordered turbulence is in reality ordered, and the organization of the butterflies movements, are well coordinated with the turbulent movements of the hurricane – the hurricanes movements depends on the movements of the butterfly. So, it is hardly applicable to conclude the absence of coordinated order and organization in biological systems (any systems) from this. If the organization of the different levels seems chaotic compared to each other, it could very well be a consequence of comparing the levels in the wrong perspective.

He also criticises the assumption that an early transfer of information to the cyto-skeleton may take place. He thinks that the cyto-skeleton could be self-organized instead. In our analysis something global and "intelligent" has to deliver the information that mediates the dynamic organization of the cyto-skelleton for the cell, for it to move coordinated. We think the positional information needed is more than the chemistry of the cytoplasm can create itself.

We believe that the development demands a great amount of information; Claus Emmeche argues that this is a problematic view (Claus Emmeche 1989, personal contact). We think the problem may be how the information is delivered to the cells, more than how many bits of information it is possible to compress with a smart compressing algorithm.

Conclusions

Our model describes the biological system as a Chinese box model of information transmitting interactions. The whole organism in parallel with the gradual development of the complexity is able to go through a radical reorganization of structure on all levels. At tissue level information can be found in a way where one complete structure replaces another, each corresponding to a morphogenetic field. We think e.g. *Hydra* makes a morphogenetic field organized by positional information that generally is accessible before and after completion of the morphogenesis.

The morphogenetic development can be explained as transfer of information through positional information that is read by the cells and their axons and dendrites (nerve connections), but it is not yet understand what mechanism that leads axons through such development. However, much perception and behaviour are embedded in the biological information indicating that the representation exists as functionally patterns in the brain, but generally not as "epi-genetic factors". We conclude that the excrescence of axons only can be described by positional information. This means that the positional information for the organization is supplied according to the position in the developing organism. We think that positional information plays a similar role in development of all nerve systems, since simple mechanisms usually are fundamental for development of more complex ones through evolution.

The information from the genes is transferred in a way making the cells able to read their position in the wholeness of a biological system and polarize into specific tissues so superior body parts, as e.g. a finger or a wing, can be formed. We think such development can only be explained by supply of positional information. Such information causes information directed self-organization of specific biological levels, which themselves are formed through information directed self-organization etc, and the supply of sufficient positional information is regulated through a feedback mechanism when more complex information is needed, defining a very sophisticated interaction between positional information and biological matter. We call such development "complex epi-genesis". The strength of the complex epi-genesis is that it allows positional information to enter into the organelles, cells and supra cellular levels simultaneously. Such interactions are a quite unexpected and seemingly, a fundamental phenomenon of living nature.

We conclude that the biological system can be described as a Chinese box system of information-transmitting interactions, where the biological system is able to develop its form and maintain this throughout life. In spite of injuries this complex system of information-transmitting interactions can always adjust all the parts of the organism in accordance to each other and the fundamental building plane. All the levels interact with each other and the parts of each level interact, mutually. Thereby, morphological information from the finely divided levels in the cells can manage to be in force at the superior levels. On the other hand, the information from the superior levels, of cause, may reach through the levels and control the cell behaviour, and thereby be able to regulate the gene expression. All these can be gathered in the following description: Through the vertebrate development a Chinese box system of morphological fields is established. Here the information-transmitting interactions between the levels and the single parts of the same level gives an enormous dynamic system, able to convey the positional information of the biological system, and maintain its form throughout the life.

The uninterrupted morphogenesis illustrates the potentials of the biological systems. The experimentally disturbed morphogenesis shows the compensational dynamics of the biological informational system (the informational redundancy). When we review the phenomena of regeneration and induction of *Hydra* and amphibians, and the higher animal's informational needs for developing their complex nervous systems, we argue – also building on the NO-GO theorem for ontogenesis as chemistry (2) - that the traditional chemical explanations of high-level informational events in ontogenesis like transmutation, regeneration, and induction, are most insufficient.

We have analyzed the informational dynamics of three embryonic compensatory reactions to different types of disturbances: 1) Transmutations of the imaginal discs of insects. 2) Regeneration after removal of embryonic tissue. 3) Embryonic induction, where two tissues that normally are separated experimentally are found to influence each other.

We found morphogenesis to be best described as a complex bifurcation, and the resulting many morphological levels of the organism as organized in a fractal manner and supported by positional information.

We suggest that some kind of real non-chemical phenomena must be taking the form in living organisms as an information-carrying dynamic fractal, causing morhogenesis and supporting the organism's morphology through time. We argue that only such a phenomena that provides information-directed self-organisation to the organism is able to explain the observed dynamic distribution of biological information through morphogenesis and the organism's ability to rejuvenation and healing.

References

[1] Ventegodt S, Hermansen TD, Flensborg-Madsen T, Nielsen ML, Clausen B, and Merrick J. Human development IV: The living cell has information-directed self-organisation. ScientificWorldJournal 2006;6:1132-1138.

[2] Ventegodt S, Hermansen TD, Flensborg-Madsen T, Nielsen ML, Clausen B, Merrick J. Human development V: Biochemistry unable to explain the emergence of biological form (morphogenesis) and therefore a new principle as source of biological information is needed. ScientificWorldJournal 2006;6:1359-67.

[3] Raven PH, Evert RF, Eichhorn SE. Biology of plants, 7 ed. New York: WH Freeman, 2005.

[4] Alberts B, Johnson A, Levis J, Raff M, Roberts K, Walker P. Molecular biology of the cell. London: Garland Science, Taylor Francis Group, 2002.

[5] Greilhuber J, Dolezel J, Lysak MA, Bennett MD. The origin, evolution and proposed stabilization of the terms 'genome size' and 'C-value' to describe nuclear DNA contents. Ann Bot (Lond) 2005;95(1): 255-60.

[6] Kutschera U, Niklas KJ. The modern theory of biological evolution: an expanded synthesis. Naturwissenschaften 2004;91(6):255-76.

[7] Kent GC. Comparative anatomy of the vertebrates. St Louis: Mosby, 1987.
[8] Thom R. Structural stability and morphogenesis. Reading, MA: Benjamin, 1975.
[9] Meinhardt H. Models of biological pattern formation. London: Academic Press, 1982.
[10] Sang HJ. Genetics and development. London: Longman, 1984.
[11] Campbell NA, Reece JB, Mitchell LG. Biology, 7 ed. San Franscisco: Benjamin Cummings, 2005.
[12] Murate M, Kishimoto Y, Sugiyama T, Fujisawa T, Takahashi-Iwanaga H, Iwanaga T. Hydra regeneration from recombined ectodermal and endodermal tissue. II. Differential stability in the ectodermal and endodermal epithelial organization. J Cell Sci 1997;110:1919-34.
[13] Mangold O. [Experimente zur Analyse der Determination und Induktion der Medullarplatte]. Wilhelm Roux' Archiv Entwicklungsmechanik des Organismen 1929;117:586-696. [German]
[14] Purves D, Lichtman JW. Principles of neural development. Sunderland, MA: Sinauer, 1985.
[15] Kandel ER, Schwartz JH, Jessel TM. Principles of neural science. New York: McGraw-Hill Medical, 2000.
[16] Horder TJ. Embryological bases of evolution. In: Goodwin BC, Holder N, Wylie CC, eds. Development and evolution. Cambridge: Cambridge University Press, 1983:315-52.
[17] Strausfeld NJ. Altas of an insect Brain. Berlin: Springer, 1976.
[18] Brody BA, Kinney HC, Kloman AS, Gilles FH. Sequence of central nervous system myelination in human infancy. I: An autopsy of myelination. J Neuropathol Exp Neurol 1987;46:283-01.
[19] Ewert JP. Neuroethology: An introduction to the neurophysiological fundamentals of behavior. Berlin: Springer, 1980.
[20] Macagno ER, Lopresti V, Levinthal C. Structure and development of neuronal connections in isogenic organisms: variations and similarities in the optic system of Daphnia magna. Proc Natl Acad Sci USA 1973;70:57-61.
[21] Levinthal F, Macagno E, Levinthal C. Anatomy and development of identified cells in isogenic organisms. Cold Spring Harb Symp Quant Biol 1976;40:321-31.
[22] Raper JA, Bastiani M, Goodman CS. Path finding by neuronal growth cones in grasshopper embryos. I. Divergent choices made by the growth cones of sibling neurons. J Neurosci 1983;3(1):20-30.
[23] Gleick J. Chaos. Making a new science. New York: Viking Penguin, 1987.
[24] Bullock A, Stallybrass O. The Harper dictionary of modern thought. London: Harper Row, 1977.

Section III:
Understanding energy and deeply structured quantum chemical fields

Chapter VII

A spiral fractal model of fine structure of physical energy

In this chapter we have made a draft of a physical fractal essence of the universe, a sketch of a new cosmology, which we believe to lay at the root of our new holistic biological paradigm. We present the fractal roomy spiralled structures and the energy-rich dancing "infinite strings" or lines of the universe that our hypothesis is based upon. The geometric language of this cosmology is symbolic and both pre-mathematical and pre-philosophical. The symbols are both text and figures, and using these we step by step explain the new model that at least to some extent is able to explain the complex informational system behind morphogenesis, ontogenesis, regeneration and healing. We suggest that it is from this highly dynamic spiralled structure that organization of cells, organs, and the wholeness of the human being including consciousness emerge. The model of "dancing fractal spirals" carries many similarities to premodern cultures descriptions of the energy of the life and universe. Examples are the Native American shamanistic descriptions of their perception of energy and the old Indian Yogis descriptions of the life-energy within the body and outside. Similar ideas of energy and matter are found in the modern superstring theories. The model of the informational system of the organism gives new meaning to Bateson's definition of information: "A difference that makes a difference", and indicates how information-directed self-organisation can exist on high structural levels in living organisms, giving birth to their subjectivity and consciousness.

Introduction

In previous chapters we have seen how it is possible to put up a general model for information in biological systems. The model shows how exchange of information between elements on each level of the organism, and between these levels to a large extent can describe how information flows to give information-directed self-organisation at all levels of the organism, and even of the supra-organismic level of the living realm. The problem is, as we also have seen, that we cannot explain this phenomenon with any known principle from either the chemical or the physical sciences. That gives us a great dilemma as what to believe.

Either we hesitate and look for our mistake – and we have done that for years actually – or we step courageously forward to identify the new and undiscovered phenomenon – a new unseen and unknown structure of the universe - that allows life to carry, store, recall, and exchange information.

Let us for a moment consider what information is: "A difference that makes a difference" as Bateson wisely put it (1) can maybe give us a fine idea. How can this be done? Only if there is receptivity in the system, a key point where a very small energy can make a very large difference. We know this principle from the "Butterfly effect" in chaotic systems: the beat of a butterfly's wing creates a cyclone in America. Most of us do not believe that it actually was the butterfly that did disturb the weather systems. The butterfly effect is an interesting phenomenon mostly met in the computer calculated world, where iterations allows a little lack of precision to be very disturbing a moment later. In life this is not seen either. You can almost stir the early embryo with a spoon without causing any noticeable difference. So living organisms are normally not perceptible for disturbances, but the question is: what is providing the information that gives all that structural stability?

Let us suggest that all the particles of a cell through many levels of interaction join into one large energetically, highly structured system, and that is what happens at the top level of this "global cell field" determines what happens everywhere in the cell – that means the cell is using "top down control". What we need is to give a model for the structure of the collective field, presumably of quantum nature. It needs to be a model that can be visualised to make sense (2). Highly complicated math will not help us much here, since we need something to see and understand.

This is therefore the effort of this paper: to present the first draft of such a fine structure of the physical energy. It should be able to control the local activities of the proteins through the delivery of a tiny energy that tips the balance between two states with a different profile of activity (cp. Bateson's concept of information) (1); it should be able to store this *information*, it should be able to pack and carry it, to recall it and to exchange it with other cells in an organism, to create multi-cellular order.

The model does not yet exist

We were not able to find such a phenomenon described in natural scientist's libraries. We found a lot of mathematical structures with fractals as the structure that looked the most "biological", but they were often quite static and did not provide us with the dynamic properties needed for a structural informational backbone of a living cell. We (SV) went to the mathematicians at the University of Copenhagen, but they could not really help. Holger Bech Nielsen knew of some structures with a slight resemblance of what we needed (called Peano curves), but they were not dynamic in three dimensions. We ended up drawing hundreds of different geometrical patterns, until one day we come up with the spiral fractal, which we present as a first solution.

To some extent, our multidimensional spiral-fractal model seems to be able to explain the informational system for the organization of cells, organs and the whole being. It seems to be able to explain central aspects of both morphogenesis, ontogenesis, and consciousness of an individual. In addition to this we believe it is also able to explain important dimensions of evolution (3), the organization of higher biological levels, such as social activity in the family, communities, whole cultures and eco systems. It does not explain everything, it is not perfect, but we believe it is a start. So please take into consideration that what we present here is ongoing research, which from our side has been going on for nearly 20 year. The reason we present it now is that we really believe that we have found something important, although the issue in its complexity in many ways is too much of a mouthful for our small research institute. We definitely need a huge international cooperation to really face this challenge in a formal and strict scientific manner. But one reason to publish it was that similar structural descriptions of energy of life and energy of the universe has been collected by anthropologists interviewing shamans from many premodern cultures on their concepts of life-energy and universal energy (4-6). Some modern experimental cosmologies also contain a concept of dynamic spirals (7).

What we found and what we continue to find surprising is that a simple geometrical "breaking up of the space-time" as the one we are proposing here, can "explain" complicated biological phenomena like enzymatic activity, cellular order, awareness, memory, and consciousness. The model we propose here is based on fractal spacious structures and bear many resemblances to the small energetic strings of modern physics (compare the superstrings) (8). We find it to some extent able to explain the distribution of information to organize the different levels and events of the morphogenesis. It is very good to explain the spontaneous self-organisation of societies and ecosystems, and it can explain how the planet is one living organism (cp. The Gaia-hypothesis) (9) and the level of collective consciousness (4).

That is if you are willing to accept a "hint from a symbolic presentation" as an explanation. We think it makes so much sense and after all it allows us not only to explain the structure and functioning of the living, but also diseases and healing. It actually allows us to believe in very complex phenomena like human adult metamorphosis, as we shall see in the following chapters.

A general model for information in biological systems

Levels and dimensions

The dimensions of space are characterized by giving degrees of freedom to a system. By thinking abstract, we can imagine these freedom-degrees of dimensions, and a new, more elastic interpretation of reality is quite conceivable to give more degrees of freedom than our well-known three-dimensional space allows. Figure 1 gives a proposal for such an organization of the space, from three dimensions to an infinite amount of "pseudo-dimensions" (not perceptionally accessible dimensions).

A spiral-fractal, that through its participation on many levels gives a big amount of curvilinear dimensions inside each other. Each of the curvilinear dimensions, have the freedom to contain its own patterns or organizations. However, these are mutually subjected to interactions from the patterns on the other levels – the curvilinear dimensions on the upper and lower levels.

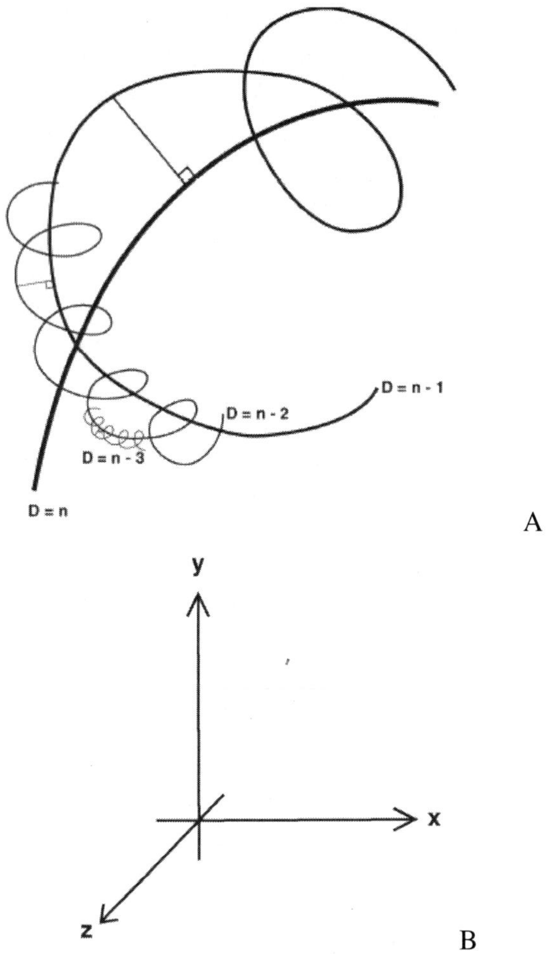

Figure 1. AB. Shows a breakdown of the three dimensional space into an infinite number of curvilinear pseudo-dimensions. A. Different fractal dimensions. B. 3-D dimensions. D =Dimensions. n – 1, n – 2, n – 3 illustrates the different levels of dimensions.

Without computer implementing of the geometry, it is not clear whether the curvilinear dimensions in the spiral fractals have these interacting qualities. If this is not the case, the geometry has to be adapted. However, it seems reasonable that the same power that organizes the spiral-fractal, should also be capable of distributing all patterns evenly through the levels of the fractal, in a such way that the fractal is realized. This could be called the fractal power of the system. This power may be opposed by the fundamental, structural differences existing on the systems different levels. In this way, the organization of particles to atoms, and atoms to molecules, organelles, cells, etc….to organisms, is equivalent to each isolated level of the fractal. The elements of this, also if they have a high degree of individualism caused by information directed self-organization that acts on each level – this is called a complex epi-genisis. Epi-genesis is the developmental process where each successive stage of normal development is built up on the foundations created by the preceding stages of development; an embryo is built up from a zygote, a seedling from an embryo, and so on. Ontogenetic development according to epigenesis is a process of increasing complexity. This contrasts with pre-formationism, where the organism is already present in the gamete(s), merely growing and unfolding during development. It does not clearly appear what riches of patterns an expansion of the fractal geometry to 3-dimentions can involve. But, if the geometry does not involve an expansion in the amount of patterns that exceeds the 2-dimentional Mandelbrot-fractal selection, and underscore the 4-dimentional Mandelbrot-fractal selection, this 3-dimentional fractal geometry has to be changed. The work represented here may not be thought of as a complete answer on the structure of the reality and the problems of the biology, but only as an attempt to establish a completely new interdisciplinary paradigm capable of explaining a lot of unanswered questions.

Energy, information, and matter

In the rational-mechanical interpretation of reality the starting point is matter represented by a particle. But in the energetic-informational interpretation of reality matter is distributed all along the spiral fractal. In this interpretation matter is not existing in a well-defined way as particles, but only as interacting organizations in the curvilinear dimensions: manifestations of energy and information. The simplest way to represent energy in the system is a consequence of destructive and constructive interference between the windings of the spiral. Destructive interference corresponds to annulment of the non-matter basic polarization, and constructive interference corresponds to an increase of the polarization (see figure 2).

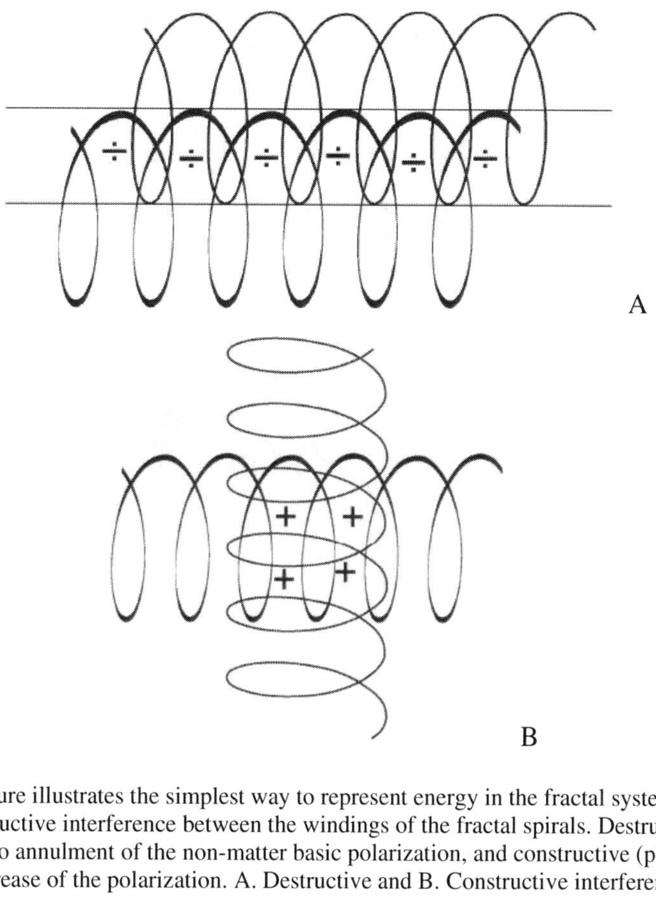

Figure 2.AB. This figure illustrates the simplest way to represent energy in the fractal system is a by destructive and constructive interference between the windings of the fractal spirals. Destructive interference (minus) corresponds to annulment of the non-matter basic polarization, and constructive (plus) interference corresponds to an increase of the polarization. A. Destructive and B. Constructive interference.

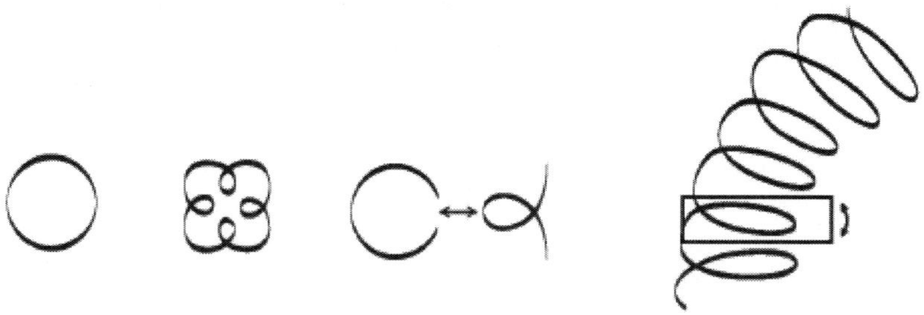

Figure 3. Particles are shown as standing waves and as stable vibrations that can be converted. An apparent particle at level n is stable oscillations at level n-1.

Information is found as patterns established by the storage in stable configurations or is moved as patterns between windings at the same level or up and down the dimensions. Particles may be formed by very stable configurations, in spirals that seem to be able to close itself in a circle; at a finer level they are simply complex oscillations around an average point (see figure 3). Systems of this type can presumably be constructed in a way, where patterns by means of interactions can be transferred between elements at the same level, and between different levels (see figure 4). The particles belonging to such a system will organize themselves according to their inner structure and spontaneously recreate the spiral fractal, when other forces do not hinder it (see figure 5). This because of their inherent lack of an independent, particular existence and their fluid nature.

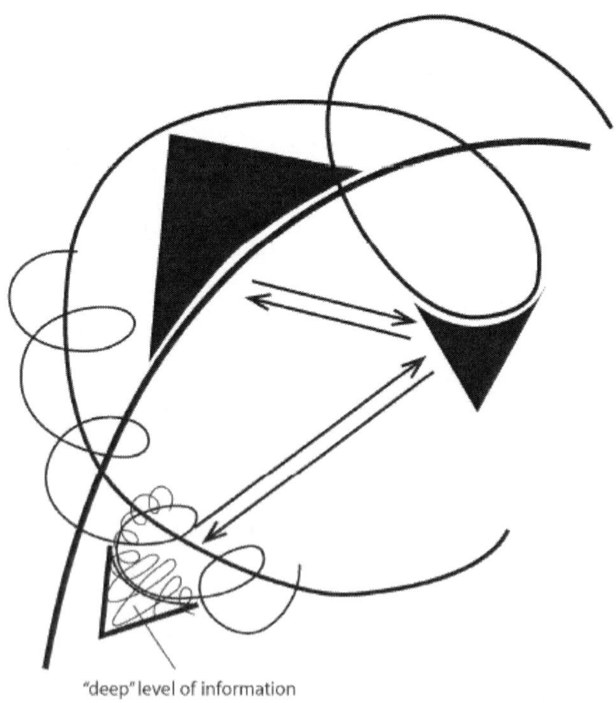

Figure 4. Illustration of information transmitting interactions. Interactions (arrows) are transferred between elements of same level, and between different levels ("the fractal croft").

Time

In the spiral-fractal time exists as dynamic patterns only. One could imagine impulses (information patterns) travelling within the fractal (around in a level or up and down through the levels) and returning with definite time intervals, giving rise to chronometers and energetic watches. We think such an energetic mechanism seems more reasonable then molecular mechanism.

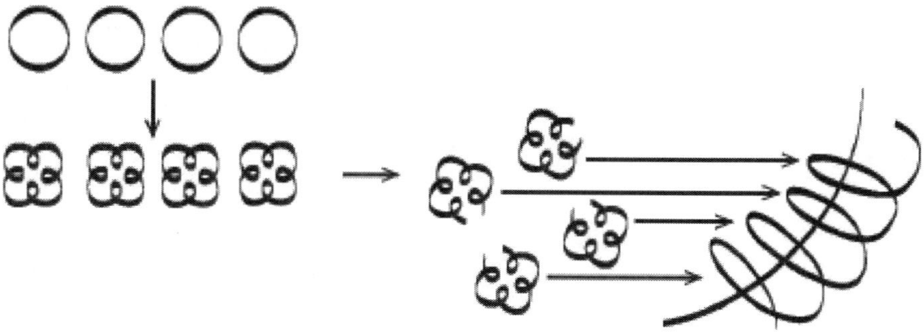

Figure 5. Self-organization of "particles" in the pattern of spiral fractals.

Discussion

The model presented should not be thought of as an attempt to provide a definite answer to the problems of biology and the structure of reality, but as an attempt to establish a new interdisciplinary direction of research, that epistemologically seems fertile. Of course there is a long journey from this energetic-informational interpretation to a closely reasoned account of the mechanisms behind the natural forces and the information transmitting interactions. But it is very likely, that a more complex breakdown of reality can explain both the information transmitting interactions and the natural forces as specific patterns of energy and information. If this is the case, it could constitute the fundament of our complex, functional reality. And yet we as human beings possibly have to accept that the heart of reality evades description. In this case understanding the world through to the emotional interpretation of reality might be the only way to real knowledge of the world.

Our model represents a holistic paradigm based on dynamic spiral-fractal spacious information carrying structures. The parts of the spiral can be compared to small energetic strings (presumably comparable to superstrings) able to explain the distribution of information necessary for the organization of the different levels of the organism during the morphogenesis, and also the higher levels of social reality from relations between biological individuals to small and large societies.

If the spiral fractal is really the backbone of the world, even black matter of the universe and the structure of the galaxies can be held by huge super-spirals penetrating the void of the physical universe. We speculate and dream, and hopefully this work will inspire other researchers and poets to speculate and dream as well, and one day we will have dreamt a new solution to the mysteries that have puzzled man for millenniums.

References

[1] Bateson G. Mind and nature: A necessary unity. New York: Dutton, 1979.

[2] Ventegodt S, Hermansen TD, Flensborg-Madsen T, Nielsen ML, Merrick J. Human development VIII: A theory of "deep" quantum chemistry and cell consciousness: Quantum chemistry controls genes and biochemistry to give cells and higher organisms consciousness and complex behavior. ScientificWorldJournal 2006;6:1441-53.

[3] Hermansen TD, Ventegodt S, Merrick J. Human development X: Explanation of macroevolution — top-down evolution materializes consciousness. The origin of metamorphosis. ScientificWorldJournal 2006;6:1656-66.

[4] Jung CG. Man and his symbols. New York: Anchor Press, 1964.

[5] Castaneda C. The teachings of Don Juan: A Yaqui way of knowledge. New York: HarperCollins, 1968.

[6] Baker D. Esoteric healing (I). London: Essendon, 1975.

[7] Edward BD, Christopher CC. Spiral dynamics: Mastering values, leadership, and change. Malden, MA: Blackwell, 1996.

[8] Davies PCW, Brown J. Superstrings: A theory of everything? Cambridge: Cambridge University Press, 1992.

[9] Lovelock J. Gaia. Copenhagen: Hovedland, 1987.

Chapter VIII

A theory of "deep" quantum chemistry and cell consciousness

Deep quantum chemistry is a theory of deeply structured quantum fields carrying the biological information of the cell, making it able to remember, intend, represent the inner and outer world for comparison, understand what it "sees", and make choices on its structure, form, behavior and division. We suggest that deep quantum chemistry gives the cell consciousness and all the qualities and abilities related to consciousness. We use geometric symbolism, which is a pre-mathematical and philosophical approach to problems that cannot yet be handled mathematically. Using Occam's razor we have started with the simplest model that works; we presume this to be a many-dimensional, spiral fractal. We suggest that all the electrons of the large biological molecules' orbitals make one huge "cell-orbital", which is structured according to the spiral fractal nature of quantum fields. Consciousness of single cells, multi cellular structures as e.g. organs, multi-cellular organisms and multi-individual colonies (like ants) and human societies can thus be explained by deep quantum chemistry. When biochemical activity is strictly controlled by the quantum-mechanical super-orbital of the cell, this orbital can deliver energetic quanta as biological information, distributed through many fractal levels of the cell to guide form and behavior of an individual single or a multi-cellular organism. The top level of information is the consciousness of the cell or organism, which controls all the biochemical processes. By this speculative work inspired by Penrose and Hameroff we hope to inspire other researchers to formulate more strict and mathematically correct hypothesis on the complex and coherence nature of matter, life and consciousness.

Introduction

"The hard problem" in biology, medicine and psychobiology is the problem of the connection between subjective and objective factors in life: how can a super-advanced, chemical machine like our body give us a conscious state? So, what is chemistry, and what is consciousness, and can these two ever meet? Recently many researchers, for example the team of Penrose and Hameroff being one of the most successful (1-4), made suggestions based on quantum mechanics, as this branch of physics has many of the qualities that is needed for a theory of consciousness, coherence and non-locality being among the most important. Penrose and Hameroff think that the coherent molecule orbitals of the microtubules and other inter- and extra cellular fibers, including the links from cell to cell of these fibers, is what is needed to give the cell consciousness.

The idea we are going to present in this chapter and the central idea in this book about human development is closely related to the Penrose-Hameroff idea. We use what we see as a holistic approach wanting to integrate all the large molecule orbitals of the cells molecules, not only the large fibers. The reason why we think this is a more constructive approach is that we want to make the consciousness of the cell continuous with all events in the cell; we want the consciousness to be connected to global information of the cell, and we want cell-consciousness to give the cell the properties of choice – the choice of moving in a direction, taking a form and size, differentiate, divide etc. We simply regard the cell as a complete little animal, and we want it to perceive and "think" very much as we are able to do it ourselves as human beings. The reason that we insist on such a weird idea is that we see the cell as carrying 99% of the complexity of the large animal; therefore we would love the cell also to provide 99% of the reason why we are conscious and feeling, why we can perceive the color red and other qualia, and why we can intuit our world. The cell is not responsible for our complicated interpretation of reality though our brain, but the colony of cells at large is responsible for the interpretation of the world, not the brain itself, as we have discussed it in our theory of the brain (5,6). So we need a really smart cell to be able to explain what is going on and we need something very creative and ingenious to explain cell intent, cell memory and all the other things we need to understand, before we will feel satisfied with our theory of holistic biology for holistic medicine.

Consciousness is a term that has been kept as a huge biological enigma in scientific history. In this paper we will try to came with a qualified guess of what consciousness is, using traditionally biological terms to explain the connections that lead to consciousness. We will discuss the term from the genes capability of keeping life going in an interaction with proteins, macromolecules, and quantum chemical equilibriums.

Scientists and religious thinkers have discussed the term consciousness in many forums. Seen in a religious perspective consciousness has been supplied with a kind of spiritual body, that the traditional thinking scientist reject without discussion. By this paper we try to build a bridge between these two different kinds of thinking and to describe consciousness as an integrated capability of the universe that is shared out to each single biological individual at the creation of it. And we do this by describing the consciousness as a function of quantum mechanical activity working in a kind of "symbiosis" by controlling the genetic and biochemical functions that is necessary for a cell or an organism to survive. This "symbiosis", that gives an organism the capability to "feel" the outer world, we call the consciousness. We describe this as a function that has progressed through evolution, and has been more and more "conscious" with the expansion of the complexity in biological organisms, which in the end has lead to the human consciousness that has been capable of developing the industrialized world.

We have already described the fractal geometry (7) and the super orbitals (7). In this chapter we will look closer at the biochemical activity that leads to the control of master genes, cell volume control, cell proliferation, cell movements, cell communication, cell interactions, etc., to be able to understand the term of cell consciousness, collectively cell consciousness (e.g. organ and tissue consciousness), and consciousness of a complete multi cellular individual.

From active proteins to energized global electron orbitals

Most of the proteins of the cell use ATP (Adenosine triphosphate) to do their job. When ATP is broken down to ADP and P, a protein is statistically forced into one allosteric shape with the proteins typical power. When it switches back and releases its build-in energy it must be recharged to work again. In this way the proteins are "paid" to dance, and the dance is typically a thing that happens in the molecule orbital, like proton pumping. The cell itself produces the ATP necessary for the cell function. ATP is produced by ATP-synthasis imbedded in the cell surface and through the energetic metabolism in the mitochondrion. This circular quantum chemical self-production of energy thereby, makes the cell able to control its own life cycles by just getting food and liquid supply from outside like a multi cellular organisms that functions like a complete body symbiosis. The connection between energetic supply from outside and quantum mechanical activity can best de illustrated by using the plants photosynthesis as a model. In plants, the sun delivers the energy that is absorbed by photo pigments. Such pigments can have a lot of different colors, e.g. red, green, yellow, blue. Each color represents a specific energetic quantum corresponding to a striked electron in a specific orbit. In this way the plant by itself, can decide the exactly quantum of energy suitable for its needs. The striked electrons then are further transported by co-factors in a process that ends up in creating nutrients for the survival of the plant. Such kind of "cyclic" quantum chemical activity guiding both the energetic building up and breaking down activities of the cell, where the cell gets its nutrients from outside and then manifests all its functions through quantum mechanical activity, is very equivalent to the function of a higher multi cellular organism, and emphasizes the whole idea in our way to think.

If the molecular orbital is connected to a higher system, this system could be energized, through the activity of co-enzymes, as well in these movements allowing the super-orbital of the cell to manage energy that can be used to control the dance of other proteins. There are many ligands modifying the activity of active proteins, and these ligands are more active or more passive as their form shifts; a system that can control the protein activity and the likelihood that a protein is in a specific allosteric configuration is then controlling the activity of the cell. The dance of the co-factors and the proteins gives much energy to the collective electron orbital; the problem is how information is handled in it. As we discussed in the paper on the cell [8], all its molecules combine to supra-molecular structures; these are often large protein molecules with extremely complex molecule orbitals (highly structured clouds of the electrons of the molecule). Interestingly, these patterns are known to be unstable; most proteins are either catalytic enzymes or structural "building block" proteins and they exist in several forms or states, often at least one active and one passive state. So basically life is regulated though the regulation of its proteins. And the proteins are all the time changing their state because of the heat-vibrations, so a small change in the likelihood for one state of the protein at one location in the cell is really what we are looking for. As we are here deeply into quantum chemistry and nobody yet have been able to calculate what is happening at these levels, all we can do is guess.

"Deep" quantum chemistry: The cell has a deeply structured integrative quantum field

It is well known from quantum mechanics that all electrons that are contained in one system are inseparable from each other. In a metal this gives the quality of an "electron gas" responsible for the metallic quality. In water the thermal movements of the water molecules disrupts so much that the collective qualities of the electrons is almost lost. In a cell the cytoplasm is a gel made of about 40% proteins, and the structure of this gel is very much like a liquid crystal; there are collective properties of the electrons but they are difficult to identify and map.

The reason why collective properties of electrons in large molecules like proteins and protein ensembles have been difficult to explore is the difficulty describing the collective high-level quantum chemical phenomena mathematically. Actually calculating the collective behavior of the electrons in molecules much larger that water has been almost impossible.

Our approach was mathematical in the beginning, then more symbolic geometrical as it is today. Geometric symbolism is a pre-mathematical and philosophical approach to problems that cannot be handled mathematically (at this moment). What we have done is to guess on the "natural" and neutral structure of electrons, if nothing disturbed them. In metal we have the crystal grid and the nuclei to cause severe disturbances of the energetically pattern a gas of completely free electrons would take.

In water we have the water molecules disturbing the pattern. In plasma, we actually see very much the same patterns as in a free gas of electrons, but strong magnetic fields necessary to hold the plasma always gives a strong artificial structure of the electron gas, and the density from liquidized nuclei also gives some deviation from a gas of free electrons.

So the way we precede is simply to guess that the pattern of the free electrons are able to give the cell its non-local qualities. The cell needs as we have seen to have a system that provides information in a dispersed non-local way. And the only place in physics we can hope to find such a phenomenon is in quantum mechanics.

Quantum mechanics also got irrational numbers (i) in its formalism [9], the same number that is a part of the mathematical structure of some of the really deep fractals like the Mandelbrot fractal. Therefore we are drawn to quantum mechanics; but the normally used formalisms of quantum mechanics do not have a structure usable for explaining living systems. So we need to find a mathematical or a symbolic geometrical language, which is both compatible with the structure of quantum mechanics and with the structure of life. Such a theory need to be a "hidden variable" theory (10) as there must be some hidden layers and structures there.

Using Occam's razor we must start with the simplest model that works; we suspect this to be the spiral fractal (7). So in this paper we investigate some of the features of cells that can be explained if we presume that: all the electrons of the molecular orbitals of all the protein molecules and co-enzymes of the cell, including all the many small molecules embedded in these large biomolecules, and co-factors transporting electrons in the mitochondrion's and mediating the photosynthesis in plants are making up one huge structure ("a cell-orbital"), which in its structure is as close to the neutral spiral fractal as possible, as this is the structure that the electron field would take if nothing were disturbing it.

Dissipative structures

A simple suggestion is by dissipative structures, which is self-organizing patterns arising in systems with a high e-flux. The problem with such patterns is that if they are to be used by biology it must be possible to build on them to make the structure of the cell form and develop.

Unfortunately all dissipative structures are bound by the lack of precision of thermal processes like Brown's movements. So they are not precise enough to build on as the natural tendency is to loose structure and information according to the laws of thermodynamics. The patterns might be there, providing the cell with a lot of creativity and a whole "menu" of patterns for use for its internal structures and its external behaviors; but the cell still needs to be highly structured levels of information in the system to direct the use, enhancement and repression of such patterns.

Iterative generators and fractal patterns

As the cell is obviously structured by some kind of fractal principle (8) it would be nice to have a phenomenon giving such fractal structures. We know that more complicated fractals are made by iteration of a sometimes extremely simple formalism, like the Mandelbrot fractal. But what in the cell can give that?

The spiral fractal needs some kind of walking impulse to function as a calculator, and if we presume that it sometimes collects this kind of impulse, or if it carries and does this in a self-reinforcing way then we can have the spiral fractal generating all kinds of such patterns.

The interesting thing of such patterns is that it really can be active at any level of detail, and in this way it can control all "local spots" in the cell, including the activity of all its protein molecules, and maybe even modify the quantum state of other, smaller molecules, like fat and sugar – and even the water of the cell.

What is cell consciousness in biochemical terms?

Cell consciousness is a very diffuse term that is hard to understand because we not yet have a usable explanation for it. Here we will try to explain the term from a biochemical point of view. To satisfy this term we need to fulfil a number of factors. The cell itself has to function as a complete "symbiosis" including all functions happening in the cell itself controlled from the quantum chemical super orbital. By changing conformation lots of proteins, guided by the energetic quanta supplied from the super orbital, implement countless processes in the cell, leading to important cell functions like e.g. volume control, control of cell cycle, control of cell proliferation, control of cell movement, etc.., but maybe most important, control of master gene and master regulatory protein activity, necessary to support all these processes. It is the gathered quantum mechanical energies standing behind and very strict controlling all this activity in a cell, gathered in a "super" functional energetic "symbiosis", that we call cell consciousness.

What is collective consciousness and consciousness of a multi cellular individual in biochemical terms?

In this chapter we will expand the term consciousness also to count for collective consciousness. Such term can be explained through the activity of fractal geometry. Through the fractal geometry (7) it is possible to gain collective control over an amount of cells that are organized in a "social society" through the development of an organ. And again the organs are organized in tissues through social activity that again organizes within a developing organism through embryogenesis and ontogenesis (8,11) forming a full developed adult organism existing of all the body cells in one complete symbiosis. If we accept that energetic quantums (12) are supported by the quantum mechanical super orbital of the individual organisms (an organism, seen as e.g. a cell, an organ, a tissue or a complete organism including all cells) to the structure of fractals, the term of collectively consciousness can be explained by gathering all super orbital energy in the fractal structures, where the fractal structure is responsible for the distribution of the energy.

In this way the energetic information (positional information) to organize the body structures as e.g. organs, tissues, whole organisms (in the principles all structures up to complete universes), can be controlled through the fractal geometry. Simply, the energetic super orbital from a single cell represents one level of the fractal. The super orbital energy from several cells delivers the information to the organization of an organ by gathering all super orbital energies from cells involved in the organ development, at the next level of the fractal and so on.

Each time a new cell enter into the organ through ontogenesis, the new cells super orbital energy is supplied to the fractal level responsible for delivering the information for the development of the specific organ. This strengthens the united energy of the fractal level, gradually as the complexity of the organism increases. This is going on until a complete adult individual has been developed, gathering all super orbital energies from the single body cells to one very strong super orbital, in humans giving the potential to human consciousness.

Consciousness involves a lot more than this. All biochemical functions and activities in all the cells of the body are involved as described above for the single cell.

Gaining collective consciousness in particular, the cell-to-cell protein interactions, hormonal and other ligands to protein interactions (cell communication) are important, as these factors are absolutely necessary for the cells of a body to communicate.

How is inter cellular control gained by the single cell?

When a protein guided by the super orbital through strict energetic control change conformation, this causes multiple activity in the cell. Some of the most important processes are listed below.

Control of Cell Volume

A very important function for the single cell to gain control over itself and thereby to exist at all is its capability to control its own volume and being able to regulate its balance of liquids. To gain this control, the cell needs energy delivered by the super orbital through co-enzyme transport and ATP. This energy is used for configuration changes of proteins as Protein Kinase C (PKC) that is placed in the cell membrane and causes transport of ions through the cell membrane (13).

Such function controls the gathered energy balance of the cell through ion balance, and is, following the holistic paradigm, a part of the united amount of quantum mechanical energy that makes up the super orbital. Such transport happens also through channels shaped by proteins changing conformation through use of ATP (14). This means that the final energetic balance is a part of and strictly controlled by the super orbital.

Control of Master Genes and Master Regulatory Proteins

How do the fractal positional information and the super orbital gain control with the DNA and the important proteins of the cell? The processes gaining control with the proteins DNA helicase and DNA topoisomerase, controls the unwinding and the super-coiling of DNA respectively (14). These processes are essential for cell activity, because it is involved both in control of cell proliferation (described under) during DNA replication, and the control of DNA transcription. Transcription of DNA leads to production of proteins of which some again are involved in control of the DNA activity (master genes and master regulatory proteins). Such processes make up the "heart" of the living cell and are under strict control from the super orbital that supports exactly the right amount of energy to make a precise regulation of a cells DNA expression. This means that involved in this process a strict control of ligands (master regulatory proteins) also have to be governed from the super orbital. This is a kind of "feedback" process, where genes produce ligands that control the DNA expression (14). Such processes are essential to gain consciousness, because DNA activity is necessary for production of new cells, and thereby the creation of several new super orbitals.

Control of the Cell Cycle

The cell cycle is essential to gain control over the cell proliferation (14,15), and thereby the amount of cells participating in the formation of a multi cellular body as e.g. an organ, a tissue or a complete body. This cycle is under strict genetic control and under strict control from master regulatory proteins as e.g. the E-box controlling proteins (16). Following the holistic paradigm these processes are again firmly controlled from the super orbital through control of protein configurations and support of ATP. This means that these processes are under stringent control from the consciousness.

Control of Cell Proliferation

Cell proliferation is a process that is firmly coupled to the cell cycle (14,15). These two processes together determine how many cells a multi cellular organism contains. Cell proliferation includes processes as control over the microtubule involved in separation of the chromosomes through cell proliferation, control over the split up of the cell membrane, the cytoplasm and the organelles, and control over the gathered cytoskeleton (15). These processes have to be governed in a very precise and strict process, where everything following the holistic paradigm is guided through the super orbital that support the precisely right amounts of energies necessary for these processes to take place. This means that this complicated cellular symbiosis is formed by lots of incredible precisely guided processes all necessary to gain cell consciousness through the production of new cells.

Control of Cytoskeleton Activity Including Cell Movements

Cell movements (8) are a consequence of the ability to control the cytoskeleton. Control of the cytoskeleton is essential for the cell to gain contact with the extra cellular world, as it is an important factor of the cell proliferation as mentioned above. The ability to move has evolved through evolution from the amoeba that crawls slowly and feels in which direction it will move (8), to cells as single cell algae that has developed flagella to control its movements (15). Control of such senses in the single organism and development of these capabilities through evolution, is following the holistic paradigm under strict control from the super orbital and is a part of the gathered qualities that makes the consciousness through quantum chemical activity.

How is intra cellular control gained between cells and multi cellular bodies?

Control of cell communication through hormone and ligand activity

To gain collective control of a multi cellular body as e.g. an organ, a tissue or a complete multi cellular organism, it is essential to be able to communicate between all cells participating in the formation of the body. To gain such control, we have already mentioned that the fractal organization of different fractal levels, where the energetic distribution of super orbitals is essential for our model to explain collectively cell communication. But this is not enough in itself.

Biochemical activity under strict control from the super orbitals of the single cells of the body, or from the super orbitals of the single bodies of a complete multi cellular organism as e.g. the human body, is following the holistic paradigm, also essential to create a collective consciousness. Such intra cellular communication is managed by genetic control through production of hormones and other ligands (15). Such proteins are secreted onto the extra cellular liquids and distributed between cells.

Hormones and other ligands bind to protein receptors of the outer cell surface, and change the conformation of these receptors (15). Such activity entrepreneurs huge amounts of activity, that both are involved in development (growth hormones), gene regulation, cell volume control, etc (15).

Such activities are essential to gain collectively consciousness of a multi cellular body, and for multi cellular bodies to gain collectively consciousness of a complete organism. In e.g. humans, the complexity of the gathered super orbital is so powerful and strong, that it has caused the development of "through" consciousness, leading to the creation of the industrialized world.

Control of cell communication through cell-to-cell adhesion

Another way for the cells to communicate directly with each other is through cell-to-cell adhesion (15). This involves protein interactions between proteins placed directly on the surface of the cell. In this way the cells are gathered in everything from clusters of two cells, to clusters of a huge amount of cells.

This kind of communication also involves the change of confirmation of the proteins involved in the cell-to-cell adhesions, and thereby the delivery of ATP energy governed from the super orbital of the single cell and from the gathered super orbital of the multi cellular body. Such processes are also following the holistic paradigm an important part of the collective body consciousness, and participate in the "gathered" consciousness of the complete multi cellular organism.

The cell's representation of the world

It is well known that the cell is attentive towards both its interior and its exterior world (15). But how is it representing the world, and how is it interpreting what it "sees"? The spiral fractal gives a simple solution to the problem of representation; because the nature of the spiral is like a spring that becomes vibrating in a specific pattern mirroring the phenomenon that impacted it. In this way the spring's cellular level will mirror both its inner and outer world.

The cell's coherence

The mutual representation of cells makes cells to cohere; they share a common reality and the whole cellular system is sharing almost the same information. Of cause will small differences like being close to the organisms borders give a little different clues to the cells that being in the middle.

The cell's remembering

Patterns wandering up and down the levels of the cell are a simple way to explain cell memory, both remembering and storing. The remembering is guided by an associative principle, as is the storage.

The cell's intent

Inherited patterns deeply embedded in the deep structure of the cells quantum field will serve as intent, giving a constant pressure on the cell to identify useful structures and situations and materializing the patterns corresponding to the inherited patterns. This is the simplest way to describe intent, it seems.

How the cell thinks

Now, can the cell think? Well, it seems that is can perform extremely complex calculations in the collective quantum field of all its electrons, judged from the symbolic geometry of the spiral fractal (7). If one imagines that the cell has an energetic impulse rotating in its border circle, then this walking impulse can be used for a kind of primitive thinking.

The cell's perception

Now, will the cell be conscious? It depends on how you define consciousness. If you define it as ability to represent and use the representation for making meaningful choices, you must ask if a global representation can be meaningful and if it can lead to choices. It seems from the symbolic geometry that the cell might even be very conscious, and that it can take decisions.

The cell's choices (to take structure, form and size, move or stay, differentiate, and divide)

A choice must be taken whenever there is several options to chose from. The cell will identify different possibilities that match its intents, and the preferred pattern must be picked. If all the "urges" of the cell is available through internal representation and all the possibilities of the external world are represented by external representation it is very simple for the cell to make its choice. It can move or stay, it can divide or not, it can differentiate or not.

The collective consciousness of cells

The deeply structured quantum field of a cell will resonate well with the quantum field of the other cells, giving the colony of cells collective consciousness. Higher order structures can then materialize in the colony making the cells building blocks of the organism. In the same way as the cell controlled its internal parts, the huge colony of cells making the organism is able to control all its organs and cells. The collective consciousness of the cell is the informational backbone of the organism, and the organism's consciousness raises at the top level of this highly ordered hierarchy.

Cell's morphogenesis

The spiral fractal and the deep quantum chemistry seem to provide the cell with many of its mysterious vital functions. It can remember and perceive, it can intent, make choices and move. This is what it takes for the cell to adapt whatever forms necessary for it self and, if social, also to serve the colony it is a part of; by using these abilities in a community of cells they can be responsible for morphogenesis of the whole organism. Many cells have morphogenesis; if cells need to change their diet, they can change their internal body structure, building a completely new body with the same molecules as before (8). Interestingly this also seems to demonstrate that cellular form is not a product of genes but of information bound to the "deep field" of the cell.

Metamorphosis

Metamorphosis seems to be the most difficult problem to explain: how can a colony of cells suddenly decide that they will shift the form of the total being? But if the morphogenesis is a part of the inherited memory, then metamorphosis will happen as soon as the possibilities of the world and the collective consciousness of the organism's cells invites this to happen. All the cells will see that now is the time to materialize this more evolved pattern, and they will chose to go that way.

How is human consciousness gained?

We think that "Metamorphous Top down" evolution (17) has evolved through consciousness, but human consciousness has developed through "Metamorphous Top down" evolution. E-box master genes that govern the circadian clock feed back loop are essential for the process of metamorphosis in all organisms from cyanobacteria to humans (16). This indicates that processes governed by the super orbital direct the process that stands behind metamorphosis in all organisms from bacteria to humans. Following the holistic paradigm such genes and other master genes involved in metamorphosis are governed by the super orbital, and thereby that consciousness stands behind metamorphosis. But through evolution, individuals so complex as humans have gained such a huge and powerful gathered super orbital, that it has given the evolution the possibility to create the human consciousness through "Metamorphous Top down" evolution.

How can bad consciousness of man lead to disease and death?

All the activity described in this paper so far, gives the feeling of a very complicated and fine regulated mechanism, leading to the term of consciousness. But, how can all these fine tuned processes always lead to fitness and survival? The fact is that they do not. Not much mismatch is necessary for malfunctions to disturb the wholeness. How then can such malfunctions arise? We think that malfunctions leading to disease and death are created by errors in the distribution of orbital energy through the fractal levels, leading to wrong positional information for the process of ontogenesis through development and to the maintenance of an adult body. This is guided by a very fine interaction between the intension of the consciousness, disturbing the patterns that shape our pictures of the reality. Thereby, we ourselves transfer a wrong or "sick" picture of the reality to the fractals that again govern a "mal" distribution of the informational energetic particles of the different fractal levels, leading to disease and death. People that suffer from such malfunctions are maybe mentally disabled and need to be guided in the right direction of a holistic medical practitioner. Following the holistic paradigm such malfunctions can be recovered by guiding a patient to look at the world in a positive way and see all the positive things in itself, face the evil inside itself and repel it. It can regulate the distribution of energetic particles of a fractal level in a way so the information for the correct household of the human body is corrected, leading to fitness and survival. In other words, we believe that fundamentally, a human is its own best medicine; just it is guided in the right direction this will lead to self-recovery through adult human metamorphosis (18).

Discussion

The presented theory of deep quantum chemistry, together with the simple structure of the spiral fractal, seems to give us a perceptive, conscious and thinking cell, creating its own world though intent and the use of historically collected information. There are many problems not solved with this model. If the structure is really infinite, why will information ever return to the here and now "surface" of reality? Why will it not just travel inwards and outwards in this structure forever and ever? How can the transport of information be so fast – almost instantaneous? We need something traveling much faster than the speed of light to explain consciousness it seems, and if everything is so speedy in this structure how can we have events in real time? And if this "highly structured collective quantum field" is really so powerful, why is it that we cannot read each other's minds much better? Actually this theory should make us so powerful that we all should be like prophets, commanding the fig tree to die, and thus just killing it or commanding the sick to stand up and be cured. As usual with such theories it is very difficult to make a powerful suggestion and not make it too powerful. The empowerment from this theory means that we should be able to remember everything that ever happened, since beginning of life on the planet; the intense coherence means that we also should know all that happens at the planet just now, and these two combined means that we should be able to remember all that ever happened on the planet. This is not our daily understanding of our human potentials; this is the reality of the wildest shaman and the wisest of Indian sages. And maybe we are all meant to know it all, to be divine, to be enlightened... Maybe this theory on deep quantum fields of information is true, and we are still evolving as humans towards our final destination. We understand fully if the reader finds our suggestion fantastic and too simplistic at the same time.

In the clinic we often have the strange experience of knowing much more about a patient that should be possible, if information was carried by our senses only. It is as if our bodies are reading the other person as this person contacts his or her old traumatic life event, and often the feeling of what happened and even the pictures and sentences of the event is clear to the therapist long before it re-surfaces into the consciousness of a patient. This sharing of information in coherent fields of humans is well known from Tibetan Buddhism, where many monks meditate on the same object, which they all visualize but from the angle of space where they are sitting. So nobody sees the same picture; all sees the same object. The Dalai Lama told us in 1996 (personally communication) that one of his friends could make a knot on a hard needle when meditating, softening the metal with his consciousness. And the Engineering Anomalies Research has for many years conducted research confirming the ability of human consciousness to interact powerfully with even dead objects like computers (19). To explain such phenomena the presented hypothesis seems also appropriate.

We are at the edge of science, the place where science meets fantasy and imagination. We have dreamt a dream of the inner mechanics of life, and it explains so much that we must share it, for other people to dream it forward to. And one day we might even be able to test the existence of deeply structured biological quantum fields in the laboratory.

It is the collective control of all the mentioned activities and all these functions of the single cell and many interacting cells in symbiosis that together involves the term consciousness regardless if it is the consciousness of a single cell, an organ, a tissue or a complete multi cellular organism. The consciousness is a symbiotic control of all energies, gathered in a wholeness, creating a superior super orbital of quantum chemical activity, regulating the whole pattern of behavioural activity necessary for survival, including all genetic and biochemical activity of a cell or a multi cellular body, whatever it applies to a single cell or a multi cellular body of higher levels during the ontogenesis, or it applies to a mature or adult organism. We think that consciousness is a function of energies that exists over all in the universe created by the energetic equilibrium that is the result of everything. The consciousness of a biological organism is only a body in the universe that has "borrowed" some of the gathered energy of the universe to gain its own consciousness gathered in its own "little" closed orbit. Seen in this perspective, all elements and components of the universe could be described as having its own consciousness. We believe that consciousness of biological organisms exists in self the most primitive organisms on earth as e.g. the bacteria. And it has existed in all organisms from the first living cell was created in the primordial soup. The consciousness of all living organisms is just borrowed from the gathered consciousness of the universe (God). We believe that consciousness in biological organisms has evolved through the evolutionary history as it gained more and more "nearness" with the surrounding world as the complexity of multi cellular organisms evolved. With the human being, this "nearness" became so revealed through the capability of independent mentality, that it was capable of developing the industrialised world. We think this development has evolved through a process we call the "Metamorphous Top down" evolution. The human being has gained so strong control over its consciousness that it is capable of focusing itself through ordering a mal-distribution of the energetic information of a fractal structure that maintains the organism in a state of sickness and psychical disturbance, in a way so it again can be fit and good for survival and capable of doing this by itself without any needs of medical products. It does this through control of master gene activity during adult human metamorphosis, that is the human's own medicine. However, sometimes it is necessary to be helped to think the right way by a holistic medical practitioner. The human has gained all these capabilities through the development of human consciousness. This means that human consciousness has been developed through "Metamorphous Top down" evolution, while "Metamorphous Top down" evolution has been implemented by consciousness through the evolutionary history.

Conclusion

We suggest that the cell has consciousness and all the qualities and abilities related to consciousness. Geometric symbolism is a pre-mathematical and philosophical approach to problems that cannot be handled mathematically (yet). What we have done is to guess on the "natural" and neutral structure of electrons if nothing disturbed them. So the way we precede is simply to guess that the pattern of the free electrons are able to give the cell its non-local qualities. Using Occam's razor we must start with the simplest model that works; we suspect this to be the spiral fractal (7). So in this paper we investigate some of the features of cells that can be explained, if we presume that all the electrons of the molecular orbitals of all the co-factors and protein molecules of the cell, including all the many small molecules embedded in these large bio-molecules are making up one huge structure ("a cell-orbital"). Its structure is as close to the neutral spiral fractal as possible, as this is the structure that the electron field would take if nothing were disturbing it.

Most of the proteins of the cell use ATP to do their job. When ATP is broken down to ADP and P, a protein is statistically forced into one allosteric shape with the proteins typical power. When it switches back and releases its build-in energy is must be recharged to work again. In this way the proteins are "paid" to dance, and the dance is typically a thing that happens in the molecule orbital, like proton pumping. Even the "payment" is administrated by the cell itself, delivered through co-factors that build up the ATP molecules through quantum chemical activity.

If the molecular orbital is connected to a higher system, this system could be energized as well in these movements allowing the super-orbital of the cell to manage energy that can be used to control the dance of other proteins. There are many ligands modifying the activity of active proteins, and these ligands are more active or more passive as their form change; a system that can control the protein activity and the likelihood that a protein is in a specific allosteric configuration is then controlling the activity of the cell. The dance of the proteins gives much energy to the collective electron orbital; the problem is now on a more detailed level how information is handled in it.

Consciousness of biological organisms is the capability of gaining full control over all aspects of life through the super orbital, by controlling all energetic activity of a cell or a multi cellular body, leading to full control of all biochemical activity in a single cell or a multi cellular body, in a way so all activity of the body interacts in a functional symbiosis. We think that consciousness is a main activity of the universe naturally assigned to all biological organisms when these are created. We believe that consciousness has evolved through the evolution from bacteria to humans, and in humans it has been so focused against the outside world that they have evolved the capability of thinking. We think this capability has evolved through "Metamorphous Top down" evolution that has given the human being the capability, through adult human metamorphosis, to gain self recovery from diseases and psychical disorders.

References

[1] Penrose R. The Emperor's new mind. Oxford: Oxford University Press, 1990.

[2] Hameroff SR, Kaszniak AW, Scott AC. Toward a science of consciousness: The first Tucson discussions and debates (complex adaptive systems). Cambridge, MA: MIT Press, 1996.

[3] Hameroff SR, Kaszniak AW, Scott AC. Toward a science of consciousness: The second Tucson discussions and debates (complex adaptive systems). Cambridge, MA: MIT Press, 1998.

[4] Hameroff SR, Kaszniak AW, Scott AC. Toward a science of consciousness: The third Tucson discussions and debates (complex adaptive systems). Cambridge, MA: MIT Press, 1996.

[5] Ventegodt S, Andersen NJ, Merrick J. Quality of life philosophy IV. The brain and consciousness. ScientificWorldJournal 2003;3:1199–1209.

[6] 6. Ventegodt S, Hermansen TD, Kandel I, Merrick J. Human development XII: a theory for the structure and function of the human brain. ScientificWorldJournal 2008;8:621-42.

[7] 7. Ventegodt S, Hermansen TD, Flensborg-Madsen T, Rald E, Nielsen ML, Merrick J. Human development VII: A spiral fractal model of fine structure of physical energy could explain central aspects of biological information, biological organization and biological creativity. ScientificWorldJournal 2006;6:1434-40.

[8] 8. Ventegodt S, Hermansen TD, Rald E, Flensborg-Madsen T, Nielsen ML, Clausen B, Merrick J. Human development IV. The living cell has information-directed self-organization. ScientificWorldJournal 2006;6:1132-8.

[9] Feynman RP, Leighton RB, Sands M. The Feynman lectures on physics: Commemorative issue Vol 1: Mainly mechanics, radiation, and heat. Redwood City, CA: Addison Wesley, 1971.

[10] Bachtold A, Schonenberger C, Strunk C, Salvetat JP, Forro L. Aharonov-Bohm oscillations in multiwalled carbon nanotubes. NANOTUBE-99 Workshop, 1999.

[11] Ventegodt S, Hermansen TD, Flensborg-Madsen T, Nielsen M, Merrick J. Human development VI: Supracellular morphogenesis. The origin of biological and cellular order. ScientificWorldJournal 2006;6:1424-33.

[12] Ventegodt S, Hermansen TD, Rald E, Flensborg-Madsen T, Nielsen ML, Clausen B, Merrick J. Human development III. Bridging brain-mind and body-mind. Itroduction to "deep" (fractal, poly-ray) cosmology. ScientificWorldJournal 2006;6:767-76.

[13] Hoffmann EK, Dunham PB. Membrane mechanisms and intracellular signalling in cell volume regulation. Int Rev Cytol 1995;161:173-262.

[14] Zubay GL. Biochemistry. Dubuque, IA: Wm Brown, 1998.

[15] Alberts B, Johnson A, Levis J, Raff M, Roberts K, Walker P. Molecular biology of the cell. London: Garland Science, Taylor Francis Group, 2002.

[16] Hardin PE. Transcription regulation within the circadian clock: the E-box and beyond. J Biol Rhythms 2004;19(5):348-60.
[17] Hermansen TD, Ventegodt S, Merrick J. Human development X: Explanation of macroevolution — top-down evolution materializes consciousness. The origin of metamorphosis. ScientificWorldJournal 2006;6:1656-66.
[18] Ventegodt S, Hermansen TD, Kandel I, Merrick J. Human development XX: A theory for accelerated, spontaneous existential healing (salutogenesis): Human adult metamorphosis. J Altern Med Res 2009;1(4):465-74.
[19] Jahn RG. The persistent paradox of psychic phenomena: An engineering perspective. Proceedings IEEE 1982;70(2):36-170.

Section IV: Understanding consciousness and human brain function

Chapter IX

Wholeness of man, his consciousness and collective consciousness

In this chapter we look at the rational and the emotional interpretation of reality in the human brain and being, and discuss the representation of the brain-mind (ego), the body-mind (Id), and the outer world in the human wholeness (the I or "soul").

Based on this we discuss a number of factors including the coherence between perception, attention and consciousness, and the relation between thought, fantasies, visions and dreams. We discuss and explain concepts as intent, will, morals and ethics. The Jungian concept of the human collective conscious and unconscious is also analysed.

We also hypothesis on the nature of intuition and consider the source of religious experience of man. These phenomena are explained based on the concept of deep quantum chemistry and infinite dancing fractal spirals making up the energetic backbone of the world. We consider man as a real wholeness and debate the concepts of subjectivity, consciousness and intent that can be deduced from such a perspective.

Introduction

Until now in this book on human development we have been looking at information transmitting interactions at the different levels of the organism. The nature of the information transmitting interactions is unknown, because we do not know the limits of their effect in time and space. Likewise, we cannot with our rational interpretation set this delimitation.

The emotional interpretation is far more suitable for this, but unfortunately far from our reason. Epistemologically we here refer to Occam's razor (1,2) and so far maintain for reasons of simplicity that the same principles that affects the lower levels also affects the higher levels. Our conclusion therefore is that the perceptual and conscious life of man may be deduced from a generalization of the effect of the information transmitting interactions to all levels.

In this chapter we discuss the representation of the wholeness of the organism and the outer world in the brain.

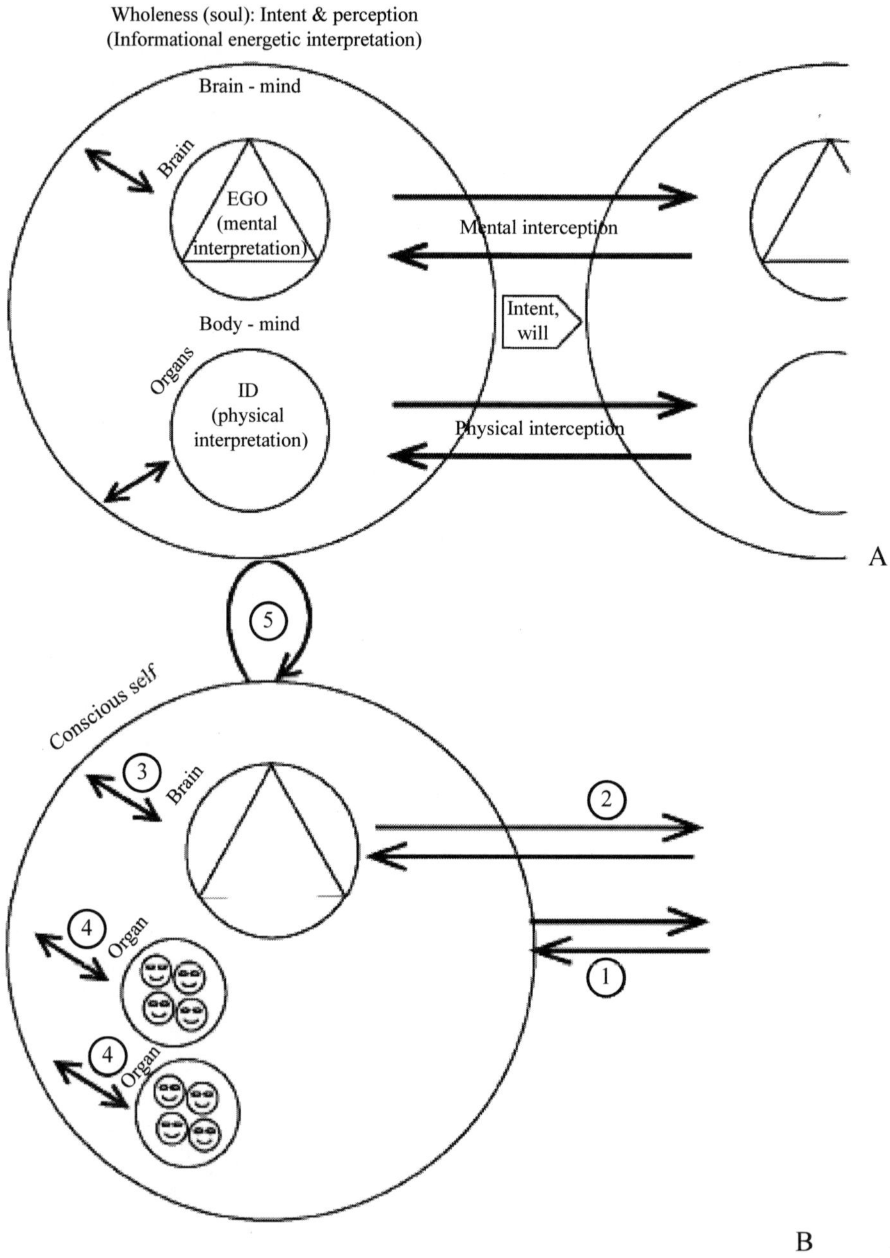

Figure 1. A model of the wholeness of man, including the functions of attention and will. The representational stages of a human being. The attention function and a survey over the representative possibilities and functional stages. This figure shows a survey of man. Man himself is part of a larger social system, to which it through interactions is adapted, and which he himself influences. The human organism can directly represent elements at its own level, humans and other living beings, and more difficultly dead things. The human organism is also able to represent lower levels, such as organ systems, and higher levels, such as the family. A. Wholeness, intention and perception. B. The 5 representational stages: 1. Directly representation of the outer world. 2. Indirectly representation of the outer world through the brain. 3. Representation of the brain in the wholeness. 4. Directly representation of the outer world. 5. The wholeness represents the wholeness.

Among all the organ systems the brain is exceptional, because it is specialized to represent the outer world through the sense organs and thereby give a simple interpretation (3) of matter with some or all elements being reduced to the same frame of description as applies to dead things. The representation of the outer world is made up as a long chain of interpretation actions corresponding with the outer world - inasmuch as the motor apparatus is under control of the brain (4).

In Figure 1 the motor apparatus is drawn below the brain, however the wholeness and the motor apparatus must be thought of as being interacting around the brain. The organism is able to represent the reality (the world) in two ways: 1) Directly through information transmitting interactions, which is the emotional way of representation, where reality is represented in the organism's wholeness (3) or 2) Indirectly through representation in the brain. This is the rational way of representing the outer world by means of the reason (3).

We believe that the brain can represent the wholeness of the organism and what goes on in it in addition to representing the outer world through the senses. This means that the organism possesses the possibility of the rational and the emotional interpretations of reality. When the simplistic representation of the world dominates in the representation of the wholeness we have the rational interpretation of reality.

When the direct representation outside the brain dominates the representation of the wholeness, we have the emotional interpretation of reality, where the brain represents the wholeness (3). In this chapter we will discuss man as a wholeness through his ability to experience the world through these two ways of interpretation.

The human as a wholeness

Attention, perception, and consciousness

The information transmitting interactions giving representation of a subject matter in the wholeness of the organism corresponds to the attention function, and the representation itself corresponds to the attention. This is in contrast to the common notion that perception happens in the brain (5). However we think that consciousness cannot be explained by brain activity alone. Consciousness is not brought about before the totality, the seat of perception, has represented the brain. Consciousness demands a representation of the subject matter in the totality of the organism, but this is not sufficient. A double, simultaneous representation is required, because the representation is a sufficient model of reality within the brain (3,6); the comparison of the two representations is necessary for being able to witness the world, in other words: to be conscious about it.

Thought, fantasies, visions, and dreams

When the totality represents the brain in a state, where important new sensory information is not entering, but where the rational interpretation of reality is sustained, the purpose of the organism of handling the outer world manifests itself through a reorganization of the representation in the brain.

The model of reality in the brain is reorganized in a way that makes it possible to retain a better handling of the world through behaviour. When the brain is working with space-time organizations this phenomenon is called thinking, and when the pictures do not represent the real world it is called visions or fantasies. In dreams, the possibility of realizing the purposes of the organism under the existing circumstances is systematically worked through.

The intent

It is evident that representation is very tightly controlled through the perception function. The patterns of the totality are tuned in a way that ensures agreement between the patterns of the totality and the patterns of a specific level or subject. This tuning is the purpose of the organism. The tuning of the totality happens, when information-transmitting interactions bring the "programs" lying as organized patterns at the deepest levels to the surface of the wholeness of the organism.

The subject matter that is represented has a given organization, which by now can be evaluated through more specific interactions between the wholeness and the information-"program". Reality is thus perceived and obtains its meaning through the purposes of the biological system.

These purposes are a collection of fundamental biological information patterns forming the basis for the unfolding of the biological system. Various aspects of wholeness are submitted to each purpose, and correspondingly different aspects of reality are perceived as a consequence of this submission.

Attention may be attracted to incidents at the various levels, and the purposes must therefore be recruited dynamically as they are needed in order for the wholeness to undergo governed transformations in a suitable way. The governing of attention must happen autonomously through shifts between the states of the totality following a genetically determined governing pattern. The realization of the more simple patterns is a condition for the more complex organizations, in the same way as in morphogenesis (7,8).

Morals and ethics

What attention captures is perceived and gives meaning in relation to the basic pattern, the purpose. The organization of the subject matter may be viewed as proper or improper – or it may be viewed as something in between. When a level or a subject matter is represented in the totality, a double evaluation follows.

On the surface the organization is perceived as suitable for the organism or unsuitable for the organism, i.e. right or wrong. Inwardly an emotional response is triggered, that evaluates the agreement between the genetic program and the structure of the subject matter in the representation of the totality.

A good agreement between the program and reality corresponds to positive emotions, while a bad agreement corresponds to negative emotions.

The will

Hitherto we have only been looking at the representation of subject matters in the totality of the organism through the attention function. It is secondary whether the representation happens directly through the totality following a purpose, resulting in an emotional interpretation of reality, or indirectly through the brain following a purpose, resulting in a rational interpretation of reality. The evaluation through emotion and perception happens in the same way.

Information transmitting interactions in principle work both ways. When purpose transmits information to the wholeness of the organism, information is also transmitted from this wholeness to the subject matter at which the attention is directed. A representation is forced upon the subject matter by the organism, and this function, which is the inverse of the attention function, is called the function of will.

The organism approaches the subject gradually. At first a fundamental adaptation is accomplished by the tuning of the totality by the purpose. In this way the basic patterns are shared between the organism and the subject. This creates a representation of the subject in the totality of the organism. Then this representation is evaluated, and the purpose is tightened, and the patterns corresponding to a certain organization of the subject are pulled up into the totality. Vice versa the desired organization is transferred to the subject.

The increasing complexity of the organism throughout life can be understood as the realization of purposes through the use of the will function. The different purposes are recruited in a succession that ensures organization of the inner levels before the outer levels. The will function is thus the organism's organizing force. It is brought about through information transmitting interactions. The purpose being a very high level of organization in the totality of the organism, forces an organization corresponding to the informational program of the organism, onto a subject with which the totality interacts. The will function can work directly upon the outer world, or indirectly through the brain. The latter leads to behaviour, and in man also to speech. Speech is an indirect mediation of interpretation of reality through words, while action is a direct mediation of interpretation of reality through the motor apparatus. It is possible to act out of an emotional interpretation of reality, because the brain is subject to the totality in this interpretation of reality [3,6]. Direct use of the will function is usually hidden from our reason and belongs to the emotional interpretation of reality.

The intuition

The direct interaction (described above) either between individuals or between an individual and the society yields a complex information transmission, but this is only available for the emotional interpretation of reality. A specific representation of the outer world focusing at certain patterns may theoretically speaking be a precise source of information, providing that patterns that reach us through information transmitting interactions can be identified and interpreted in a precise and sober-minded way.

The difficulty of obtaining this precision and sober-mindedness has led to a pronounced and legitimate scepticism against the value of this data source, because people who claim to have intuition rarely have developed this gift. Intuition in its developed form is the ability to let one's attention float in the complex dynamics consisting of the wholeness and all the individuals of the society. When the wholeness represents the outer world in a general way, one obtains experiences of unity with nature and the like, that belongs to the class of experiences that collide with an undeveloped reason.

The collective (un-)consciousness

When you start thinking in this strange way, that reality is more a hidden order that what meets the eye; the reality of a collective mind or collective consciousness is not a farfetched idea. If the world is created like a huge many-leveled fractal, we might all be a part of such a huge meta-structure, which connects us and collects our perceptions and experiences to bring them to the next level for all to enjoy. This is in a way scary because we often prefer that our consciousness is private; the Jungian concept of a common consciousness is basically born our of the opposite idea: that consciousness cannot be private; it is always public although most people luckily for us do knot know how to approach the collective consciousness and "crack the code" to get admission to our innermost hidden secrets and unspoken emotions.

The religious experience

Last we will discuss the religious experience. In our culture one of the least known qualities of man's wholeness is man's ability to represent himself, to be aware of himself. This is an experience obtained either by chance or, more commonly, after having exercised the control of attention to a level, where he can exclude all other representations. It is usually described as the religious experience: the experience of God, of the divine principle enlightenment etc. In spite of the evidently positive, beautiful and real aspects of this experience, it demands much soberness to be able to use this kind of experiences in a positive way. Very often people lose their way in the overwhelming experiences, or they use them for self-asserting purposes, a fact that has resulted in a prevailing and justified skepticism against "religious" people.

Conclusions

Many aspects are involved when we look at the man as wholeness, including a suitable combination of the rational and emotional interpretation of reality. These aspects are summarised and discussed in this paper:

- We think that consciousness cannot be explained by brain activity alone. A double, simultaneous representation is required. The comparison of these two representations is necessary to be conscious about the world.
- The phenomenon of thinking is when the brain is working with space-time organizations leading to pictures. When these do not represent the real world it is called visions or fantasies. In dreams the possibility of realizing the purposes of the organism is systematically worked through.

- The patterns of the totality are tuned in a way that ensures agreement between the patterns of the totality and the patterns of a specific level or subject. This tuning is the organism's intent.
- The organization of a subject matter may be viewed as proper or improper or something in between. A good agreement between the genetic program and reality corresponds to positive emotions, while a poor agreement corresponds to negative emotions. Tuning of these functions is called moral and ethics.
- A representation is forced upon a subject matter by the organism. This function is called the function of will and can work directly upon the outer world, or indirectly through the brain. The latter leads to behaviour, and in man also to speech. Speech is an indirect mediation of interpretation of reality through words, while action is a direct mediation of interpretation of reality through the motor apparatus.
- The direct interaction between individuals or between an individual and the social levels yields a complex information transmission available for the intuition. In its developed form intuition is the ability to let one's attention float in the complex dynamics consisting of the wholeness and all the individuals of the society.
- In our culture one of the least known qualities of man's wholeness is man's ability to represent himself, to be aware of himself. This is an experience obtained either by chance or after having exercised the control of attention to a level, where he can exclude all other representations. It is usually described as the religious experience.

References

[1] Thorburn WM. Occam's razor. Mind 1915;24:287-8.
[2] Thorburn WM. The myth of Occam's razor. Mind 1918;27:345-53.
[3] Ventegodt S, Andersen NJ, Merrick J. The life mission theory V. A theory of the anti-self and explaining the evil side of man. ScientificWorldJournal 2003;3:1302-13.
[4] Kandel ER, Schwartz JH. Principles of neural science. Amsterdam: Elsevier, 2000.
[5] Hubel DH, Wiesel TN. Brain and visual perception: The story of a 25-year collaboration. Oxford: Oxford University Press, 2004.
[6] Ventegodt S, Hermansen TD, Kandel I, Merrick J. Human development XII: a theory for the structure and function of the human brain. ScientificWorldJournal 2008;8:621-42.
[7] Ventegodt S, Hermansen TD, Flensborg-Madsen T, Nielsen ML, Merrick J. Human development IV: The living cell has information-directed self-organisation. ScientificWorldJournal 2006;6:1132-8.
[8] Ventegodt S, Hermansen TD, Flensborg-Madsen T, Nielsen M, Merrick J. Human development VI: Supracellular morphogenesis. The origin of biological and cellular order. ScientificWorldJournal 2006;6:1424-33.

Chapter X

The origin of metamorphosis

In this chapter we first give a short discussion of the macro-evolution viewing life as information directed, complex, dynamic systems. On this basis we give our explanation of the origin of life and discuss the top down evolution of molecules, proteins and the macro-evolution. We discuss these subjects according to our new holistic biological paradigm. In the view of this we discuss the macro-evolution of the organism, the macro-evolution of the species, the biosphere, and the macro-evolution of human society. After this we discuss the shift in evolution from natural selection to a new proposed process of nature called the "metamorphous top down" evolution. We discuss the capability of the evolutionary shift to govern some of the processes that lead to the formation of new species. We discuss the mechanisms we think is behind this proposed shift in evolution and conclude that this event is able to explain the huge biological diversity of nature in combination with evolutionary naturally selection. We also discuss this event of nature as an isolated, but integrated part of the universe. We propose the most important genetic and biochemical processes that we think is behind the evolutionary shift as a complicated symbiosis of mechanisms leading to metamorphosis in all biological individuals, from bacteria to humans. The energetic super orbital that manifests the consciousness govern all these processes through quantum chemical activity. This is the key to evolutionary shift through the consciousness and this process we propose to call "adult human metamorphosis".

Introduction

In this chapter we will discuss the capability of nature for creating the huge amounts of biological diversity that has been formed through evolution. As this is a theoretical discussion, we only guess and suggest how nature governs this incredible ability to create new species. Therefore the written text maybe here and there give the expression that we think this is the definite truth, but our purpose is only to suggest a new explanation of these functions of nature, because we have not found that the existing knowledge is able to explain these things in details. Therefore we have developed a new way to explain events of the nature that has not been explained like this before. We believe it is a step towards the truth and believe this explanation can work until something better has been developed. It is not our intention here to summarize the modern evolutionary biology's theories of the relation between form and function; this has been done elsewhere (1-4).

Awareness, intention, and will are the driving forces behind morphogenesis. When it comes to the description of man, it is evident that the biological information that is buried deeply in the lower levels of the biological system manifests itself by means of organization of the complex dynamics of the organism as a whole. This organization corresponds to a certain intention and creates a specific interaction between the wholeness and a certain item with a corresponding pattern (5). If the pattern of the wholeness corresponds with organs and organ systems, will power as the organizing force is directed towards these and the organization of the organism is enforced. This is the morphogenesis interpreted through the intension. The interpretation is particularly relevant, because it implies that the organization of the organism is functioning caused by an intention to do so (5). In principle there is no difference between an intension directed towards outer or inner organization. In this paper we discuss the macro-evolution on earth from the evolution of an organism to the evolution of the human society.

The evolutionary activity of nature usually is described as happening through natural selection (6). We think that such evolutionary events are not capable of describing the great diversity of biological organisms developed through evolution of life and existing at the surface of earth today. Therefore, the need for a new evolutionary theory has been pressing on for the evolutionary scientists to be able to explain the evolution of diversity in greater details. A new evolutionary hypothesis capable of satisfying these needs is a new "fly-wing theory" (21). Following this the fly developed its wings through metamorphosis in a process called the "Metamorphous Top down" evolution.

The capability of nature to create new species and thereby increase the biodiversity has been a main problem of evolutionary science until Charles Darwin (1809-1882) in 1859 published his book "The origin of species" (6). He gave scientists a tool that was powerful and partially able to explain the huge biodiversity that ever had existed on earth. But to fulfil the needs for the nature to create this huge biodiversity, we think that something else has to interact with the Darwinian "naturally selection". Therefore we have proposed the "Metamorphous Top down" evolution as a tool for nature to create biological diversity in a tight interaction with the natural selection proposed by Darwin (in the fly-wing theory in preparation for publication).

Charles Darwin based his evolutionary theory concerning "natural selection" on a "bottom up" development, where the evolution went from varieties to taxa, based on the idea that a variety that represents the trait that makes an individual most fit for survival, will eliminate the others through natural selection (6) (already Darwin knew that insects have sperm from 20 males to choose between, and all individuals choose, especially the females). On the other hand, the principle of the "Metamorphous Top down" evolution is based on the capability of an individual and consciousness to go through a process of metamorphosis that develops a specific trait necessary for the individual to stay fit and survive in its habitat or micro habitat, and thereby give a great contribution to the diversity of living organisms existing on the surface earth.

Below, we will first discuss the macro-evolution as a consequence of the evolution, and afterwards some of the mechanisms behind the great amount of diversity developed through evolution. We discuss the capability of the "Metamorphous Top down" evolution to force the creation of new species, and we will do this by looking at it as a local event that is an integrated part of the whole universe, but acts as its own "tiny" (read local) universe, as a small part of the wholeness. This is not a religious explanation of how God interacted with nature and created the species, but as a tool for scientists to understand some of the mechanisms of nature that leads to the creation of new species, and try to see this "tiny" event in a greater perspective by looking at it in a fractal manner compared to the universe as a wholeness. By doing this, we will try to make readers understand that this only is a little isolated event of the universe, but it demands huge amounts of coordination and powers to be able to happen.

Macro-evolution of the biosphere and its biological life

Concerning the origin of life, it is unlikely that the morphological information of the organism is situated in the DNA base sequences, because of 1) the C-value paradox, e.g. (7), 2) normal self-organization of proteins is unable to explain the general morphological features (8) and 3) form can apparently evolve without corresponding molecular evolution (4,9). Supposing that the information giving form lies in the fine structure of energy, the amount of information is not a problem, and likewise the information transport becomes understandable. We believe, that the origin of life can be given a reasonable explanation. According to our hypothesis, the fine structure of energy has an inherent ability to drive the self-organization of all levels, from the smallest units to prokayotes (bacteria and the like), unicellular to multi cellular eukaryotes. We believe that there is a creative principle in nature in which evolutionary selection works. We do not think this is caused by mutations in DNA, because proteins do not give rise to form in itself (9). Therefore, we believe it makes perfect sense if the fine structure of energy works in a creative way and organizes evolution. The evolutionary leaps, that hitherto have lacked an explanation, e.g. the evolution from unicellular to multi-cellular organism can also be explained through our hypothesis, since the creative principle works through creation of still higher organization levels. The more or less constant mutation rates of DNA are called micro-evolution in contrast to the non-molecular evolution, which is usually called macro-evolution. In this paper we discuss the macro-evolution in the view of our new holistic biological paradigm.

Top down evolution of molecules

The evolution of function in the first era of life on earth, presumably lasting the first couple of billion years, was strongly dependent on molecular evolution greatly assisted by the ongoing evolution of shape and function. Molecular evolution was we believe, also guided by high-level organizational forces. If modern day bacteria mirror the early bacteria by recombining DNA (9) they were able to produce new proteins and if the informational system was as smart as we envision it, it could even tell immediately if a protein would be useful for a certain intended task. This is very much the way bacteria gain resistance towards antibiotics today. The recombination part of molecular evolution might be a random process, as we today do not believe that the cell controls the small molecules like DNA fragments, which are subject to strong Brown's stochastic movements. We also believe that molecular evolution was also strongly influenced by natural selection.

Macro-evolution of the organism

The macro-evolution of the organism can be divided into following parts: evolution of the inner organs (10) and evolution of the brain and the limbic apparatus (11). The evolution of the structure of the inner organs was narrowly tied to their function. The function of an organ can be thought of as a result of its structure, however, the function have to act upon the structure through the system's evaluation and later modification of the structure. Structure may be an evolutionary product of function instead of the other way around. This demands that the biological system has a double representation of its structure and its function. This co-representation may be the foundation of the evolution of intension in the biological system.

The macro-evolution of the species and the biosphere

The macro evolution of the limbic apparatus and the brain as we recognize it from the vertebrates that all have the same muscles, bones, brain building plane, etc (4), but different macroscopic shape of all its parts, counts in a very high degree upon adaptation after the surroundings and specific circumstances (6). This can be understood by looking at all individuals within a species through specifically, information transforming interactions in its biosphere. Such interactions make it possible for each individual to interact with the others within the same species, and the same counts for other living individuals in their milieu.

Now, we will have a greater look at the interaction that all organisms in a biosphere may have with each other. All living being seems to have the ability to represent all the other organisms as well as higher levels of organization in its ecological niche. The functions of will and attention is understood as the capability to transform the representation of own organization to other life and represent other life in oneself, that recognizes biological systems in general (5). This counts also for plants, sponge, and single celled animals. This was exactly the original premise for the formation of multi-cellular organisms. All ecosystems are developed through long time of evolution, where all life forms have arranged themselves in accordance to each other and may have influenced each other's evolution through the capability to force a representation. An example of governing of the development of life has appeared through our house animals. An example is the dog. Dogs have been breed for almost all purposes: gundogs, watchdogs, lap dogs, sheep dogs, etc.

Normally it is accepted that the development happens through natural selection. The consequence of the description of life as complex dynamics is that the development in the same degree happens through arrangement to the other life in the biosphere. Exactly the human being has to a very high degree disengaged itself from this arrangement through the use of its reason and through technical development. Conforming to, counts under the direct use of the will in the highest degree to the emotional interpretation of reality.

Life is developed through common arrangement, through an ongoing development of the capability to use the physical circumstances. Behind the evolution is a driving, organizing power as is in the structure of the energy itself (compare with the spiral fractal).

Macro-evolution of the human society

Information transmitting interdependencies between man as an individual and the higher organizational levels of the human communities are a necessity for the social function of the individual. The functionality of the family is completely dependent on the ability of the participants to adapt to a community. Human society should be viewed as a biological structure in itself, which is under evolutionary development. The process of democratisation is a beautiful expression of the evolution of mankind towards integration into a greater society. The human society is a very complex biological structure, much more complex than an anthill, however, in principle there is no difference. When mankind learns to respect the whole, it will at last be able to function in the world in a constructive way.

What drive in evolution has created the diversity on earth?

What has driven evolutionary development in this direction? What forces of nature are capable of creating a world so diverse that a biosphere can cover a huge symbiotic ecosystem as the main ecosystem of earth? We think that powers like naturally selection and "metamorphous top down" evolution interacts with each other, to be able to create a biodiversity necessary for the ecosystem of the biosphere to survive. Below we will describe what we think makes the evolutionary drive for biodiversity. We will discuss the mechanism in nature that drive the evolutionary shift from naturally selection to "metamorphous top down" evolution. This will be discussed in details from the advanced biochemical and genetic processes involved in this shift, to the perspective of an isolated event of nature compared to the great wholeness of the universe.

The force that drive the shift in evolution is governed by quantum energy

The shift in evolution, from our point of view, between the Darwinist's naturally selection and the "metamorphous top down" evolution are governed by quantum chemical activity directed from the super orbital (12) of nature combined with the super orbitals of the individual organisms involved in the evolutionary event, through their consciousness. The individual's consciousness is driven by the needs for survival in a specific habitat or microhabitat. The natural consciousness is driven by the need for the nature to keep the biological diversity on the surface earth (GAIA) as enormous as possible at all. We imagine that the super orbitals of the involved parts gain a kind of resonance with each other, and this resonance through quantum mechanical activity activates the energy necessary for satisfying the needs of the individuals for survival in their habitat and the needs of the nature to gain a more complex diversity. Such activity forms a gathered super orbital for the involved parts that drive the development against formation of new species. In this way new species can be formed that is fit to survive in all kinds of milieus and fitting all habitats and microhabitats. Such mechanism is very practical for the nature, because the temperature of the nature keeping species alive are very dependent on an extremely complex diversity of living organisms on the surface earth (13). This means that the shift in development is a kind of feedback mechanism that is turned on to satisfy needs of earth for more diversity to gain survival as a huge symbiotic ecosystem and to satisfy the individual organism's needs to adjust for survival, when the "natural selection" fails to satisfy these needs. This means that nature has created the "metamorphous top down" evolution as a mechanism that is able to interact with the "natural selection", when it is necessary for the single individual to adjust itself, because the "natural selection" is not able to fulfil these needs, and the individual has got the capability to drive this event by its consciousness. Likewise, the nature itself is by its consciousness capable of driving this force, when natural selection fails to create a diversity of living organisms necessary for survival of the huge symbiotic ecosystem that is manifested in the biosphere of the earth.

We think that such powers are working all over in nature, because it is the way for nature to guard itself against destruction. The main power of nature is interaction between energies that shape a super orbital for each single component in nature. Combined they will shape a huge super orbital gathering all super orbital energies in universal consciousness, capable of organizing the whole universe (nature) in one huge symbiotic "organism". This "symbiosis" represents the highest fractal level, and could in religious terms cover the term, God, but in traditional scientific thinking it covers the capability of the universe to coordinate all functions within the universe in each tiny little detail. It is the same kind of powers that drive the shift in evolution, but on a lower fractal level. The gathered super orbital of the nature and the individuals involved in such evolutionary event makes their own "tiny" universe, governed by the super orbital that delivers the quantum mechanical energy for driving the shift in evolution.

If we look at the evolutionary shift from "natural selection" to "metamorphous top down" evolution as an isolated event, then it is obvious to imagine that this in itself is a very complicated and sensitive process that involves immensely many complicated processes that has to co-operate in a very fine timed interaction. These interactions involve biochemical activity and master gene control of the single individual organisms involved in the evolutionary event in order to direct the process of metamorphosis in a given direction. This way nature gets its needs for more complex biodiversity satisfied and the individual organisms gets their needs for developing a peculiar trait necessary for survival in a specific habitat or micro habitat.

Below we will describe some of the processes that we think is involved in "metamorphous top down" evolution, governed in following sequences: chemical bombing rate → biochemical reactions → genetic control → morphogenesis → manifested mirror image → genetic regime control → parallel metamorphosis of several individuals leading to a new species. We think these processes are tightly regulated by the orbital mastering the processes, and we will try to describe our thoughts concerning the interactions between quantum mechanical activity of the super orbital, the biochemical and regulatory processes happening in the single biological organism through metamorphosis.

Biochemical and genetic interactions

When nature shifts from evolutionary naturally selection to "metamorphous top down" evolution a very huge and complicated apparatus involving a huge amount of strict and coordinated processes of genetic and biochemical origin has to happen for these powerful processes to function as the "symbiotic" wholeness they are. Below we will try to give our explanation for how this coordination can take place by going through a number of genetic and biochemical activities that we think are important candidates behind the shift in evolution involved in increasing the biodiversity of the earth.

Chemical bombing rate

The first chemical signals for forcing the event of "metamorphous top down" evolution, is a chemical bombing rate. This is governed by the quantum chemical energy from the super orbital. This initiates the activity of genes involved in production of pheromones and other airborne ligands necessary for initiating a biochemical reaction in individuals having the needs for the same peculiar trait for survival as the donor organism. This can be individuals in the same habitat, or because the ligands are airborne, individuals from other populations around the world in need for the same trait of survival in their habitats. Such ligands trigger the biochemical activity leading to genetic control by getting absorbed through the mucous membranes of the nose or through membranes of other olfactory organs. Through the body liquids they are transported to the surface of the body cells, where they interacts with protein receptors that changes conformation and thereby are strictly controlled from the consciousness arisen from the super orbital (12). Such activity initiates multiple processes and interactions in the cell.

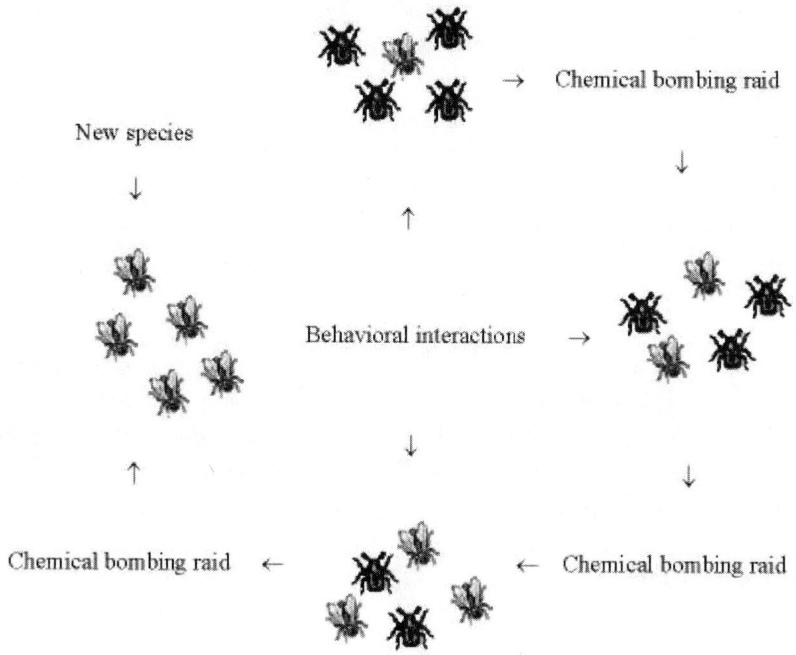

Figure 1. When there in a population is a breeding ground for the development of a specific trait as e.g. the development of wings, this will gain behavioural patterns that favour the development of wings in the population through activation of biochemical reactions. At the same time, chemical substances are transferred between the individuals, also initiating biochemical reactions. Such biochemical reactions insure the control over the genetic regime, so more and more individualsdevelop the wings. The more individuals that have got the wings, the greater evolutionary pressure there is in the population in favour for wing development. In the end, this will involve the creation of a new species.

Biochemical reactions

The conformation change of the cell surface receptors (9,12) is the first biochemical reaction in the sequence of events leading to metamorphosis. All processes initiated in the cell by this action are of biochemical and genetic character. Such actions gives the cell the possibility to adjust itself to processes leading to organ and tissue production, and in the end the capability of forming a complete adult individual organism with its own consciousness. Changing of protein conformations of the cell surface inform the super orbital (consciousness) (12) about initiation of metamorphosis (14). Thereby the super orbital adjusts its energetic activity to coordinate a long row of biochemical activity leading to control of master genes involved in the morphogenesis. This includes a row of feed back mechanisms involving genetic activity for production of master regulatory proteins that regulates the expression of the metamorphous master genes. It also leads to control of the processes of cell cycle control and control of cell proliferation. These processes also include huge amounts of biochemical activity and control of master genes and master regulatory proteins. From this it is easy to imagine the huge "symbiosis" of processes and mechanisms lying behind the united process that initiates the "metamorphous top down" evolution.

Control of master genes and master regulatory proteins

The metamorphous process itself is governed through activity of master genes and master regulatory proteins. A lot of master genes have been identified that are involved in the process of metamorphosis through evolution (a summary of some of the most important of these are in preparation for publication). Such genes produce proteins that are involved in the regulation of expression of all other genes of the single individual's genome, called master regulatory proteins (9). This means that the super orbital by consciousness, through strict and precisely energetic control of the quantum chemical energies needed for expression of genes necessary for all developmental processes of a forming body to process, is able to control each tiny little biochemical process needed for "metamorphous top down" evolution. This includes production of hormones and other ligands needed for configuration changes of proteins, membrane proteins involved in cell to cell adhesion and transport of ions through the cell membrane, ATPase and ATPsynthetase involved in production and degradation of ATP (the main energy resource of the cell) (9,15). But also production of proteins involved in regulation of mitochondria regulatory proteins and thereby involved in the main production of quantum chemical ATP energy of the cell (15), that again is under strict control from the super orbital. And this amazing symbiosis of processes controlled from the super orbital, gives only a humble overview over a little corner of the biochemical processes of nature, and may be considered as much more complex and complicated in reality, because scientists have only been able to map a bit of the great whole of processes and mechanisms involved.

Morphogenesis

The control of master genes and master regulatory proteins as mentioned above, leads to the metamorphous processes necessary for the nature and the individual organisms to satisfy their needs for fitness and survival. These processes make the gathered super orbital of the individual organisms and nature (their joint consciousness) capable of, through influence of the quantum mechanical orbital energy needed for positional information on the different fractal levels, organizing its own evolution of such traits through developmental enforcement of the traits necessary for survival in a specific habitat or microhabitat. Concerning the nature, such processes make it capable of, through the gathered super orbital, to force the single individuals involved in the processes to go through the "metamorphous top down" evolution for satisfying the natural need to multiply the diversity of species necessary for keeping the huge gathered ecosystem of the earth fit and stable (13).

Manifested mirror image

Following the hypothesis of "metamorphous top down" evolution this process is manifested as a mirror image influencing the whole species genetic regime, and through the airborne diffusion of chemicals during chemical bombing rate such mirror image can be manifested in isolated habitats needing the same peculiar traits for survival and thereby giving the capability of developing different species with e.g. wings, as the species of the class Diptera: the fly, gnats and mosquito. Such functions are governed through the diffusion of pheromones carried by the wind, and other airborne proteins secreted by an individual through chemical bombing rates as mentioned above and through changes in behavioural patterns, that again implement a huge amounts of biochemical reactions governed from the super orbital. Such processes implement the same kind of biochemical actions governed from the super orbital, as mentioned under the section "Chemical bombing rate". But in the connection with "manifested mirror image", the behavioural patterns that are implemented between the individuals within a habitat are of major importance. It is those that make the differences between the manifested mirror developments in the different populations. Thereby, they force the different populations against different development and by this forces the development of an even more complex diversity of species. This means that in each isolated habitat, through internal chemical bombing rate and behavioural activity in combination, the development goes through the evolution of a new species (see Figure 1), but in different isolated habitats such development leads to different species (not identical but "of the same kind"), as for instance gnats, mosquito and the fly of Diptera. This means that different behavioural patterns involved in the implementation of a manifested mirror image, is the reason why some isolated populations develops to gnats, others to flies and even others to mosquitoes. Such developmental differences are caused by very small differences in the positional information of the different fractal levels (16), leading to small changes in the evolution of a population through the ontogenesis of the adult individual, and are under tight control from the super orbital gathered between nature and the involved individuals. This means that these use their gathered consciousness to gain their needs for survival.

Genetic regime control

The manifested mirror image leads to "genetic regime control". This means that the gathered super orbital in this way gets the control over the genes of all biological organisms involved in the development of a specific species in a way so it is capable to coordinate the genetic activity towards developing "morphological identical" individuals within the new species. Such control is governed by the super orbital (12) through parallel organization of the distribution of positional information at the different fractal levels (16). In this way, the gathered super orbital is able to guide the development of a new species. Thereby, both the nature and the individual organisms governed from the super orbital, has got their needs covered. The nature has expanded the diversity of species, and the individuals gotten their need for survival satisfied.

The process of adult human metamorphosis

”Metamorphous top down” evolution is driven by the gathered super orbital. The super orbital exists of all energies involved in all processes involved in the evolutionary shift. This is also our definition of consciousness (12). Therefore, the shift in evolution is governed by the consciousness. By this interpretation, the consciousness is involved in the development of a huge amount of species and peculiar traits of individual biological organisms through evolution from bacteria to humans. We think that such evolutionary mechanism as the “metamorphous top down” evolution has been interacting with the mechanisms of natural selection through evolution, and these two mechanisms together are able to explain the huge diversity evolved through evolution. This process of ”metamorphous top down” evolution has also been able to force the development of a more realistic consciousness with the development of more complex organisms. Also here we have a kind of feedback function. Consciousness force the ”metamorphous top down” evolution to happen, and the ”metamorphous top down” evolution evolve a more complex consciousness leading to human consciousness as, until now, a temporary end of the evolution. What comes next? We do not think the evolution has stopped yet. As a very “advanced” result of the ”metamorphous top down” evolution humans have evolved the capability to go through an adult human metamorphosis leading to recovery from illnesses and distortion in order to gain fitness and survival. The adult human metamorphosis, is a delicate kind of ”metamorphous top down” evolution and is probably capable of leading the development of humans to new heights of evolution.

Discussion

This chapter discusses the mechanisms behind the evolutionary shift between “natural selection” and “metamorphous top dawn” evolution. We see the processes involved in this evolutionary event as a function of consciousness. The consciousness drives the evolution to make its shift by a tight and strict regulation of quantum chemical energies that are governed by, and a part of the super orbital. The nature uses the fractal structural levels to govern the distribution of positional information in a way that leads to the creation of new species. In this way the shift in evolution is able to satisfy both the earth’s needs for creation of new species, and the single individual need for fitness and survival. In this way nature is able to keep on and maintain its biological nature as a huge symbiotic ecosystem covering the whole biosphere of the earth, through consciousness.

We imagine the main power of nature (read universe) as one complete symbiosis of energies gathered in one huge super orbital, and isolated events in nature as the shift in evolution, as a local tiny “universe” acting in itself, but as a part of the great wholeness of nature. Such local “universes” are the building stone of nature (the universe), and can be manifested at all fractal levels independent of how huge or tiny they are as events compared to the wholeness. This means that the event of “metamorphous top down” evolution is only a little part of the wholeness that includes everything from the smallest parts of an atom to the universe itself. But as an isolated event, the evolutionary shift draws our attention, because it is a main function of nature, which through evolution has lead to the development of us as human beings. This is one of the main reasons why we have drawn our attention specifically to this particular event in this paper.

However, we think the shift in evolution is governed by the super orbital that is formed by gathering all super orbitals from all parts involved in the evolutionary shift, (compared to the absolute universe) forming a gathered local "universal" super orbital. This very powerful local super orbital governs the shift in evolution, from the evolutionary natural selection to the "metamorphous top down" evolution, through a huge serial of biochemical and genetic processes (listed over), that is under tight control from the super orbital (read consciousness) through quantum mechanical activity. All this activity through "metamorphous top down" evolution has lead to the development of the human being and its capability for gaining spontaneously healing and self-recovery through adult human metamorphosis (14). We do not think this is the end of evolution, but through adult human metamorphosis in interaction with evolutionary naturally selection, nature has the capability to drive the evolution of the human beings to even higher evolutionary goals. By such evolutionary activity the evolution pushes the formation of new peculiarly traits necessary for survival for the single human being. Human evolution is still going on, and the most constructive perspective must be that we are on our way to our evolutionary destiny as conscious, healthy, and happy people fully able to using all talents and create optimal value to our world.

Conclusion

It is unlikely that the morphological information of the organism is situated in the DNA base sequences, because of the C-value paradox, and because the normal self-organization of proteins is unable to explain the general morphology and because form apparently can evolve without corresponding molecular evolution. According to our hypothesis, the fine structure of energy has an inherent ability to drive the self-organization of all evolutionary levels, from the smallest units to prokaryotes. We believe that there is a creative principle in nature where evolutionary selection works. We believe it makes perfect sense if the fine structure of energy works in a creative way and organizes evolution. The evolution from unicellular to multi cellular organisms has also become explainable through our hypothesis, since the creative principle works through creation of still higher organization levels.

We think the "metamorphous top down" evolution evolved from the nature to satisfy its own needs for creation of huge biological diversity for its survival as a huge symbiotic ecosystem in the biosphere of the earth. We also think that the "metamorphous top down" evolution evolved as a result of the single biological individual needs for fitness and survival in a specific habitat or microhabitat. We think this shift in evolution is guided from the super orbital gathered by the super orbitals from all parts involved in the evolutionary event. We guess that the super orbital governs all genetic and biochemical processes involved in the whole serial of events involved in "metamorphous top down" evolution. Through this we think that nature by use of its consciousness has initiated the creation of the great diversity developed through evolution and living on earth today. Furthermore, we think that the gathered super orbital (consciousness) that governs the evolutionary shift is only a local part of the huge super orbital (the main consciousness) of the universe.

We think that "metamorphous top down" evolution is driven by the consciousness in all organisms from bacteria to humans, but the human consciousness has developed from the "metamorphous top down" evolution and given the human being the capability to make spontaneous self healing from mental and physical diseases through a process we call "adult human metamorphosis. This process is a sophisticated kind of "metamorphous top down" evolution that makes the human capable of, in interaction with nature, an even higher level of evolution.

References

[1] Futuyma DJ. Evolutionary biology. Sunderland, MA: Sinauer Associates, 1997.
[2] Futuyma DJ. Evolution. Sunderland, MA: Sinauer Associates, 2005.
[3] Anapol F, German RZ, Jablonski NG. Shaping primate evolution: Form, function, and behaviour. Cambridge studies in biological and evolutionary anthropology). Cambridge: Cambridge University Press, 2004.
[4] Minelli A. The development of animal form: Ontogeny, morphology, and evolution. Cambridge: Cambridge University Press, 2003.
[5] Ventegodt S, Andersen NJ, Merrick J. QOL philosophy V: Seizing the meaning of life and getting well again. ScientificWorldJournal 2003;3:1210-29.
[6] Darwin C. The origin of species. New York: Gramercy, 1995.
[7] Gregory TR. Coincidence, coevolution, or causation? DNA content, cell size, and the C-value enigma. Biol Rev 2001;76:65-101.
[8] Ventegodt S, Hermansen TD, Flensborg-Madsen T, Nielsen ML, Clausen B, Merrick J. Human development V: Biochemistry unable to explain the emergence of biological form (morphogenesis) and therefore a new principle as source of biological information is needed. ScientificWorldJournal 2006;6:1359-67.
[9] Alberts B, Johnson A, Levis J, Raff M, Roberts K, Walker P. Molecular biology of the cell, Fourth edition. London: Garland Science, Taylor Francis Group, 2002.
[10] Diogo R. Morphological evolution, aptations, homoplasies, constraints, and evolutionary trends: Catfishes as a case study on general phylogeny and macroevolution. Enfield: Science Publishers, 2004.
[11] Striedter GF. Principles of brain evolution. Sunderland, MA: Sinauer Associates, 2004.
[12] Ventegodt S, Hermansen TD, Flensborg-Madsen T, Nielsen ML, Merrick J. Human development VIII: A theory of "deep" quantum chemistry and cell consciousness: Quantum chemistry controls genes and biochemistry to give cells and higher organisms consciousness and complex behaviour. ScientificWorldJournal 2006;6:1441-53.
[13] Lovelock J. Gaia: A new look at life on earth. Oxford: Oxford University Press, 2000.
[14] Ventegodt S, Hermansen TD, Kandel I, Merrick J. Human development XX: A theory for accelerated, spontaneous existential healing (salutogenesis): Human adult metamorphosis. J Altern Med Res 2009;1(4):465-74.
[15] Zubay G. Biochemistry. Dubuque, IA: William C Brown, 1999.
[16] Ventegodt, S., Hermansen, T.D., Flensborg-Madsen, T., Rald, E., Nielsen, M.L., Clausen, B. and Merrick, J. (2006) Human development VII: A spiral fractal model of fine structure of physical energy could explain central aspects of biological information, biological organisation, and biological creativity. ScientificWorldJournal 6, 1434-1440.
[17] Ventegodt S, Hermansen TD, Rald E, Flensborg-Madsen T, Nielsen ML, Clausen B, Merrick J. Human development IV. The living cell has information-directed self-organization. ScientificWorldJournal 2006;6:1132-8.

[18] Ventegodt S, Hermansen TD, Flensborg-Madsen T, Nielsen M, Merrick J. Human development VI: Supracellular morphogenesis. The origin of biological and cellular order. ScientificWorldJournal 2006;6:1424-33.

Chapter XI

The structure of the cerebral cortex

The structure of human consciousness is thought to be closely connected to the structure of cerebral cortex. One of the most appreciated concepts in this regard is the Szanthagothei model of a modular building of neo-cortex. The modules are believed to organize brain activity pretty much like a computer. We looked at examples in the literature and argue that there is no significant evidence that supports Szanthagothei's model. We discuss the use of the limited genetic information, the corticocortical afferents termination and the columns in primary sensory cortex as arguments for the existence of the cortex-module. Further, we discuss the results of experiments with Luminization Microscopy (LM) colouration of myalinized fibres, in which vertical bundles of afferent/efferent fibres that could support the cortex module are identified. We conclude that sensory maps seem not to be an expression for simple specific connectivity, but rather to be functional defined. We also conclude that evidence for the existence of the postulated module or column does not exist in the discussed material. This opens up for an important discussion of the brain as functionally directed by biological information (information-directed self-organisation), and for consciousness being closely linked to the structure of the universe at large. Consciousness is thus not a local phenomena limited to the brain, but a much more global phenomena connected to the wholeness of the world.

Introduction

In our quest for a new understanding of the mysterious connection between life, matter and consciousness (1-10), we have now turned our interest towards the human brain. The structure held responsible for the structure and quality of human consciousness is the neo-cortex, the only structure that discriminates us qualitatively from the other vertebrates. Our motivation for the exploration of the life-matter-consciousness-link is our finding that quality of life, health and ability primarily is determined by our consciousness. This discovery has lead us to the interesting possibility that guided shifts in the state of consciousness and the development of self-insight with a more positive, responsible and constructive philosophy of life can be used as medicine to help cure patients. Based on this strategy we have made a series of papers setting the strategy for alleviating healing many different physical, mental, existential, and sexual diseases and health problems (11-52). Another motivation for a deep exploration of the matter-life-consciousness link is the strange fact that "holistic healing" - the healing of the patients total life and existence, including body, mind and spirit – seems to be closely connected to the patient recovering his experience of "sense of coherence" (SOC) (53-59). SOC is actually the experience that one conscious being is connected to the whole universe though our physical and mental existence (59); to bring the patient back to be an integrated part of the world seems to be the fundamental idea of all medicine, the tradition going all the way back to Hippocrates and his students (60).

Many theories concerning the neo-cortical function are based on the possible existence of discrete modules in the neo-cortex as suggested by Szentagothai (1). It is tempting to do this, because the function of neo-cortex is a lot easier to understand and modulate in a mathematical way. Unfortunately, evidence does not seem to exist concerning modules in Szentagothai's understanding, as approximately 300 μ large, discrete columns that are able to make the "building stones" of association-cortex. In this chapter, we discuss the possibility of the existence of Szentagothai's cortical modules.

Does evidence for Szentagothai's cortex module exist?

Szentagothai (2), wrote "It was then (1974) that the assembly of larger tissue complexes from repetitive units of similar build - not unlike the integrated circuits in electronics technology - became an attractive conceptual model to explain how such an immense complexity in "wiring" might be put together without having to make unrealistic demands on the genetic apparatus responsible for this feat in system engineering". Szentagothai had both the genes and the microchips in his mind, when he introduced the modular concept into the association cortex. He realized that repetition of a structure coded by the genes as basis of the cortex development would economize the specific amount of information delivered by the genes. Even if we do not know of any molecular mechanism of such kind, this is a good idea, because such copy mechanism does not deal with DNA replication. Instead it deals with supra cellular patterns. Therefore, we have gradually gotten confident with the description of such phenomena through the conception of positional information. Analogue examples such as the development of testine-villi may be described in this way and therefore modules are an attractive idea. Furthermore, genes can be thought to deliver "continuous modules" (as in visual cortex, see later), and maybe create connectivity in a closely netted, unlimited, roomy way.

About an experiment (1,3) invented to prove that association cortex is constructed by discrete modules Szentagothai (1) wrote, "Here at last is unambiguous evidence for really columnar cortical architecture". In this experiment, proteins containing injected 3H-amino acids were transported to cortical areas of the brain. After subsequent auto-photography, these experiments unveiled the finest 300 μ broad vertical columns in all layers of cortex (2-4). Szentagothai (2) wrote, "It is most interesting to compare the overall size of the columns with the arborisation pattern of individual cortico-cortical afferents as it appears in the Golgi picture. Majorossy shows that both the size and the shape of the arborisation space correspond very closely to the columns, as delineated by the degeneration or autoradiographic tracing techniques". Consequently, this column is analogues to the termination column of the corticocortical afferent. The big question then, is if the identified 300 μ broad termination column represents the module Szentagothai tries to prove the existence of. Seemingly, he has not found evidence enough to support such kind of structure (1,2).

Szentagothai supposed that this area had a great convergence. But Goldman and Rakic (3) said: "Such clear modular pictures as those of the figure (3) are not often seen, because the injections necessarily must cover many modules that result in agglomerations of the labelled modules, often in strips". This indicates an overlap between the termination columns and why fully discrete modules seem not to exist. But, of cause, a "half discrete" module represented by the largest concentration of afferents, cannot be excluded on the basis of this.

In cortex about 2,500,000,000 corticocortical afferents have been identified by Kandel et al (5). But cortex has not room for more than approx. 3,000,000 columns of 300 μ in diameter, only enough to supply approx. 100 afferents to each column. If then a discrete module exists, a great amount of afferents may terminate within this. Therefore, it does not seem reasonable to take the single afferents termination, as indication for the existence of superior modules.

Szentagothai supports his opinion, that the primary sensory cortex has a columnar structure, on evidences from early research (2,6,7). He emphasized two examples from respectively the somatic sensory cortex, 1), and the visual cortex, 2), that he thinks corresponds to the potential modules of the association cortex. However, he realized that the modules are not evident in association cortex.

1. *Barrels:* In a rodent's brain, "barrels" exists in the fourth layer of the somatic sensory cortex. Each barrel contains 2500 neurons arranged around a hollow centre. 100 somatic afferent fibres lead to these barrels from receptors around the barrels. If a barrel is removed, the other barrels are regularly distributed and fill out the area where the barrel is missing. This arrangement reminds us of the primates somatic sensory cortex, where inputs also are recorded in the fourth layer. If some nerve treads are damaged in this layer, a corresponding re-arrangement to that of the barrels can be realized in this layer, see (5). Because a rodent only has 50-100 barrels each delivering huge amounts of information, it seems reasonable that each barrel has an individual discrete representation in cortex. This is a unique exception of the organization of the somatic sensory cortex (5). Such kind of organization of the representation only exists in the fourth layer of the somatic sensory cortex, why it cannot be explained as a cortical column. Because the amount of barrels can vary independent of the area of the modality, it can be stated that if the amount of barrels determines the size of cortex, a single barrel cannot be a structural module.
2. *Hyper columns*: The hyper columns are found in primary visual cortex (5). In one direction these are organized in ocular dominance columns, and in the other direction they are organized in orientation columns. However, we do not think the expression column fits here, because the ocular dominance column is defined by the termination of the afferent in the fourth layer where they do not make discrete columns, but instead alternate stripes from right and left eye. Perpendicular to these, the orientation columns can be defined as serial prisms that separately can be adjusted so their angle is a bit displaced compared with the prism. But nothing indicates that the

line detector is not continuous. Also, the hyper column is arbitrary defined as a 1-2 mm large part of the cortex that makes up a right and a left stripe on the one side, and a complete round of the line detector on the other. The colour modality are handled in "blobs" in layer two and three of cortex (8). Therefore, it is not likely that they represent the cortical module. Generally, for example concerning the receptive field [5] (5), all cells are identical in a very thin layer of the somatic sensory cortex. Possible, this field corresponds to the "micro-columns" described by (2,9), of approx. 10-30 μ, but not to the expected "micro column".

Maybe, cortex is constructed by continuous "modules" as the hyper column that very well could be a repeated structure coded by the genes, and seemingly, Szentagothai did not find any support for his discrete modules in this possibility, because the primary sensory cortex did not show any characteristics of a discrete modular structure. Also in the primary motor cortex the evidence for the presence of columns is weak (Hultborn, Personally communication), because the structure of stripes can only be distinguished here, where the organization seems much more to look like the visual cortex, analogue to the ocular dominance stripes (10).

Other data that could support the existence of a cortex module

In colouration experiments with myalinized fibres, vertical bundles of afferent- and/or efferent-fibres are often seen many places in the cortex as reviewed by Williams and Warwick (11). In LM, the distance between these is approx. 2-3 pyramid cell diameter (20-150 μ) and approx. 10 μ in thickness, and seemingly 10-20 of these myalinized axons are seen to loose the myalinization throughout the cortex (own observations). These "weigot-columns" are hardly Szentagothai's modules, because 1) the distance between these is too small and 2) each of the large modules should receive 1000 afferents and deliver 1,000 efferents, and not 10-20. But maybe these10-20 myalinized axons correspond to "micro-columns" of 5 μ thickness described by Mountcastle (12).

It seems like the outspread termination of mono-aminergic fibres in neo-cortex can be observed as a gap of around 100 μ (13). This distance is too little to fit with modules of 300 μ. But other researchers think that cortex contains non-modular "overlapping columns" of 800 μ (12).

Sensory maps seems not to be an expression for simple specific connectivity

The somatotope map shows immediately reorganization (5), when the supplying nerve is cut. This shows that the afferents terminations happen through a large area, and do not indicate that they are "hard wired" in a simple way. We can guess that such phenomena are caused by complex connectivity, and find it obvious to set the organization of the map in connection with the attention and the positional information that manage the complex dynamics of the morphogenesis.

In the same way, because the thalamus-cortical afferents terminates opposite large areas, the ocular dominance columns seem to make up a kind of functional organizations (4). Szentagothai (1) realized this and proposed that inhibitory inter-neurons adjust the afferents. But such arrangement does not support the existence of structural discrete columns in the sensory cortex.

Discussion

Millennia ago, around 300 BC, at the island of Cos in old Greece, the students of the famous physician Hippocrates (60) worked to help their patients to step into character, get direction in life, and use their human talents for the benefit of their surrounding world. For all that we know this approach was extremely efficient medicine that helped the patients to recover health, quality of life, and ability, which resulted in Hippocrates attaining great fame. For more than 2000 years this was what medicine was about in most of Europe.

On other continents similar medical systems were developed. The medicine wheel of the native Americans, the African Sangoma culture, the Samic Shamans of northern Europe, the healers of the Australian Aboriginals, the ayurvedic doctors of India, the acupuncturists of China, and the herbal doctors of Tibet all seems to be fundamentally character medicine, re-connecting man to his world. Interestingly, the Hippocratic and the transcultural medical traditions gave birth to two succesfull movements in the last century: psychoanalysis (75) and psychodynamic therapy (76,77), developed further into today's clinical holistic medicine, integrating "short-term psychodynamic psychotherapy" (STPP) (78,79) with the many traditions of bodywork developed further into emotionally realizing bodywork by Reich (80), Lowen (81) and Rosen (82) and existential work, today often much inspired by Antonovsky and the concept of SOC (sense of coherence) (83,84). What we have learned by following this long journey of medicine that the grand medical heritage from the planet's different cultures teach us to work on body, mind and spirit at the same time; medicine men of all kinds have always combined talking, touching, and praying. The fundamental problem of understanding the connection of human consciousness and health is this: how do we understand the experience of true connectedness with the universe of a human being, if the brain is just a computer generating the consciousness? The module concept reduces the brain from being a pattern-formatting organ under information directed control from the wholeness of the being and the wholeness of the world, into just being a computer with a pretty isolated function.

The importance of the module concept is seen in the attractiveness of this concept, which gives a feeling of understanding the brain, at least to a certain level, as a computer, or network of computers - the modules, in many models working very much like central processing units (CPUs). If there are not modules and not real structure of the cerebral cortex, only billions of brain-cells firing at their own will, a highly structured consciousness really is a true mystery. This mystery opens up for the possibility that our consciousness is structured true our complex interaction with the surrounding world on an informational level. And this is fundamentally what we need consciousness to be, to understand the extreme importance of the concept of "sense of coherence" for the art and science of holistic healing.

Conclusion

When small 30 μ cortical columns are taken from a random spot of cortex, no significant difference in the amount of cells can be identified. When we look at corticocortical myelinized afferents, no organization of 300 μ can be identified. When many afferents are coloured at the same time, we can see no superior columns. This means that the well known 300 μ termination columns from corticocortical afferents overlaps. Such structure does not support the existence of the discrete module proposed by Szentagothai. "The "barrels" of the rodents somatosensory cortex does not seem to be modules, and discrete modules in visual- or somatosensory cortex, does not seem to be discrete. At the cell level, the large basket cell seems to be independent of, and opposite to, a 300 μ module. The thalamus cortical afferents are shown to terminate in the fourth layer, independent of the corticocortical afferents termination column. Besides this, the most corticocortical afferents are shown to terminate laminar (so the fibres from respectively layer 7 and layer 5 of neo-cortex projects to the same layer). These considerations do not seam directly to permit any modules. Therefore, we conclude that evidence for the existence of the postulated module or column does not exist in the literature. There is no evidence for an app. 300 μ large vertically placed structure that could be able to make up for the cortical ground-unit proposed by Szentagothai.

References

[1] Hermansen TD, Ventegodt S, Rald E, Clausen B, Nielsen ML, Merrick J. Human development I: twenty fundamental problems of biology, medicine, and neuro-psychology related to biological information. ScientificWorldJournal 2006;6:747-59.

[2] Ventegodt S, Hermansen TD, Nielsen ML, Clausen B, Merrick J. Human development II: we need an integrated theory for matter, life and consciousness to understand life and healing. ScientificWorldJournal 2006;6:760-6.

[3] Ventegodt S, Hermansen TD, Rald E, Flensborg-Madsen T, Nielsen ML, Clausen B, Merrick J. Human development III: bridging brain-mind and body-mind. introduction to "deep" (fractal, poly-ray) cosmology. ScientificWorldJournal 2006;6:767-76.

[4] Ventegodt S, Hermansen TD, Flensborg-Madsen T, Nielsen ML, Clausen B, Merrick J.(2006). Human development IV: the living cell has information-directed self-organisation. ScientificWorldJournal 6, 1132-1138.

[5] Ventegodt S, Hermansen TD, Flensborg-Madsen T, Nielsen ML, Clausen B, Merrick J. Human development V: biochemistry unable to explain the emergence of biological form (morphogenesis) and therefore a new principle as source of biological information is needed. ScientificWorldJournal 2006;6:1359-67.

[6] Ventegodt S, Hermansen TD, Flensborg-Madsen T, Nielsen M, Merrick J. Human development VI: Supracellular morphogenesis. The origin of biological and cellular order. ScientificWorldJournal 2006;6:1424-33.

[7] Ventegodt S, Hermansen TD, Flensborg-Madsen T, Rald E, Nielsen ML, Merrick J. Human development VII: A spiral fractal model of fine structure of physical energy could explain central aspects of biological information, biological organization and biological creativity. ScientificWorldJournal 2006;6:1434-40.

[8] Ventegodt S, Hermansen TD, Flensborg-Madsen T, Nielsen ML, Merrick J. Human development VIII: A theory of "deep" quantum chemistry and cell consciousness: Quantum chemistry controls genes and

biochemistry to give cells and higher organisms consciousness and complex behavior. ScientificWorldJournal 2006;6:1441-53.
[9] Ventegodt S, Hermansen TD, Flensborg-Madsen T, Rald E, Nielsen ML, Merrick J. Human development IX: A model of the wholeness of man, his consciousness and collective consciousness. ScientificWorldJournal 2006;6:1454-9.
[10] Hermansen TD, Ventegodt S, Merrick J. Human development X: Explanation of macroevolution — top-down evolution materializes consciousness. The origin of metamorphosis. ScientificWorldJournal 2006;6:1656-66.
[11] Ventegodt S, Flensborg-Madsen T, Andersen NJ, Nielsen M, Mohammed M, Merrick J. Global quality of life (QOL), health and ability are primarily determined by our consciousness. Research findings from Denmark 1991-2004. Soc Indicator Res 2005;71:87-122.
[12] Ventegodt S, Merrick J. Clinical holistic medicine: Applied consciousness-based medicine. ScientificWorldJournal 2004;4:96-9.
[13] Ventegodt S, Morad M, Merrick J. Clinical holistic medicine: Classic art of healing or the therapeutic touch. ScientificWorldJournal 2004;4:134-47.
[14] Ventegodt S, Morad M, Merrick J. Clinical holistic medicine: The "new medicine", the multi-paradigmatic physician and the medical record. ScientificWorldJournal 2004;4:273-85.
[15] Ventegodt S, Morad M, Merrick J. Clinical holistic medicine: Holistic pelvic examination and holistic treatment of infertility. ScientificWorldJournal 2004;4:148-58.
[16] Ventegodt S, Morad M, Hyam E, Merrick J. Clinical holistic medicine: Use and limitations of the biomedical paradigm ScientificWorldJournal 2004;4:295-306.
[17] Ventegodt S, Morad M, Kandel I, Merrick J. Clinical holistic medicine: Social problems disguised as illness. ScientificWorldJournal 2004;4:286-94.
[18] Ventegodt S, Morad M, Andersen NJ, Merrick J. Clinical holistic medicine Tools for a medical science based on consciousness. ScientificWorldJournal 2004;4:347-61.
[19] Ventegodt S, Morad M, Merrick J. Clinical holistic medicine: Prevention through healthy lifestyle and quality of life. Oral Health Prev Dent 2004;1:239-45.
[20] Ventegodt S, Morad M, Hyam E, Merrick J. Clinical holistic medicine: When biomedicine is inadequate. ScientificWorldJournal 2004;4:333-46.
[21] Ventegodt S, Morad M, Merrick J. Clinical holistic medicine: Holistic treatment of children. ScientificWorldJournal 2004;4:581-8.
[22] Ventegodt S, Morad M, Merrick J. Clinical holistic medicine: Problems in sex and living together. ScientificWorldJournal 2004;4:562-70.
[23] Ventegodt S, Morad M, Hyam E, Merrick J. Clinical holistic medicine: Holistic sexology and treatment of vulvodynia through existential therapy and acceptance through touch. ScientificWorldJournal 2004;4:571-80.
[24] Ventegodt S, Flensborg-Madsen T, Andersen NJ, Morad M, Merrick J. Clinical holistic medicine: A Pilot on HIV and Quality of Life and a Suggested treatment of HIV and AIDS. ScientificWorldJournal 2004;4:264-72.
[25] Ventegodt S, Morad M, Merrick J. Clinical holistic medicine: Induction of Spontaneous Remission of Cancer by Recovery of the Human Character and the Purpose of Life (the Life Mission). ScientificWorldJournal 2004;4:362-77.
[26] Ventegodt S, Morad M, Kandel I, Merrick J. Clinical holistic medicine: Treatment of physical health problems without a known cause, exemplified by hypertension and tinnitus. ScientificWorldJournal 2004;4:716-24.
[27] Ventegodt S, Morad M, Merrick J. Clinical holistic medicine: Developing from asthma, allergy and eczema. ScientificWorldJournal 2004;4:936-42.
[28] Ventegodt S, Morad M, Press J, Merrick J, Shek D. Clinical holistic medicine: Holistic adolescent medicine. ScientificWorldJournal 2004;4:551-61.
[29] Ventegodt S, Solheim E, Saunte ME, Morad M, Kandel I, Merrick J. Clinical holistic medicine: Metastatic cancer. ScientificWorldJournal 2004;4:913-35.
[30] Ventegodt S, Morad M, Kandel I, Merrick J. Clinical holistic medicine: a psychological theory of dependency to improve quality of life. ScientificWorldJournal 2004;4:638-48.

[31] Ventegodt S, Merrick J. Clinical holistic medicine: Chronic infections and autoimmune diseases. ScientificWorldJournal 2005;5:155-64.
[32] Ventegodt S, Kandel I, Neikrug S, Merrick J. Clinical holistic medicine: Holistic treatment of rape and incest traumas. ScientificWorldJournal 2005;5:288-97.
[33] Ventegodt S, Morad M, Merrick J. Clinical holistic medicine: Chronic pain in the locomotor system. ScientificWorldJournal 2005;5:165-72.
[34] Ventegodt S, Merrick J. Clinical holistic medicine: Chronic pain in internal organs. ScientificWorldJournal 2005;5:205-10.
[35] Ventegodt S, Kandel I, Neikrug S, Merrick J. Clinical holistic medicine: The existential crisis – life crisis, stress and burnout. ScientificWorldJournal 2005;5:300-12.
[36] Ventegodt S, Gringols G, Merrick J. Clinical holistic medicine: Holistic rehabilitation. ScientificWorldJournal 2005;5:280-7.
[37] Ventegodt S, Andersen NJ, Neikrug S, Kandel I, Merrick J. Clinical holistic medicine: Mental disorders in a holistic perspective. ScientificWorldJournal 2005;5:313-23.
[38] Ventegodt S, Andersen NJ, Neikrug S, Kandel I, Merrick J. Clinical Holistic Medicine: Holistic Treatment of Mental Disorders. ScientificWorldJournal 2005;5:427-45.
[39] Ventegodt S, Merrick J. Clinical holistic medicine: The patient with multiple diseases. ScientificWorldJournal 2005;5:324-39.
[40] Ventegodt S, Clausen B, Nielsen ML, Merrick J. Advanced tools for holistic medicine. ScientificWorldJournal 2006;6:2048-65.
[41] Ventegodt S, Clausen B, Merrick J. Clinical holistic medicine: The case story of Anna: I. Long term effect of child sexual abuse and incest with a treatment approach. ScientificWorldJournal 2006;6:1965-76.
[42] Ventegodt S, Morad M, Merrick J. Clinical holistic medicine: the case story of Anna. II. Patient diary as a tool in treatment. ScientificWorldJournal 2006;6:2006-34.
[43] Ventegodt, S., Morad, M., and Merrick, J. (2006) Clinical holistic medicine: The case story of Anna. III. Rehabilitation of philosophy of life during holistic existential therapy for childhood sexual abuse. ScientificWorldJournal 6, 2080-2091.
[44] Ventegodt S, Merrick J. Suicide from a holistic point of view. ScientificWorldJournal 2005;5:759-66.
[45] Ventegodt S, Clausen B, Merrick J. Clinical holistic medicine: Holistic sexology and acupressure through the vagina (Hippocratic pelvic massage). ScientificWorldJournal 2006;6:2066-79.
[46] Ventegodt S, Clausen B, Merrick J. Clinical holistic medicine: Pilot study on the effect of vaginal acupressure (Hippocratic pelvic massage). ScientificWorldJournal 2006;6:2100-16.
[47] Ventegodt S, Thegler S, Andreasen T, Struve F, Enevoldsen L, Bassaine L, et al. Clinical holistic medicine: Psychodynamic short-time therapy complemented with bodywork. A clinical follow-up study of 109 patients. ScientificWorldJournal 2006;6:2220-38.
[48] Ventegodt S, Thegler S, Andreasen T, Struve F, Enevoldsen L, Bassaine L, et al. Clinical holistic medicine (mindful, short-term psychodynamic psychotherapy complemented with bodywork) in the treatment of experienced impaired sexual functioning. ScientificWorldJournal 2007;7:324-9.
[49] Ventegodt S, Thegler S, Andreasen T, Struve F, Enevoldsen L, Bassaine L, et al. Clinical holistic medicine (mindful, short-term psychodynamic psychotherapy complemented with bodywork) improves quality of life, health, and ability by induction of Antonovsky-salutogenesis. ScientificWorldJournal 2007;7:317-23.
[50] Ventegodt S, Thegler S, Andreasen T, Struve F, Enevoldsen L, Bassaine L, et al. Clinical holistic medicine (mindful, short-term psychodynamic psychotherapy complemented with bodywork) in the treatment of experienced physical illness and chronic pain. ScientificWorldJournal 2007;7:310-6.
[51] Ventegodt S, Thegler S, Andreasen T, Struve F, Enevoldsen L, Bassaine L, et al. Clinical holistic medicine (mindful, short-term psychodynamic psychotherapy complemented with bodywork) in the treatment of experienced mental illness. ScientificWorldJournal 2007;7:306-9.
[52] Ventegodt S, Thegler S, Andreasen T, Struve F, Enevoldsen L, Bassaine L, et al. Self-reported low self-esteem. Intervention and follow-up in a clinical setting. ScientificWorldJournal 2007;7:299-305.
[53] Flensborg-Madsen T, Ventegodt S, Merrick J. Sense of coherence and physical health. A Review of previous findings. ScientificWorldJournal 2005;5:665-73.

[54] Flensborg-Madsen T, Ventegodt S, Merrick J. Why is Antonovsky's sense of coherence not correlated to physical health? Analysing Antonovsky's 29-item sense of coherence scale (SOCS). ScientificWorldJournal 2005;5:767-76.
[55] Flensborg-Madsen T, Ventegodt S, Merrick J. Sense of coherence and health. The construction of an amendment to Antonovsky's sense of coherence scale (SOC II). ScientificWorldJournal 2006;6:2133-9.
[56] Flensborg-Madsen T, Ventegodt S, Merrick J. Sense of coherence and physical health. A cross-sectional study using a new SOC scale (SOC II). ScientificWorldJournal 2006;6:2200-21.
[57] Flensborg-Madsen T, Ventegodt S, Merrick J. Sense of coherence and physical health. Testing Antonovsky's theory. ScientificWorldJournal 2006;6:2212-9.
[58] Flensborg-Madsen T, Ventegodt S, Merrick J. Sense of coherence and health. The emotional sense of coherence (SOC-E) was found to be the best-known predictor of physical health. ScientificWorldJournal 2006;6:2147-57.
[59] Ventegodt S, Flensborg-Madsen T, Andersen NJ, Merrick J. Life mission theory VII: Theory of existential (Antonovsky) coherence: a theory of quality of life, health and ability for use in holistic medicine. ScientificWorldJournal 2005;5:377-89.
[60] Jones WHS. Hippocrates. Vol I–IV. London: William Heinemann, 1923-1931.
[61] Szentagothai J. The modular architectonic principle of neural centers. Rev Physiol Biochem. Pharmacol 1983;98:11-61.
[62] Szentagothai J. The neuron network of the cerebral cortex: A functional interpretation. Proc R Soc Lond B 1978;201:219-48.
[63] Eccles JC, Jones EG, Peters A. Cerebral cortex, vol. 2. New York: Plenum Press, 1984.
[64] Gilbert CD, Wiesel TN. Functional organization of the visual Cortex. Nature 1979;280:120-5.
[65] Kandel ER, Schwartz JH. Principles of neural science. New York: Elsevier, 2000.
[66] Mountcastle VB. Modality and topographic properties of single neurons of cat's somatic sensory cortex. J Neurophysiol 1957;20(4):408-34.
[67] Hubel DH, Wiesel TN. Receptive fields of single neurones in the cat's striate cortex. J Physiol 1959;48:574-91.
[68] Livingstone MS. Art, illusion and the visual system. Sci Am 1988;258(1):78-85.
[69] Szentagothai J. The "module concept" in cerebral cortex architecture. Brain Res 1975;95:475-96.
[70] Goldman-Rakic PS. Modular organization of the prefrontal cortex. Trends Neurosci 1984;7:419-24.
[71] Williams PL, Warwick R. Grays anatomy. Edinburgh: Churchill Livingstone, 1980.
[72] Mountcastle VB. An organizing principle for cerebral function: the unit module and the distributed system. In: Edelman GM, Mountcatle VB, eds. Mindful brain. Cambridge, MA: MIT Press, 1978:7-50.
[73] Cook ND. The brain code. London: Methuen, 1986.
[74] Rockel AJ, Hions RW, Powell TPS. The basic uniformity in the structure of the neocortex. Brain 1980;103:221–4.
[75] Jones E. The life and works of Sigmund Freud. New York: Basic Books, 1961.
[76] Jung CG. Man and his symbols. New York: Anchor Press, 1964.
[77] Jung CG. Psychology and alchemy. Collected Works of CG Jung. Vol. 12. Princeton, NJ: Princeton University Press, 1968.
[78] Leichsenring F, Rabung S, Leibing E. The efficacy of short-term psychodynamic psychotherapy in specific psychiatric disorders: a meta-analysis. Arch Gen Psychiatry 2004;61(12):1208-16.
[79] Leichsenring F. Are psychodynamic and psychoanalytic therapies effective? A review of empirical data. Int J Psychoanal 2005;86(Pt 3):841-68.
[80] Reich W. [Die Function des Orgasmus]. Köln: Kiepenheuer Witsch, 1969. [German]
[81] Lowen A. Honoring the body. Alachua, FL: Bioenergetics Press, 2004.
[82] Rosen M, Brenner S. Rosen method bodywork. Accessing the unconscious through touch. Berkeley, CA: North Atlantic Books, 2003.
[83] Antonovsky A. Health, stress and coping. London: Jossey-Bass, 1985.
[84] Antonovsky A. Unravelling the mystery of health. How people manage stress and stay well. San Franscisco: Jossey-Bass, 1987.

Chapter XII

Structure and function of the human brain

The human brain is probably the most complicated single structure in the biological universe. The cerebral cortex that traditionally is connected with the consciousness is extremely complex. The brain contains approximately 1,000,000 km of nerve fibres indicating its enormous complexity that makes it difficult for scientists to reveal the function of the brain. In this chapter we propose a new model for brain functions: information-guided self-organization of neural patterns, where information is provided from the abstract wholeness of the biophysical system of an organism (often called the true self, or the "soul"). We present a number of arguments in favour of this model that provide self-conscious control over the thought process or cognition. Our arguments arise from analysing experimental data from different research fields: histology, anatomy, electro-encephalography (EEG), cerebral blood flow, neuropsychology, evolutionary studies and mathematics. We criticize the popular network theories as a consequence of a simplistic, mechanical interpretation of reality (philosophical materialism) applied to the brain. We demonstrate how viewing the brain functions as information-guided self-organization of neural patterns can explain the structure of conscious mentation; we seem to have a dual hierarchical representation in the cerebral cortex: one for sensation-perception and one for will-action. The model explains many of our unique mental abilities to think, memorize, associate, discriminate, and make abstractions. The presented model of conscious brain seems also able to explain the function of the simpler brains like that of insects and hydra.

Introduction

In our previous publications (1-11), we have discussed the nature of biology, cell communication and deeply structured quantum fields, the structure of the neocortex and human consciousness (12-20). With such a background in the literature, you should be able to understand our motivation for the development of this new theory of the brain as we need a strong scientific foundation for holistic medicine (21-24). Holistic medicine can provide us with new treatments for many physical, mental, sexual, and existential human sufferings (25-60). Our former philosophical work on life, brain, consciousness, biological information, order, health, and disease have lead to successful treatment of patients with many different health problems (61-66). It is our hope that a more profound model of the human brain and consciousness can facilitate further progress. Our work is based on many publications by gifted researchers like Freud, Jung, Reich, Lowen, Rosen, Anand and Antonovsky (67-75) and many others, who created the foundation for the psychodynamic psychotherapy (76-78).

The present work is neurophilosophical, not traditional neuroscientific. It is based on many different sources of contemporary thoughts and is thus highly interdisciplinary. Our core ideas or *axioms* are:

3. Everything has a solid particle and an energetic wave aspect, according to the laws of quantum physics.
4. Everything is thus an aspect energy. We live in a quantum world where everything, when it comes down to it, is interfering, non-local energy fields of quantum nature (8). These fields are structured and can carry information that can be used by the living organisms (1-10).
5. Everything is thus an aspect of matter, like atoms and molecules giving rise to biochemistry that can be used by the living organisms.
6. A living organism has two sides, a subjective and an objective side; matter and consciousness that are bridged by biological mechanics, including according to our understanding both biochemical and informational quantum mechanics. Our hypothesis is that all organisms, even bacteria, have consciousness and a sense of subjectivity. We doubt that a virus has awareness or subjectivity, but then again we need to define what we mean by "subjectivity" and "consciousness". Those two are difficult and thorny terms with many meanings and connotations; so by consciousness we mean an entities' ability to represent in its informational field the inner and outer world in a meaningful way. Meaning points of purpose carried by the entity. If a virus can sense its surrounding world without sheer mechanical stimuli, then it has awareness and consciousness by our definition. Awareness is thus the sub-atomic ("quantum") quality derived from the entities' wholeness (the informational field) that "senses" (represents) the inner and outer world.
7. When two particles of any kind are contained in the same system they share a common quantum state and thus stops being two separate entities; their common quantum state can be used to coordinate the living system's parts with its wholeness, and this is true for life at all levels: from molecular ensembles, organelles, organelle-systems, cells, organs, organ-systems, organism and the levels of the outer ecosystem in which the organism participates. What is life in this sense? Is it not a supernatural vital force, a cause, an agent, or an independent being like the Indian or Christian concept of the soul? No, but rather a materialisation of an extremely complex and mysterious quality or aspect of the universal energy, providing the living being through billions of years of evolution with autonomy, light, meaning, joy, purpose, will and choice.
8. The level of separateness defines the extent to which any part of a global ecosystem (living totality) has autonomy; the level of the organism has a rather high degree of autonomy giving rise to the organism's sense of self. Every part of the whole has

thus to some extent autonomy; parts that merges completely (i.e. cells that merges into muscle fibres) surrender autonomy to the system. All parts are partly autonomous and partly controlled by the levels above and below them.

9. We are well aware of the depths of the mathematical structures rising from chaos theory, fractal geometry, and complex dynamics (79,80), but we want to emphasis that we do not believe that self-organising patterns can control neither consciousness nor behaviour. Quite on the contrary, we find the mathematical structures derived from the new mathematic (81) to be rather far from biological patterns, so it is quite obvious that we have not yet found the key to a profound scientific understanding of life.

This paper is our first presentation with interpretations based on these fundamental axioms of the current knowledge of the human brain. We are thus taking an absolutely opposite stand-point than normally done in neuroscience, where the nerve-cell is seen as the mechanical unit of the brain, under control of genes, hormones, neurotransmitters, neuropeptides, and functional stimuli. Our hope is to be able to understand the nature and structure of consciousness and especially the functional relationship between consciousness, mind, and the physical (or cellular) body. Our hope is that such a model may allow us to understand emotions and psychosomatics – and in the end, existential healing (23,24,26) and salutogenesis (defined by Aaron Antonovsky (1923-1994) as healing of physical and mental illnesses trough the rehabilitation of "the sense of coherence")(74,75).

The idea that the cell is conscious might seem strange to many people, but that was a conclusion that Sir Roger Penrose and other researchers reached at the SOL-meeting at the Niels Bohr Institute in Copenhagen 1996 (82). Since this meeting this has been our understanding of living organisms: that they always carry consciousness. This means that every cell in our body to some modest extent has an independent consciousness and subjectivity. It is this phenomenon that allows cells to develop into cancer cells, and cells to be cultivated in the laboratory, also after the dead of the multi-cellular organism. We know that this position will be hard to accept in modern neuroscience: that every nerve cell think for it self and make its own independent decisions, which make it so much more than just a small "mechanical computer". Sir Roger Penrose has published a similar hypothesis in his book "Shadows of the mind" (83). It could be argued that we should draw a line somewhere between those unicellular organisms that may have primitive consciousness versus the cells of our nails or hair (84), but we still find it most likely and in accordance with the philosophical principle of Occam's raiser that consciousness is a trait carried by all living beings, but on different levels of complexity.

Does a zygote have a sense of self?

This leaves us to the natural question: Does a zygote have a sense of self and consciousness? Based on empirical research with people re-experiencing their conception, the answer definitely seems to be affirmative, but the objectivity of such studies has been disputed (84). If you think this is a little to farfetched we must tell you that in spite of being quite sceptical ourselves, we as scientists often see patients spontaneously regress all the way into the womb in intensive holistic therapy. The first author have after 20 years of consecutive therapy suddenly regressed during holotropic breastwork (86) all the way back to the zygotic state; according to this experience there was indeed a conscious "I" from the very beginning of life (87), but then again, when you become very experienced late in life, you start to relax your scepticism a little and believe in all sorts of things, and then you are not really reliable any more, are you? To keep it simple, which is a must in order to comply with Occam's raiser, the sense of self (being an independent soul) does not develop gradually from the zygote – embryo – foetus – newborn – infant – child – adult – old age; this feeling of being an autonomous creatures is with us from the very beginning as an innate trait of the person; what is developed though time and experience is the level of complexity of the mind (brain-mind), and perhaps also the complexity of the being.

Many scientists believe that the human brain is a kind of independent computer able to understand almost everything in this universe, but at the same time it seems clear that we human beings never fully will be able to comprehend consciousness and the brain itself. This paradox may be a consequence of a narrow, rationalistic and materialistic interpretation of reality (1-10).

If the interpretation of reality gets more complex and less naive this might give us a chance of developing a more transcendent and deeper understanding of the brain and consciousness. The description of life as a complex dynamic and information-directed self-organizing system (4) is an example of this point of view and it can settle the above-mentioned paradox. From this perspective, consciousness is seen as a characteristic of all living things. Because of the relative high degree of outer separateness self-consciousness is especially prevalent in the human brain, which has a large degree of autonomy; the abilities of the human mental self (Ego) are complex and advanced. The body is the next fairly independent biological entity, carrying as Freud noticed correctly, the consciousness of the Id (and the body-mind). Finally the human wholeness also carries its representation of us called our "spiritual self", "true self" or "soul". This gives man three dominant representations of self: The wholeness-related self (below called the Soul), the brain-related self (below called the Ego), and the body-related self (below called the Id). And then we have the "I", the integrative self that emerges from an organic synthesis of body, mind and spirit. This "I" is often in spiritual literature and poetry called the "heart" (Comp. the lyrics of Madonna: "When you heart is open"). In Freud's work the super-ego is synonymous with the Soul; Freud often said that everything good in man comes from the Superego (67).

The interesting question, "Doesn't the self operate through the brain?", leads us to an immediate "of course", and a secondary remorse on deeper reflection, that in the end brings the realisation that our thoughts and behaviour might be controlled by our self, but feelings, sexuality and spirituality are as well controlled by our self, so are we really living in the brain? This is most definitely not our experience, if we must be honest. Emotions are felt in the body; and our sexual feelings are definitely felt primarily in our sexual organs. And maybe this is not merely a joke, but human biological reality that must be respected, as Freud insisted.

We know that there seems to be an interpretation of our feelings in the limbic system, and without this happening we could not be mentally aware that we were emotionally hurt or sexually exited. But to insist that feelings and being is merely a brain-thing, that is insisting on something that obviously conflicts with our common sense (senses communis). As therapists we are emotionally oriented people, and we really like to place the "I" in our hearts much more that in our brains. This gives a much more human contact and a much richer emotional life. On the other hand we must agree that mental consciousness obviously is focused in the middle of the brain (in what the Indians have called the "3rd eye" for 7,000 years). We like to use our experiences as basis for understanding; not vice versa. And this is the true reason for making this chapter : We need to stop thinking so much and start sensing and explore our inner self to really understand what is going on. We suspect that many neuroscientists often miss the obvious truth, because they do not sincerely feel what is happening inside and reflect upon it. Basically this is a question of using subjectivity in research, and putting sufficient emphasis on the qualitative aspect of science. In this chapter we want to analyse the implications of this for our understanding of the brain.

Consciousness

All the cells of the organism carry consciousness. Human consciousness is basically embedded in a quantum field arising from the combination of all individual cells' consciousnesses (4-9,83). This is highly debatable, we must admit, but we humbly ask you to play with this idea before you decide to dismiss it. This structure of Soul, Ego and Id, all carrying semi-individual consciousness means that the whole organism (the "heart" or "I") can "see" and analyse all its lower-level conscious functions; thus both brain-mind functions, body-mind and even wholeness-related "spiritual intelligence" functions are observed and interacted with by the "I". It is very usual in therapy that a person suddenly observes an inner process of cancer, or observers his own autistic side, and this could not be done without the ability of such a "meta-perspective" provided by the abstract I. This is somewhat related to the strange observations that patients during surgery can observe the surgical theatre from above, from a view-point where they obviously feel out of the body. Most interesting this shift of perspective also often happens in intensive holistic therapy. So it is hard not to believe for us as therapists witnessing this happen all the time. Not that we understand it, or even like it. Human consciousness is way too mysterious to be cosy.

The "I" or "heart" can see the brain from a perspective "outside" the brain, but inside the organism. And the body can see the brain and the brain can see the body – energetically, or if you prefer by direct transference of information from one "wholeness" to the other. Is the wholeness of the body different from the wholeness of the brain? Does an organism have a wholeness? Does an organism behave as one entity? The unreflected answer is that the organism carries the wholeness and its organs do not. But after deeper reflection and meditation on ants and corrals and after that on human societies and the way consciousness is manipulated in a society collectively i.e. by the media, we must admit that the organisms are not that free to think, feel and act as we would like them to be. After thorough studies on sexuality and human unconsciousness (un-integrated traits of brain-mind and body-mind) we must admit that our organs are pretty powerful actors in their own right. So the picture is much more complicated and in the end every level of us has some degree dependence and freedom.

The "I" can use the bodies' emotional intelligence – the faculty of intelligence connected to the body-cells' collective consciousness field (and of course the wisdom of a mature brain's cortico-limbic system) – to get a clear picture of the mental processes and the basic machinery that creates its own mind. This is the typical perspective of the Tibetan Buddhists

Yogis reached in deep meditation, but only little acknowledged in the west until recently (87). You could argue that these are subjective reports of their mental states and experiences and thus they are probably not a good argument in support of our thesis in a scientific journal. But our direct experiences of the world might be as real as our mental reflections upon it; we would actually argue: more real!

The brain is represented in the organism's wholeness, in the same way as the organism is represented in the brain. You could argue that you have never seen your own brain, but if you can sense where you mental activity is centred– right behind your eyes – then your wholeness has already acknowledged that you can sense you own brain, or at least its energy and quality. So what we basically claim is that the brain cannot be understood without understanding the "I" and its consciousness, and Id, and the Soul, and their individual consciousnesses.

In a well-integrated organism the self rules ("I" am in power); from its placement at the top-level of the organism it can strongly impact what is going on at all the lower levels, including in the brain. The model we are going to develop further was originally made by the psychoanalysts to allow us to understand how I-born consciousness (as in "living by heart") can be causal in our life (67-72). Interestingly this model seems to be the normal understanding of man in most pre-modern societies (9), and most interestingly it opens up to an explanation of collective consciousness (9,68,69) that is normal in pre-modern societies, but almost forgotten or neglected in our culture in spite of Jung (68,69), Grof (85,86) and other prominent psychic researchers stressing its meaning and importance to us. It seems that it actually came from the pre-modern cultures into psychoanalysis especially by Jung. The model we present is thus not purely based on theory, but on lots of practical experience, and it is also in accordance with the philosophy of life that has arisen from our research in quality of life and health during the last two decades.

Understanding the complex patterns of brain activity

To understand the various functions of the brain we need an integrative theory for brain function that accounts for the control of the mental functions on the highest level of the brain. The multi-dimensional connectivity that follows the extreme con- and divergence in the architecture of the human neocortex, the results of countless EEG-measurements, and the measurements of the high and almost constant brain-energy-usage indicates that the cortex cerebri is a machinery that almost continuously delivers a huge selection of patterns that floats into each other (79,80,88-91). Mathematical analyses (90-92) have indicated that the cerebral cortex cannot organize the patterns by itself in a meaningful way, leaving us with the most fundamental problem of how the brain is controlled.

It may be assumed that sensory inputs to the brain temporarily can stabilize its chaotic neural self-organising patterns, creating a sensory perception of simple information-directed self-organisation. But this does not explain much – thinking, understanding, perceiving etc. Therefore, another much more efficient and innate organizing system may exist, that makes the brain function as it does. Morphological and evolutionary data (93,94) seem to show that the nervous system is developed and functions through an intense communication with self – or in more scientific language: the brain is totally imbedded in the complex informational dynamics of the organismic wholeness. The organismic wholeness entirely depends on a well functioning, living brain as expressed by John Zachary Young (1907-1997), professor of anatomy at the University College London: "No brain, no mind, no nothing."

Actually the brain is completely absorbed in the organism, and has, as we see it, no completely independent function of its own, in spite of it being in many structural and physiological ways a highly autonomous organ. Even the slightest action on the brain is in some way influenced by the totally and intentions of the being who owns it.

One can ask at what stage of embryological development of an organism, like a bacterium or a human being, is there a communication between the brain and self? Little do we know about how bacteria process information, but they obviously do, and the close distance between the genes and their global level tells us that there must be an intense inner representation of the bacteria's wholeness. Do we have a self, when we are at the single cell level of development viz. the zygote? Yes, we seemingly do. Do we have self even before conception? No, at least not in our philosophy. Then again it is wise to remember that half the population of the planet would disagree here.

It is normally believed in neuroscience that a lot of the cortical-subcortical activity is spontaneous, involuntary, automatic and subconscious. This is often presented as a fact in many standard textbooks of neurobiology, neuroscience and neurophysiology; but it is worthwhile remembering that we never have seen a brain keeping these "automatic" functions on its own (in vitro). And if we count the consciousness of the Self and the Body, this might not be the case at all. The interesting thought-experiment to do here is to imagine a person's brain isolated and fed in a jar: Will this brain still be able to think and feel like the person it came from? Will it still function at all? We know of course that in vitro developing neurons still fire. But will this activity be able to create any collective meaning without the informational guidance from the Self and the body? The consequence of our thinking is that this is not possible. Unfortunately we do not know of any experiments that can decide this for us.

On the contrary, we know of many experiments that indicate that the functional order of the brain is highly fluent and rapidly reorganised. The cortical representation of all sensory and motor functions in neural maps and their well known and quite mysterious, momentary reorganization[88] seems to confirm that brain function is controlled from "a level above", because the representations are not fixed in the physical brain – the maps are not hardwired. Studies of blood flow and lesions show a hierarchically ordered structure of representations in cortical networks (88,93,94). These examples are in accordance with the psychological developmental studies showing that the consciousness has its starting point in the functions of the body with the primary sensory and motor areas creating the foundation for the hierarchy. The higher integrative areas create the intermediate, and the highest integrative areas – especially prefrontal cortex – create the top of the hierarchy (88).

It is well known that some brain lesions like that of the Brocka's area are followed by expressive aphasia. So there are lots of hard-wired solutions in the brain also, of course, on lower levels of the brain. But all high level, consciousness-related organisation seems to be informationally directed. And we believe that the Self provides this information, and that the Self thus controls its brain. This link between Self and brain is crucial for our understanding of i.e. the altered brain function and physiology in illnesses like depression and schizophrenia.

In appendix 1, we give a short evaluation of the value of central experimental data for our understanding of the brain with a special attention to the function of the cortex cerebri. We will now discuss how the brain functions according to our holistic understanding. The fundamental structure of the neocortex (95,96) was reviewed in (11) and will not be repeated here.

A theory for brain function

What does the brain do? Basically the brain connects Ego, I, Soul, and Id to the outer world. It carries the organism's rational interpretation of the world and allows it to realise the intentions of Self and Id through plans carried out in time and space. The brain creates mental perception, rational understanding and visual, auditory, somatosensory interpretation of the inner and outer world, and proper, rational actions and inner adjustments from the Ego's, Soul's, Id's, and I's intent.

One could ask if I, Soul, Ego and Id are generated from a well-functioning brain, and this could very well be so. The different inner personalities that we label all these names could easily rise from a less than perfect integrity of the brain. So one cannot judge just by thinking, if there is any rationale behind the more complex model of human reality that we propose; we do it because we respect our "common sense", our direct experience of life; it is this sense that allows us to help our patients in holistic therapy, and provides us with the power of healing. So we cannot just give it up. And when we take our experience to meet neuroscience, things do not fit. What we feel and experience is simply not compatible with the mechanical interpretation of reality we find in contemporary neurophysiology books; even profound books like Principles of Neural science[88] does not reflect on the quality of existential depth, joy, light, and innate wisdom that we sense is connected to consciousness. Frankly the concept of consciousness is hardly addressed in neurophysiology of today. The "hard" problem (how subjective consciousness is produced from chemistry and physics) has not been solved, and it is much too often just ignored.

This is done through intensive mapping of the inner and outer world. The brain is in touch with the outside world through the senses and the apparatus of movement, and in touch with the person's "inner world" through feelings, intuitions, finer sensations, intentions, states of consciousness and being: dreams are very much the materialisation of this inward contact.

Throughout life a more and more detailed model of reality is build up, using the fundamental dimensions of space and time. The brain and mind are harvesting experiences through the presence of sensory qualitative units called "qualia" - like the colour red - that is solely produced by the nervous system and the organism itself; a nervous signal can in principle not be read (unless you accept that it can carry a more subtle and finer level of information, i.e. quantum level information). Qualia are combined through time and space into elements that can be perceived and manipulated, giving birth to the phenomenological world. The intensity of qualia is established, and its location in time and space is noticed, and all these neural measurements are integrated into a dynamic perception. Mental elements can be static or dynamic, corresponding to nouns and verbs in the language (see Chomsky's famous concept of "deep structures" of language, and Piaget's model of development of human consciousness) (97-99). The nonverbal mental phenomena like pure visual images, touch, taste and emotions are also based on qualia, but they are often just taken for granted and not abstracted to higher logical levels. But they can be, as in the Indian and Tibetan art of erotic tantra, were sexual feelings and unified sexual poles are abstracted to oneness ("sunya") (73).

The mind is thus basically nothing but a highly dynamic model of reality constructed in this way, just by combining the elements on higher and higher, more and more abstract levels through experience and memory, and the mental faculty of abstraction and concretisation. The model is organised though association and dissociation. Logic and sets are used for giving rational structure.

The brain is constantly preparing a row of behavioural and perceptual strategies to meet the intentions of the whole organism (the "I" motivated by "Soul, Id, and Ego" according to psychodynamic theory). The brain interprets all experiences and sensory inputs from the outer world in agreement with the organism's intentions, accumulating concrete strategies for action and for perceptual and intellectual analysis. These strategies are gradually revised as new goals immerge through a changing life. When the resistance is too big – when realising ones dreams and intentions are too difficult and painful - the goals are replaced in resignation with smaller, more obtainable goals. Such events result in the degeneration of the intent (life mission) (12-20,50,51,54-57) and the personal character, which sometimes even lead to mental illness like depression and schizophrenia, as suggested by Bleuler, Freud, Jung and others.

The only interpretation of the brain that is in accordance with all the collected data from all the sources mentioned above is that the brain is a pattern-machine, that continuously produces concrete and abstract patterns combining into "sensori-motor-pictures"; it seems that this process is guided by the organism's abstract high-level perceptual faculty of finding meaning in chaos, and a similarly abstract faculty of intent. It thus seems that it is the self (the organism's wholeness) that guides the brain in its making plans for achievement and realization of the abstract and concrete goals of the human being. The reality is interpreted in agreement with the intention and is represented for the conscious wholeness, where it is evaluated and processed. This seems to be in accordance with Arthur Schopenhauer (1788-1860) a German philosopher, who believed that the will to live is the fundamental reality.

Information guided self-organization of the neural patterns

We have analysed the process of morphogenesis and found that it is happening through information-directed self-organization of cells and tissue; all cell movements and differentiations are initiated and directed through information-directed self-organization of molecules and organelles[4]. It seems reasonable to suggest that the brain functions in a completely similar way through information-directed self-organisation of complex, dynamic, hierarchically organised, neural patterns.

The neural connectivity patterns are specified through information-transferring interactions on many levels of the living organism. This means that both the patterns of connectivity and the functional neural patterns of the working brain interact with the information-bearing, complex dynamic processes of the biological system; please recall that we have found this to be a real phenomenon existing in the organism at a quantum level (8). The functional neural patterns are different from the structural. But when structures and functions are developed in parallel through evolution, it must be that the functional patterns also interact with the informational level of the organism; a fine example of this is the *Hydra*; in the Hydra the neural network is constantly updated by the organism; if the body of hydra is reshaped by cutting, the neural information reconstructs it with no hesitation.

We can look and make sense of even the most complicated of patterns like for example at turbulent water flowing, growing plants and ecosystems, or computer produced fractals; from this it is clear that our brain that has the capability to form extremely complex patterns for perceptual use – much more complex and complicated than those structures that are before our eyes. The brain is formatting extremely complex patterns. This neural "modelling medium" (or matrix) can be organized either by sense input or by the consciousness and intention of the organism.

Since our senses are always flooded by information, it is obvious that a considerable and continuous selection of the incoming data is happening at all times. The intent of the Self determines what is interesting for the being from an existential perspective, and therefore defines the contents of its perception. Therefore it is correct to state that the content of consciousness it actually coursed by the Self. This is an extremely interesting conclusion as it makes the perception of the mentally ill including delusions, hallucinations, emotional flattening understandable (17, 50, 51). Most interestingly, intent is connected to the philosophy of life, and revising one's philosophy of life seems to completely alter the realm of perceptions; the patient in deep philosophical exploration is often "travelling from heaven to hell and vice versa". Also the perception of body, sexuality, the partner etc is completely mouldable by the person's philosophy of life.

The huge mass of data not found to be relevant for the realization of the organism's intentions are selected by the materialisation of a "pyramid of consciousness" from the abstract intentions to the concrete perceptions and behaviours; the intentions materialise "concrete plans" that support the specific sense-impressions. For the organism's wholeness intention materialises existential relevant experiences; in the brain intentions materialise the physical and well know description of the world used for everyday-living: the interpretation of the reality.

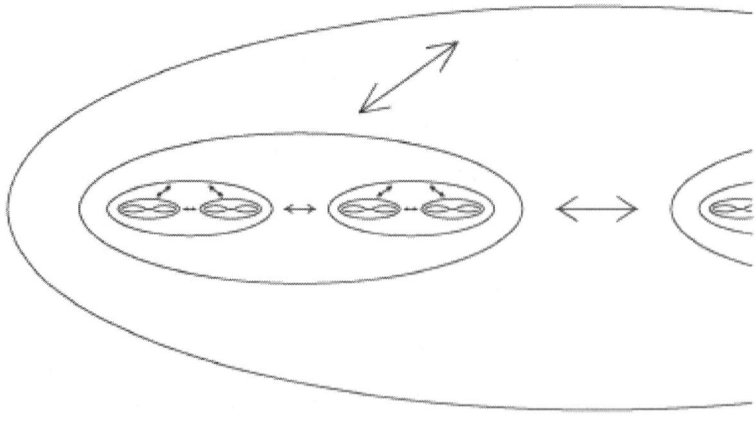

Figure 1. The living organism seen as a fractal structure with interacting parts on many levels. The arrows illustrate the information-transmitting interactions and are based on generalized empiric results. The figure illustrates how biologic informational systems are structured as a fractal "Chinese box"; the information transmitting interactions is seen between the different parts of the same level and between levels.

On a mechanistic, informational level the intentions must come to the brain in the form of superior matrixes of guiding patterns – these patterns correspond to those of the organism's wholeness; they organize the top-level patterns selected by the functioning brain, and the brain's self organising nature takes care of the rest. The organism's plans and strategies for self-realization are carried out in agreement with the rational interpretation of the organism's sense impressions. We know that in humans dreams are very important in this process.

A double hierarchical model for the reality representation of cortex cerebri

In order to simplify this description, we will only concentrate on the largest structures of the prosencephalon. The subject for the discussion will be the integrative structures such as the limbic system of cortex cerebri, the basal ganglia, and thalamus. The last two of these areas will be considered as equally connected with the areas of cortex cerebri. We think these structures represent a person's "emotionally close social relationships in the group" (the limbic system), groupings in the basal ganglia corresponding to the motor and verbal activities related to the outside world, and the whole perceived and cognised reality model in the thalamus and the related cortex. In the following, these structures are included in the term, cortex cerebri.

In cortex cerebri, two hierarchies seemingly exist, one for perception and one for action. Together, these stretch out the reality model, in the brain. One of these is a variant of the Chinese box system (see Figure 1) existing in all biological systems. In this, the brain as a whole and cortex cerebri is build up by substructures somewhat similar to the cortical areas. These again are separated in supra cellular structures – for instance feature-detectors in visual cortex – that again are build up by cells, etc as discussed above. The second hierarchy, roughly speaking, goes lengthwise through the brain.

The primary sensory input from vision, hearing, and somatic senses, give representations of these sense-spaces, in particular places. Seemingly, these areas converge to superior integrative areas that again converge to the highest integrative areas. The last ones are directly related to motor cortex, from where movement are controlled. However, the somato-sensory cortex is also placed next to somato-motor cortex. Figure 2 shows these hierarchies separated and together.

Explanation of the ability to associate, discriminate and abstract

The cerebral areas represent a separate level and use a time-space hierarchy (see discussion above). But it also has to interact with the superior wholeness of the brain. Data corresponding to everything that has happened on each level, could possibly be stored in each of the cortical areas from the complex dynamic at the lower levels. Recall that the storage of data may happen in an extremely controlled way, because each area contains unimaginable amount of information.

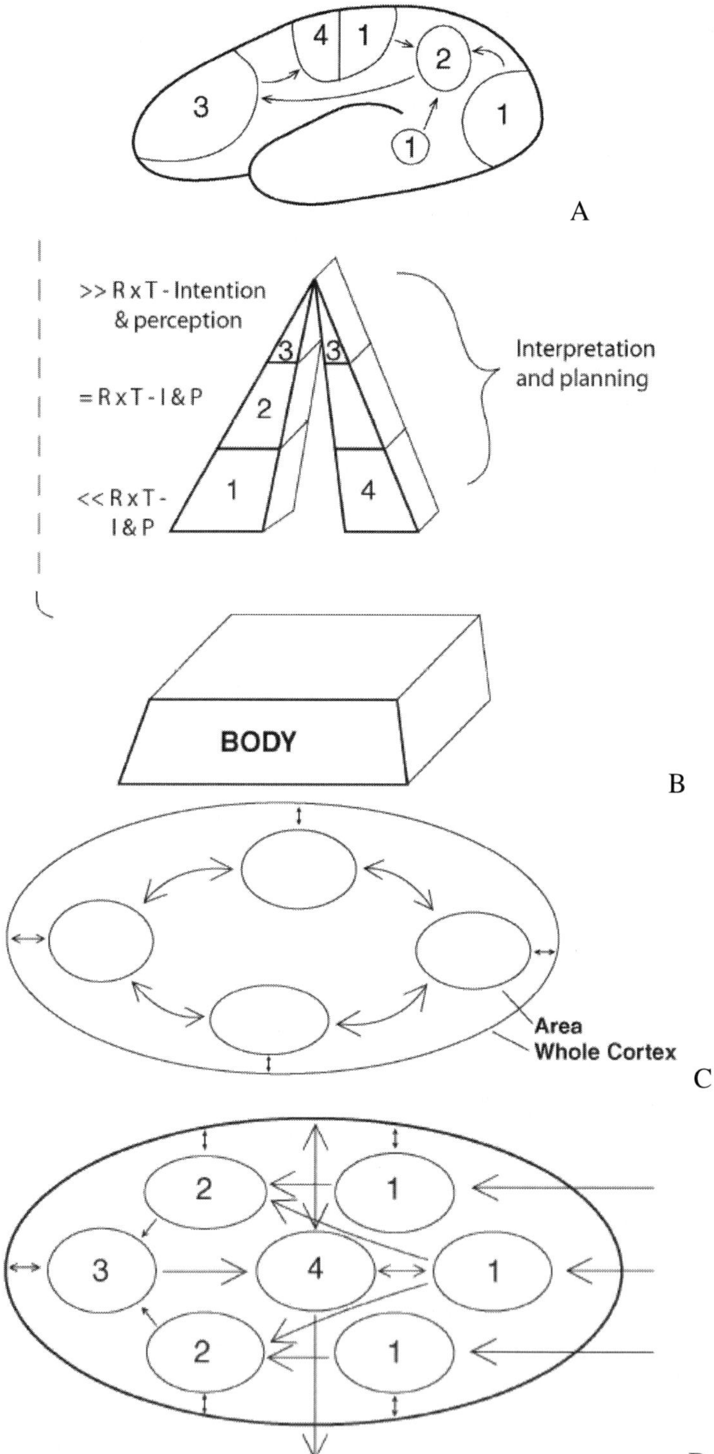

Figure 2. ABCD. A representation of the model of reality in the (human) brain as a consequence of a double hierarchy. A illustrates parts of the brain that interacts by each other. B illustrates the hierarchy of quantified qualia organised in space and time (I = Intention, P = Perception.). C illustrates the Chinese box hierarchy of interactions in the brain. D illustrates the double hierarchical representation of reality in the the brain.

A consequence of the ongoing information transmitting interactions on all levels of the brain is that every recall of data happens in an associated way. This means that only the data that passes into the larger pattern is retrieved, and it is the most superior pattern that organizes the complete recall, which in this connection is the intention. The ability to discriminate is presumably achieved through the structure of the functional patterns themselves. The ability to abstract and generalize follow from a co-representation of many elements having common traits in an n-dimensional, self-organizing, associative room (7) that will make up an informational body which forms exactly as the generalization or abstraction of all represented elements in this area. Such bodies (or sets) that are built by big amounts of smaller bodies will in a similar way correspond to higher levels of abstraction.

A proposal for the generation of the reality model in the brain

Figure 3 shows how a four-levels model of how reality can be created from simple repetitive co-groupings of elements of quantified qualia in space and time. First, sense impressions are analysed to meaning units in meaning unit analysers (this maybe too speculative like the "grandmother" cell and the mechanism might be energetic and not mechanic at all; comp. the visual feature detectors). The distribution of these "meaning units" in space and time is determined in the representation as positional information (compare the traditional use of this concept in ontogenesis (5)– the quantifying of the qualities. To realize the relationship between space and time in these groupings at different levels must be imagined. Thus the fundamental structure of space-time must be fully understood for us to fully comprehend consciousness.

The first grouping level causes the creation of sense elements corresponding to the perceptual level of qualia localised in space and time (this process takes milliseconds). On the motor side a corresponding grouping to motor-elements exists. The second grouping level is co-grouping of the different sense-modalities with sense elements in bigger space and time (many milliseconds or seconds) to perceptions. These cover the whole sensory-perceptual space. In the same way, motor-elements are grouped together to create concrete movements.

The third grouping level is higher space and time (seconds or minutes) groupings of perceptions and behaviour. This gives a reason-effect, where relationships between perceptions and behaviour, as possibilities, and those between behaviour and perceptions, as consequences. In the last case, the possibilities of a perception for action, and actions for perceptual consequences, is getting clear. On the third level of grouping, the co-grouping of the perceptions with behaviour in space and time, gives the reality models concrete functional elements – as for instance a cup – together with those processes that move and transform the elements into each other (the cup for example, can break in pieces when it is dropped on the floor). A child's limited intellectuality and its ability to interpret the concrete objects and possibilities of the surrounding world, has been build up at this point. The language with its nouns and verbs is introduced at this stage of life. Here, it is central to understand that the reality of the child is developed in identity with the biological intensions and needs.

The fourth group level replies to the groupings of possibilities and consequences, on higher space-time levels (from min. to years) that fit with the more complicated intentions of the human beings. In the reasonability of interpretation, this corresponds to the intellectual development, because the reality interpretation here is lifted from the concrete world into the world of ideas, abstractions, principles, hypotheses, lawfulness, and logic. In the emotional reality interpretation, this corresponds to the recognition of extremely complicated situations in the reality, outside the body. The hierarchical representation through cortex cerebri fits with the existing data of the brain. This representation also makes good sense. Since this fits with co-groupings of sense data, and data concerning motor functions, in a number of levels that, in the end, represents the human conceptions, ideas, and intentions.

The purpose by the first concrete grouping levels is to realise the concrete intentions of the child as eating, drinking, playing, etc. On the other hand, the abstract groupings serve the purpose of realizing the long-term intentions of the adult individual, through huge spaces and times.

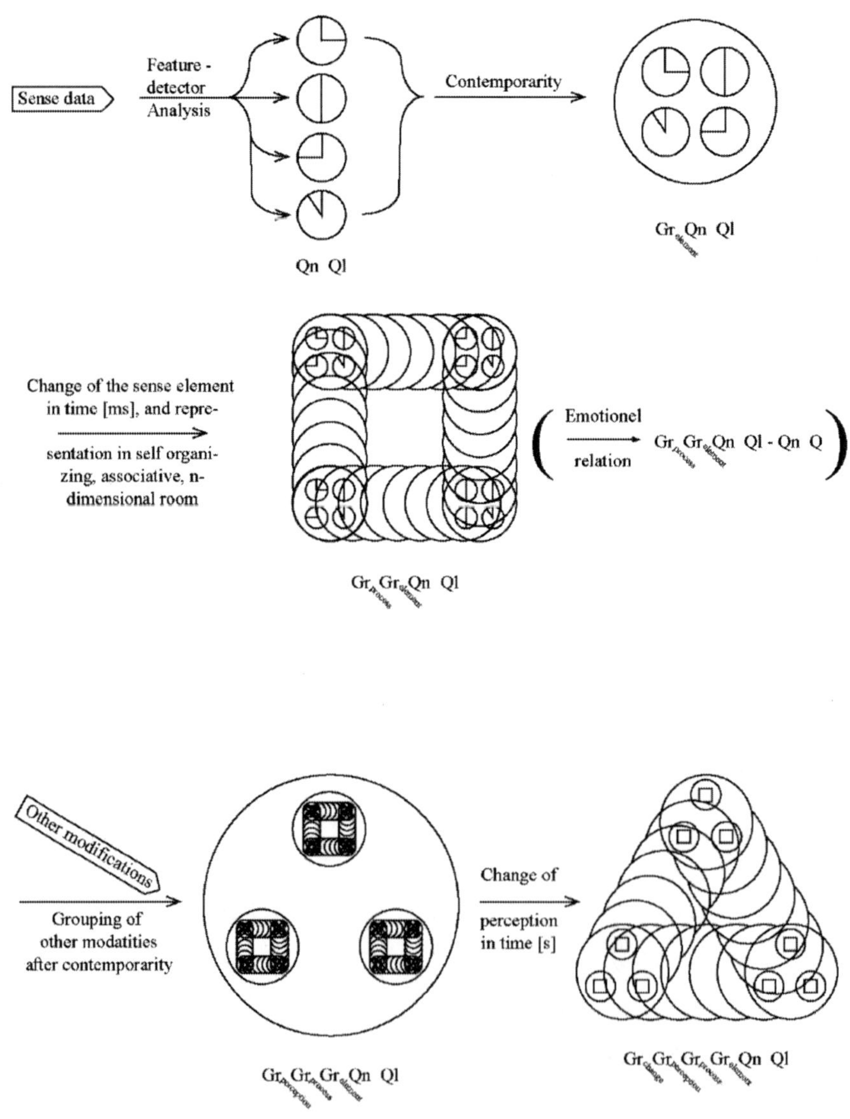

Figure 3. A model of the representation of reality in the human brain (see text).

For the brain, it is the case, that the intentions are superior patterns that organize the highest level of co-grouping of the elements of the reality model. A specific intention, in this way, results in a selection of specific superior consequence groupings, and these again, correspond to a plan or strategy. This strategy is realised through a de-grouping to sensory-motor elements, completely analogous to the superstructure of the reality model (see Figure 3). The intention, therefore, is the innermost craft that organizes the reality model in the brain. But, the co-ordinated learning acts directly on the intention. Also, a great part of the parent's reality interpretation is transferred in a direct way, through information directed interactions to the child as learned ideas with value-ladings attached to these. Therefore, the reality model is often filled with glaring conflicts, misty ideas and direct contradictions. This results in a really complicated structure caused by the demands of social interactions.

Discussion

We do not believe that the brain is a "neural network" that has a conscious activity of its own; neither do we believe that such an "isolated brain" could have an independent activity corresponding to thoughts and dreams. We believe that models for self-organizing associative memory have demonstrated a fundamental disability in thinking and perceiving; as the computerised "neural network" of an isolated brain can't have an independent activity like thinking. We believe the brain is a highly complex pattern machine that continuously produces concrete and abstract "sensory-motor-pictures" guided by the intention of the wholeness. The mechanism by which this is done by the brain is by using information-directed dissipative neural patterns, the information coming from both the senses and the organism's wholeness. The brains central job is to make meaningful plans for achievement and realization of goals presented to the brain at its top-level by the human wholeness (the "I"), often in dreams. The reality is interpreted in agreement with the intention and is through the brain's highest level represented for the organism's wholeness in which it is evaluated emotionally. The presented model is built on a number of axiomatic statements derived from the former papers in this series. It is quite surprising to us that it is possible to get such a clear picture of human brain function that in so many ways is helpful i.e. in explaining the perceptive distortions of the mentally ill from the degeneration of intent. The model seems highly helpful in relation to clinical holistic therapy, where these distorted perceptions in the form of transferences and projections are happening at all the times. The clear understanding of their neural basis will presumably make it much easier to deal with the mentally ill in therapy, and increase the number that can be helped by scientific holistic medicine. The problem of the proposed model is that you need a holistic philosophy admitting the individual cell consciousness etc. to use it.

Conclusions

We propose a theory for the function of the (human) brain claiming that it on a mechanical level works through information-directed self-organization of neurally produced extremely complex patterns, which only add up to meaningful perceptions and actions because of the fundamental will, or intentions, of the individual. We assume likewise that the morphogenesis of the brain is happening through the information directed self-organization of cells and tissues, and that this informational link is active throughout the individual life, securing an extremely close informational connection between mind and Self (the wholeness of the organism). All growth, absorption and modifications of nerve cells, axons, and dendrites are guided through information-directed self-organization of the molecules and organelles (1-10).

When we look at the most complicated, visual patterns as for example turbulent water flow, growing plants or computer produced fractals (101), and understand how fast and direct the brain interprets even the most complicated of visions, we find it obvious that a "patterned medium" exists in our brain that immediately forms even extremely complex patterns; these extremely complex patterns are organized in many hierarchical levels to create the well known model of reality in the human brain. A possible consequence of the information-transmitting biological interactions in the "deep quantum field" (8) is that recalls of information happens in an associative way. Only memorised patterns that resonate with actually activated patterns are recalled. This ability to discriminate is achieved through the structure of the functional patterns themselves. The ability to abstract and generalize follows from a co-representation of many elements having common traits in an n-dimensional, self-organizing, associative space.

In cortex cerebri two hierarchies stretch out the reality model in the brain. One of these is a variant of the Chinese box system. In this, the brain is build by structures in the cortical areas separated in supra-cellular structures that in turn are build by cells. In the second hierarchy, the primary sensory input from vision, hearing, and somatic feeling, gives representations of these sensory-spaces. The last hierarchy is directly related to motor cortex, the part of the brain from where movement is controlled. This sum up our proposal for the generation of the In our model of reality in the (human) brain, the. The many-levelled organising biological (positional) information organises the distribution of the qualia (meaningful units) in space and time. We suggest that the intentions of the whole individual are represented in the brain as superior patterns that organize the highest levels of the model of reality in the brain; this high-level co-grouping of all the elements of consciousness by the intention is the innermost craft that organizes the perceptual and acting powers of the brain. The brain is by evolution structured to transform all modalities of sensation and intention into its integrated perception of reality, in a form that empowers it for optimal action to achieve self-actualisation – the full realisation of the abstract "I".

Appendix 1: The value of existing sources of knowledge of the brain for understanding its function

Sources of experimental data about the structure and function of the brain: Data concerning the brain come from different sources, but yet they are not - even when pooled together - adequate to fully explain consciousness and the mind. To postulate that the functions of the brain are understood, we need the collected data to make sense and be able to explain all aspects of the mind; but most of the data does not make much sense – as the EEG and data definitely do not sum up to a nice understanding of what is going on. Some essential sources of data on the brain are mentioned below.

Anatomically the brain can be divided in a forebrain, a midbrain and a hindbrain that again are separated in several structures – as for example cortex cerebri separated in sub structures of cerebral cortex. These sub structures are split in supra cellular structures in the primary visual cortex – meaning analysers/feature detectors – that seemingly analyse incoming counts nerve cell signals.

The morphogenesis of the brain is uniform to that of other organs and can be essentially separated from those in reverse order, when axons and dendrites make the connectivity between nerve cells. This connects the different parts of the brain with each other so sensory inputs from receptors in the sense organs and body can be associated with motor outputs of the apparatus of movement.

Histological studies (79) have shown that pyramid cells (that count approximately half of all cells in cerebral cortex) converge to about 1,000 of the 3 million cells in cerebral cortex positioned only 3-4 cells apart from each other. Electron microscopically measurements of the distribution of sound impressions in cortex, give the same kinds of results (88,94).

Connectivity: The cortex cerebri can be understood as a surface having a 1,000-dimentional connectivity. The neuron length is about 1,000,000 km indicating the tremendous complexity of the brain. The existence of meaning-unity analysers/feature detectors, is most evidently in the primary visual cortex, but probably exists in all primary cortices. It has been proposed (95,96), that cerebral association-cortex, in general, should be organized in functionally modules of 300 μ. But this is not supported by evidence in the literature, because the existing columns only seem to be caused by corticocortical termination from pyramid cells (11).

Physiological studies have shown an intense interaction between all structures of the brain. Thalamus, fore instance, may be assumed to play a central role as a regulator of input to the cerebral cortex. Motor outputs that happen through motor cortex, seemingly is essentially influenced by the limbic structures and the basic ganglions, and also the patterns of movement (and perceptional patterns) from the cerebellum are commissioned. Generally, all parts of the brain are mutually connected with each other, but the total brain function is not yet understood.

Electro physiological studies have shown that inputs are well-arranged in cortex in two-dimensional maps repeated many times throughout the different brain areas. Somatotopic maps in somatosensory and motor areas, represents the different sense-receptor types and motor aspects of the body. Visio type maps represent the visual field in the visual areas, and tonotropic maps represent the tone scale. The organization of the smell sense in cortex is not yet understood. Trials with maps show that these cortical representations almost instantaneously can be organised, why they hardly are associated directly with axons and dendrites (88).

Electron microscopic tests of complex cells and groups of cells show, most clearly in primary visual cortex, an organization that is used to analyse meaning-unities of visual impressions.

Data from electro encephalogram (EEG) is very difficult to interpret. Simultaneously, they reflect the activity from cells in the cortical surface. Normally the electronic voltage arises and decreases 8-30 times per second when awake. The most important result is that a high amount of patterns integrates constantly while the brain works (88).

Studies of the blood flow in cerebral cortex (4,5) and of cortical lesions, have given a rather detailed map of the localization of the cerebral cortex's function. We know that data from the primary sensory areas as prefrontal cortex are dispositioned in areas of higher levels.

Psychological studies (93-94) have shown that the consciousness is build of sense-motor activities, and that the development through childhood follows the development of the body organization. The consciousness consists of sense and motor functions created by sense-motor elements. Later in the development, the child learns to think in abstractions and then talk. In this way, the consciousness gets freedom to use the body in more complex ways.

Philosophy. What we think gives the consciousness meaning. This interpretation is fundamentally connected with the body. The development of the consciousness as an objective matter of interpretation, results in hiding the original physical beginning where the connection between body and mind is being obscure.

Evolutionary studies reveal that simple nerve systems such as *Hydras*, that is thought to be the first organism with a nerve network, could function through information-transformed interactions between nerve system and health. A quite simple nerve network consisting of neurons, where the nerve impulses can move along in all directions from the stimulus point (100), gives *Hydra* a possibility to carry out a very complex behaviour as for example catching of prey, swimming, and somersaults. Seemingly this nerve network does not need any practise to function.

Mathematical analysis; cortex cerebri as a model machine: Systems with elements that exitatorically and inhibitorically influences each other, such as neurons do, show a stabile self-organization. However, mathematical analysis show that when the connectivity is huge, as it is in cortex, the ability to stabilize the self-organization break down (90,91), and all patterns become liquid and non-durable.

Metabolism. About 20% of our energy is used by the brain that only makes up 2% of the body weight. It is estimated that this huge consumption of energy, corresponds with an almost constant activity of impulses in all the brains nerve cells (N. A. Lassen, personally communication 1990). However, seemingly through an extreme "mental burden" this metabolism does not increase considerably.

Discussion of the significance of the above listed data sources

How does the brain work? Throughout time, there have been a lot of proposals on how to explain the functions of the brain. However, nobody has been able to explain "higher psychic functions" and existing explanations seems not to support what we know about the brain. For example, it is often proposed that the brain process its data in the same way as computers. Indeed computer implemented "neural network" with learning specific "synapse strengths" can be trained to process input patterns into specific output patterns.

Coordinated learning through synapse strength is known from lower level animals (88), but as mentioned, research on the cortical maps have documented that cortical representations are not hard-wired. We are therefore forced to conclude that the brain is not using "neural networks" of any known type. A neural network is basically a slave without creativity, and do not have an independent activity corresponding to thoughts and dreams, which is an other major problem if you try to explain the brain as a neural net.

Also, when it comes to consciousness, the non-local key quality of consciousness it very difficult to produce even in theory with a neural network; they are popular because they are easy to produce in silicon, but please remember that the electric activity of a computer is extremely local!!! Having dispersed information into a network is not helping us getting non-local consciousness as data now is stuck to certain positions in the network "Convergent groupings of feature-detectors" on an number of levels can theoretically give us a single cell that represents any object knows from our world - the famous idea of a "grandmother cell" that fires only when we see our grandmother! But the problem is of course how such the activity in a cell should be able to give us the consciousness of seeing our grandmother – how will this single-cell activity be clearly represented in out global brain activity? We believe that models for self-organizing associative memory like the famous Kohonen model (92) have the same disability as the other "neural network" and cannot have an independent activity like thinking.

Most interestingly there have never been constructed a truly self-learning neural network; all existing neural networks are in some way controlled from the outside though the programming of the net. Nothing artificial has yet been created that function like the brain, not even most superficially. This in itself should make people think about the difference between machines of silicon and living beings.

The most famous problem in brain research is the psychophysical problem. The materialistic version of this problem goes: how can consciousness emerge from biochemistry? How can any dead back of inorganic atoms end up feeling and living? Most people intuitively agrees that no matter how many balls you put in the bag, or how ingeniously you combine them with electromagnetic springs, they still do not live and laugh. As to the dualistic proposals to find a solution of the psychophysical problem (102), these proposals of the brain function seem to be an expression for philosophic resignation: The problem simply is to hard, so instead of solving it we are cheating.

Most researches that haves worked with this problem has actually in the end given up explaining consciousness, perception, intention etc (102). Brain research has turned completely mechanical, into molecular research; as if the brain researchers all has come to believe that just by finding smarter molecules they will some pretty day be able to solve the fundamental problems of brain and consciousness. (Maybe that is what is going to happen in the end when we finally get to understand proteins and find "intelligent molecules" that can read the extremely small energies of the directive collective quantum field of the biological system)(4-8).

It is not that we do not believe in science. But we need to acknowledge reality and stop moving away from the fundamental problems we need to solve – just avoiding the pain of not getting anywhere will not give us a new scientific understanding of the brain. We cannot understand the wholeness as a sum of its parts; the wholeness is always more. That is one of the most central understandings of holistic philosophy (1-11).

The brain (neocortrec) is making lots and lots of extremely complex patterns; that's pretty much what we can tell from all the brain research done. The nature of these patterns; the geometry they must be described in, the way they interact with other patterns, sensory stimuli and the inner biological informational system of the organism, remains most unfortunately still in the dark.

References

[1] Hermansen TD, Ventegodt S, Rald E, Clausen B, Nielsen ML, Merrick J. Human development I. Twenty fundamental problems of biology, medicine and neuropsychology related to biological information. ScientificWorldJournal 2006;6:747-59.

[2] Ventegodt S, Hermansen TD, Nielsen ML, Clausen B, Merrick J. Human development II. We need an integrated theory for matter, life and consciousness to understand life and healing. ScientificWorldJournal 2006;6:760-6.

[3] Ventegodt S, Hermansen TD, Rald E, Flensborg-Madsen T, Nielsen ML, Clausen B, Merrick J. Human development III. Bridging brain-mind and body-mind. Itroduction to "deep" (fractal, poly-ray) cosmology. ScientificWorldJournal 2006;6:767-76.

[4] Ventegodt S, Hermansen TD, Rald E, Flensborg-Madsen T, Nielsen ML, Clausen B, Merrick J. Human development IV. The living cell has information-directed self-organization. ScientificWorldJournal 2006;6:1132-8.

[5] Ventegodt S, Hermansen TD, Flensborg-Madsen T, Nielsen ML, Clausen B, Merrick J. Human development V: Biochemistry unable to explain the emergence of biological form (morphogenesis) and therefore a new principle as source of biological information is needed. ScientificWorldJournal 2006;6:1359-67.

[6] Ventegodt S, Hermansen TD, Flensborg-Madsen T, Nielsen M, Merrick J. Human development VI: Supracellular morphogenesis. The origin of biological and cellular order. ScientificWorldJournal 2006;6:1424-33.

[7] Ventegodt S, Hermansen TD, Flensborg-Madsen T, Rald E, Nielsen ML, Merrick J. Human development VII: A spiral fractal model of fine structure of physical energy could explain central aspects of biological information, biological organization and biological creativity. ScientificWorldJournal 2006;6:1434-40.

[8] Ventegodt S, Hermansen TD, Flensborg-Madsen T, Nielsen ML, Merrick J. Human development VIII: A theory of "deep" quantum chemistry and cell consciousness: Quantum chemistry controls genes and biochemistry to give cells and higher organisms consciousness and complex behavior. ScientificWorldJournal 2006;6:1441-53.

[9] Ventegodt S, Hermansen TD, Flensborg-Madsen T, Rald E, Nielsen ML, Merrick J. Human development IX: A model of the wholeness of man, his consciousness and collective consciousness. ScientificWorldJournal 2006;6:1454-9.

[10] Hermansen TD, Ventegodt S, Merrick J. Human development X: Explanation of macroevolution — top-down evolution materializes consciousness. The origin of metamorphosis. ScientificWorldJournal 2006;6:1656-66.

[11] Hermansen TD, Ventegodt S, Kandel I. Human development XI: The structure of the cerebral cortex. Are there really modules in the brain? ScientificWorldJournal 2007;7:1922-9.

[12] Ventegodt S, Andersen NJ, Merrick J. Editorial: Five theories of human existence. ScientificWorldJournal 2003;3:1272-6.

[13] Ventegodt S. The life mission theory: A theory for a consciousness-based medicine. Int J Adolesc Med Health 2003;15(1):89-91.

[14] Ventegodt S, Andersen NJ, Merrick J. The life mission theory II: The structure of the life purpose and the ego. ScientificWorldJournal 2003;3:1277-85.

[15] Ventegodt S, Andersen NJ, Merrick J. The life mission theory III: Theory of talent. ScientificWorldJournal 2003;3:1286-93.

[16] Ventegodt S, Merrick J. The life mission theory IV. A theory of child development. ScientificWorldJournal 2003;3:1294-1301.

[17] Ventegodt S, Andersen NJ, Merrick J. The life mission theory V. A theory of the anti-self and explaining the evil side of man. ScientificWorldJournal 2003;3:1302-13.

[18] Ventegodt S, Andersen NJ, Merrick J. The life mission theory VI: A theory for the human character. ScientificWorldJournal 2004;4:859-880.

[19] Ventegodt S, Flensborg-Madsen T, Andersen NJ, Merrick J. Life mission theory VII: Theory of existential (Antonovsky) coherence: a theory of quality of life, health and ability for use in holistic medicine. ScientificWorldJournal 2005;5:377-89.

[20] Ventegodt S, Merrick J. Life mission theory VIII: A theory for pain. J Pain Manage 2008;1(1):5-10.

[21] Ventegodt S, Andersen NJ, Merrick J. Holistic medicine: Scientific challenges. ScientificWorldJournal 2003;3:1108-16.

[22] Ventegodt S, Andersen NJ, Merrick J. Holistic Medicine II: The square-curve paradigm for research in alternative, complementary and holistic medicine: A cost-effective, easy and scientifically valid design for evidence based medicine. ScientificWorldJournal 2003;3: 1117-27.

[23] Ventegodt S, Andersen NJ, Merrick J. Holistic Medicine III: The holistic process theory of healing. ScientificWorldJournal 2003;3:1138-46.

[24] Ventegodt S, Andersen NJ, Merrick J. Holistic Medicine IV: Principles of the holistic process of healing in a group setting. ScientificWorldJournal 2003;3:1294-1301.

[25] Ventegodt S, Merrick J. Clinical holistic medicine: Applied consciousness-based medicine. ScientificWorldJournal 2004;4:96-9.

[26] Ventegodt S, Morad M, Merrick J. Clinical holistic medicine: Classic art of healing or the therapeutic touch. ScientificWorldJournal 2004;4:134-47.

[27] Ventegodt S, Morad M, Merrick J. Clinical holistic medicine: The "new medicine", the multi-paradigmatic physician and the medical record. ScientificWorldJournal 2004;4:273-85.

[28] Ventegodt S, Morad M, Merrick J. Clinical holistic medicine: Holistic pelvic examination and holistic treatment of infertility. ScientificWorldJournal 2004;4:148-58.

[29] Ventegodt S, Morad M, Hyam E, Merrick J. Clinical holistic medicine: Use and limitations of the biomedical paradigm. ScientificWorldJournal 2004;4:295-306.

[30] Ventegodt S, Morad M, Kandel I, Merrick J. Clinical holistic medicine: Social problems disguised as illness. ScientificWorldJournal 2004;4:286-94.

[31] Ventegodt S, Morad M, Andersen NJ, Merrick J. Clinical holistic medicine Tools for a medical science based on consciousness. ScientificWorldJournal 2004;4:347-61.

[32] Ventegodt S, Morad M, Merrick J. Clinical holistic medicine: Prevention through healthy lifestyle and quality of life. Oral Health Prev Dent 2004;1:239-45.

[33] Ventegodt S, Morad M, Hyam E, Merrick J. Clinical holistic medicine: When biomedicine is inadequate. ScientificWorldJournal 2004;4:333-46.

[34] Ventegodt S, Morad M, Merrick J. Clinical holistic medicine: Holistic treatment of children. ScientificWorldJournal 2004;4:581-8.

[35] Ventegodt S, Morad M, Merrick J. Clinical holistic medicine: Problems in sex and living together. ScientificWorldJournal 2004;4:562-70.

[36] Ventegodt S, Morad M, Hyam E, Merrick J. Clinical holistic medicine: Holistic sexology and treatment of vulvodynia through existential therapy and acceptance through touch. ScientificWorldJournal 2004;4:571-80.

[37] Ventegodt S, Flensborg-Madsen T, Andersen NJ, Morad M, Merrick J. Clinical holistic medicine: A Pilot on HIV and Quality of Life and a Suggested treatment of HIV and AIDS. ScientificWorldJournal 2004;4:264-72.

[38] Ventegodt S, Morad M, Merrick J. Clinical holistic medicine: Induction of Spontaneous Remission of Cancer by Recovery of the Human Character and the Purpose of Life (the Life Mission). ScientificWorldJournal 2004;4:362-77.

[39] Ventegodt S, Morad M, Kandel I, Merrick J. Clinical holistic medicine: Treatment of physical health problems without a known cause, exemplified by hypertension and tinnitus. ScientificWorldJournal 2004;4:716-24.

[40] Ventegodt S, Morad M, Merrick J. Clinical holistic medicine: Developing from asthma, allergy and eczema. ScientificWorldJournal 2004;4:936-42.

[41] Ventegodt S, Morad M, Press J, Merrick J, Shek DTL. Clinical holistic medicine: Holistic adolescent medicine. ScientificWorldJournal 2004;4:551-61.

[42] Ventegodt S, Solheim E, Saunte ME, Morad M, Kandel I, Merrick J. Clinical holistic medicine: Metastatic cancer. ScientificWorldJournal 2004;4:913-35.

[43] Ventegodt S, Morad M, Kandel I, Merrick J. Clinical holistic medicine: a psychological theory of dependency to improve quality of life. ScientificWorldJournal 2004;4:638-48.
[44] Ventegodt S, Merrick J. Clinical holistic medicine: Chronic infections and autoimmune diseases. ScientificWorldJournal 2005;5:155-64.
[45] Ventegodt S, Kandel I, Neikrug S, Merrick J. Clinical holistic medicine: Holistic treatment of rape and incest traumas. ScientificWorldJournal 2005;5:288-97.
[46] Ventegodt S, Morad M, Merrick J. Clinical holistic medicine: Chronic pain in the locomotor system. ScientificWorldJournal 2005;5:165-72.
[47] Ventegodt S, Merrick J. Clinical holistic medicine: Chronic pain in internal organs. ScientificWorldJournal 2005;5:205-10.
[48] Ventegodt S, Kandel I, Neikrug S, Merrick J. Clinical holistic medicine: The existential crisis – life crisis, stress and burnout. ScientificWorldJournal 2005;5:300-12.
[49] Ventegodt S, Gringols G, Merrick J. Clinical holistic medicine: Holistic rehabilitation. ScientificWorldJournal 2005;5:280-7.
[50] Ventegodt S, Andersen NJ, Neikrug S, Kandel I, Merrick J. Clinical holistic medicine: Mental disorders in a holistic perspective. ScientificWorldJournal 2005;5:313-23.
[51] Ventegodt S, Andersen NJ, Neikrug S, Kandel I, Merrick J. Clinical Holistic Medicine: Holistic Treatment of Mental Disorders. ScientificWorldJournal 2005;5:427-45.
[52] Ventegodt S, Merrick J. Clinical holistic medicine: The patient with multiple diseases. ScientificWorldJournal 2005;5:324-39.
[53] Ventegodt S, Clausen B, Nielsen ML, Merrick J. Advanced tools for holistic medicine. ScientificWorldJournal 2006;6:2048-65.
[54] Ventegodt S, Clausen B, Merrick J. Clinical holistic medicine: The case story of Anna: I. Long term effect of child sexual abuse and incest with a treatment approach. ScientificWorldJournal 2006;6:1965-76.
[55] Ventegodt S, Clausen B, Merrick J. Clinical holistic medicine: the case story of Anna. II. Patient diary as a tool in treatment. ScientificWorldJournal 2006;6:2006-34.
[56] Ventegodt S, Clausen B, Merrick J. Clinical holistic medicine: The case story of Anna. III. Rehabilitation of philosophy of life during holistic existential therapy for childhood sexual abuse. ScientificWorldJournal 2006;6:2080-91.
[57] Ventegodt S, Merrick J. Suicide from a holistic point of view. ScientificWorldJournal 2005;5:759-66.
[58] Ventegodt S, Clausen B, Omar HA, Merrick J. Clinical holistic medicine: Holistic sexology and acupressure through the vagina (Hippocratic pelvic massage ScientificWorldJournal 2006;6:2066-79.
[59] Ventegodt S, Clausen B, Merrick J. Clinical holistic medicine: Pilot study on the effect of vaginal acupressure (Hippocratic pelvic massage). ScientificWorldJournal 2006;6:2100-16.
[60] Ventegodt S. [Min brug af vaginal akupressur]. My use of acupressure. Ugeskr Laeger 2006;168(7):715-6. [Danish]
[61] Ventegodt S, Thegler S, Andreasen T, Struve F, Enevoldsen L, Bassaine L, et al. Clinical holistic medicine: Psychodynamic short-time therapy complemented with bodywork. A clinical follow-up study of 109 patients. ScientificWorldJournal 2006;6:2220-38.
[62] Ventegodt S, Thegler S, Andreasen T, Struve F, Enevoldsen L, Bassaine L, et al. Clinical holistic medicine (mindful, short-term psychodynamic psychotherapy complemented with bodywork) in the treatment of experienced impaired sexual functioning. ScientificWorldJournal 2007;7:324-9.
[63] Ventegodt S, Thegler S, Andreasen T, Struve F, Enevoldsen L, Bassaine L, et al. Clinical holistic medicine (mindful, short-term psychodynamic psychotherapy complemented with bodywork) improves quality of life, health, and ability by induction of Antonovsky-salutogenesis. ScientificWorldJournal 2007;7:317-23.
[64] Ventegodt S, Thegler S, Andreasen T, Struve F, Enevoldsen L, Bassaine L, et al. Clinical holistic medicine (mindful, short-term psychodynamic psychotherapy complemented with bodywork) in the treatment of experienced physical illness and chronic pain. ScientificWorldJournal 2007;7:310-6.
[65] Ventegodt S, Thegler S, Andreasen T, Struve F, Enevoldsen L, Bassaine L, et al. Clinical holistic medicine (mindful, short-term psychodynamic psychotherapy complemented with bodywork) in the treatment of experienced mental illness. ScientificWorldJournal 2007;7: 306-9.

[66] Ventegodt S, Thegler S, Andreasen T, Struve F, Enevoldsen L, Bassaine L, et al. Self-reported low self-esteem. Intervention and follow-up in a clinical setting. ScientificWorldJournal 2007;7:299-305.
[67] Jones E. The life and works of Sigmund Freud. New York: Basic Books, 1961.
[68] Jung CG. Man and his symbols. New York: Anchor Press, 1964.
[69] Jung CG. Psychology and alchemy. Collected Works of CG Jung. Vol. 12. Princeton, NJ: Princeton University Press, 1968.
[70] Reich W. [Die Function des Orgasmus]. Köln: Kiepenheuer Witsch, 1969. [German]
[71] Lowen A. Honoring the body. Alachua, FL: Bioenergetics Press, 2004.
[72] Rosen M, Brenner S. Rosen method bodywork. Accessing the unconscious through touch. Berkeley, CA: North Atlantic Books, 2003.
[73] Anand M. The art of sexual ecstasy. The path of sacred sexuality for Western lovers. New York: Jeremy P Tarcher/Putnam, 1989.
[74] Antonovsky A. Health, stress and coping. London: Jossey-Bass, 1985.
[75] Antonovsky A. Unravelling the mystery of health. How people manage stress and stay well. San Franscisco: Jossey-Bass, 1987.
[76] Leichsenring F, Rabung S, Leibing E. The efficacy of short-term psychodynamic psychotherapy in specific psychiatric disorders: a meta-analysis. Arch Gen Psychiatry 2004;61(12):1208-16.
[77] Leichsenring F. Are psychodynamic and psychoanalytic therapies effective? A review of empirical data. Int J Psychoanal 2005;86(Pt 3):841-8.
[78] Leichsenring F, Leibing E. Psychodynamic psychotherapy: a systematic review of techniques, indications and empirical evidence. Psychol Psychother 2007;80(Pt 2):,217-28.
[79] Kelso JAS. Dynamic patterns: The self-organization of brain and behavior. London: MIT Press, 1995.
[80] Kelso JAS, Engstrøm DA. The complementary nature. Cambridge, MA: MIT Press, 2006.
[81] Wolfram S. A new kind of science. Champaign, IL: Wolfram Media, 2002.
[82] Ventegodt S. Philosophy of life that heals. [Livsfilosofi der helbreder.] Copenhagen: Forskningscenterets Forlag, 1999. [Danish]
[83] Penrose R. Shadows of the mind. Oxford: Oxford University Press, 1996.
[84] Hofstadter DR. What is it like to be a strange loop? In: Kriegel U, Williford K, eds. Self-representational approaches to consciousnesses. Cambridge, MA: MIT press, 2006:465-515.
[85] Grof S. LSD Psychotherapy: Exploring the frontiers of the hidden mind. Alameda, CA: Hunter House, 1980
[86] Grof S. Implications of modern consciousness research for psychology: Holotropic experiences and their healing and heuristic potential. Humanistic Psychol 2003;31(2-3):50-85.
[87] Goleman D. Healing emotions: Conversations with the Dalai Lama on the mindfulness, emotions, and health. Boston, MA: Mind Life Inst, 1997.
[88] Kandel ER, Schwartz JH. Principles of neural science. Amsterdam: Elsevier, 2000.
[89] Freeman WJ. Analysis of cerebral cortex by use of control systems theory. Logistics Rev 1967;3:5-40.
[90] Babloyantz A. Self-organization phenomena resulting from cell-cell contact. J Theor Biol 1977;68(4):551-61.
[91] Babloyantz A, Kaczmarek LK. Self-organization in biological systems with multiple cellular contacts. Bull Math Biol 1979;41(2):193-201.
[92] Kohonen T. Selforganization and associative memory. Berlin: Springer Verlag, 1984.
[93] Lassen NA, Roland PE, Larsen B, Melamed E, Soh K. Mapping of human cerebral functions: a study of the regional cerebral blood flow pattern during rest, its reproducibility and the activations seen during basic sensory and motor functions. Acta Neurol Scand Suppl 1977;64:262-3, 274-5.
[94] Friberg L, Olsen TS, Roland PE, Paulson OB, Lassen NA. Focal increase of blood flow in the cerebral cortex of man during vestibular stimulation. Brain 1985;8(3):609-23.
[95] Szentagothai J. The "module concept" in cerebral cortex architecture. Brain Res 1975;95:475-96.
[96] Mountcastle VB. An organizing principle for cerebral function: The unit module and the distributed system. In: Edelman GM, Mountcastle VB, eds. The mindful brain: Cortical organization and group-selective theory of higher brain function. Cambridge, MA: MIT Press, 1978:7-34.
[97] Piaget J. The construction of reality in the child. New York: Basic Books, 1954.

[98] Piaget J. The growth of logical thinking from childhood to adolescence. New York: Basic Books, 1958.
[99] Piaget J. The psychology of the child. New York: Basic Books, 1969.
[100] Campbell NA, Reece JB, Mitchell LG. Biology. Menlo Park, CA: Addison Wesley Longman, 2002.
[101] Peitgen H-O. Beauty of fractals: Images of complex dynamical systems. Berlin: Springer, 1986.
[102] McGinn C. The mysterious flame. New York: Basic Books, 2000.

Chapter XIII

Sound, music and our understanding of the human brain

The functioning brain behaves like one highly structured, coherent informational field. It can be popularly described as a "coherent ball of energy", making the idea of a local highly structured quantum field carrying the consciousness very appealing. If that is so, the structure of the experience of music might be a quite unique window into a hidden quantum reality of the brain, and even of life itself. The structure of music is then a mirror of a much more complex, but similar, structure of the energetic field of the working brain. This paper discusses how the perception of music is organised in the human brain with respect to the known tone scales of major and minor. The patterns used by the brain seems to be similar to the overtones of vibrating matter giving a positive experience of harmonies in major; but we also like the minor scale, which can explain brain-patterns being fractal-like, giving a symmetric "downward reflections" of the major scale into the minor scale. We analyse the implication of beautiful, ugly tones and harmonies for the model. We conclude that when it comes to simple perception of harmonies, the most simple is the most beautiful, and the most complex is the most ugly; but in music even the most disharmonic harmony can be beautiful, if experienced as a part of a dynamical release of musical tension. This can be taken as a general metaphor of painful yet meaningful and developing experiences in human life.

Introduction

Music, a complex cultural phenomenon of sounds, tones, harmonies, and rhythm, must be understood by the way it is created, and the way it influence the human being listening to it, in the context it is used. Many other vertebrates, however, besides the human being (e.g. songbirds and whales) use a language of tones and rhythms, when they communicate. Therefore, in this paper we will describe music as a phenomenon of nature, closely linked to biology and of course especially the nerve system of vertebrates.

As music has a highly mathematic structure we suggest that it is mirroring a mathematical functional aspect of the vertebrate brain. Most interestingly music seems to have a simple, mathematical fundament independent on the vertebrate morphology and brain size, which might give us a clue to understanding the functioning brain, and more specifically a new way to understand the universal language of rhythms and tones. In this paper we will investigate the elements of mathematics and morphology that creates the foundation of music.

Music can be analysed into its elements: rhythm, tone and sound; harmony, melody and harmonic tension; and words, meaning and intended effect. In this paper we will primarily discuss the principles behind the formation of tones and harmonies. But it is very important for the understanding of music as a whole, that we also discuss the principles behind rhythm and sound, because these, like the language of tones, also are important and products of nature. Therefore, to understand some of the basic problems concerning the music we begin with investigating what sound and rhythm really are.

What it sound?

Sound is formed by swinging bodies' rhythmic movements of the air. Such movements influence the liquids of the inner ear to move caused by vibrations in the eardrum spread through the hammer, anvil and stirrup (1). This movement makes a "permanent" wave in the cochlea (1). Membrane basilaris is affected and swings around its normal main position by which hair cells are activated and sends off impulses to the ganglion tympani (1). From there the signals are sent to the cortex cerebri for processing and interpretation (1). But simultaneously, the cortex cerebri is sending signals to the ganglion tympani – that functions as its own "tiny brain" of the ear – that via these signals modifies the activity of the brain in a way that is highly complicated, and not fully understood (2).

Thus we cannot look at the ear as an object of measurement that is objective, but instead as an instrument with a high degree of selectivity, that only register impulses of interest for the brain. However, it is also possible, that the ears neural system to a certain degree "by itself" is able to produce the specific signal needed by the brain, but to find out about this, more experimental research is needed. This means that a fundamental trait of sound is an interpretation of external signals reaching the ear through the molecules of the air.

What is rhythm?

Noise is a sound that cannot be organized by the brain. The rhythmic organization of the noise is a splitting up in beats. The beat is a pulse of sound separated by a silence, noise or sound of different intensity or quality. Rhythm is an organization of the time by a serial of beats, partitioning the time in intervals (3). If we accelerate beat we will get a transition from rhythm to tone. This transition happens at 10 Hz. Formerly it was perceived that this transition happens at 20 Hz (4), but our own experiences in the laboratory indicate 10 Hz as the transition border.

A really interesting phenomenon that can be observed by this is transition between quantity and quality. This transition illustrates how our brain interprets reality: where quantities grow in intensity and finally gets incomprehensibly large, we find the interface between the consciousness and the sub-consciousness. In this way the phenomena will appear in a new way not any more available for the analytical mind, but still available as experiential phenomena through the senses.

What is a tone?

A body that vibrates between 10 and 20,000 Hz makes movements in the air that forms the tones. The body vibrates as a sinus function representing the simplest form something is able to vibrate. Such tone is ideally seen absolutely pure, because it is not contaminated by other tones vibrating simultaneously with it (5).

Overtones and the vibrating string: Figure 1 shows a vibrating string, just as the ancient Greeks used as a starting point for understanding the tonal side of the music. The ideal vibration, the sinus curve, forms when the string vibrates between maximum and minimum. But bodies that vibrate acts more advanced, because it is not only the end of the string that stands more or less still when it vibrates. Here and there, on the nodes along the string, the string stands more still then it does other places along the string. In Figure 1-3 only the activity around the nodes are shown.

What is sound?

To understand a component that makes sound we most understand: 1) the attack and the specific form trough witch the sound is created, and 2) the quality of the created sound. Sound is created by the specific form of the instrument, and reflects the materials of the instrument, which has its own specific spectrum of overtones. In theory all overtones are always present, but each swinging body strengthens and individually hampers the spectrum of overtones.

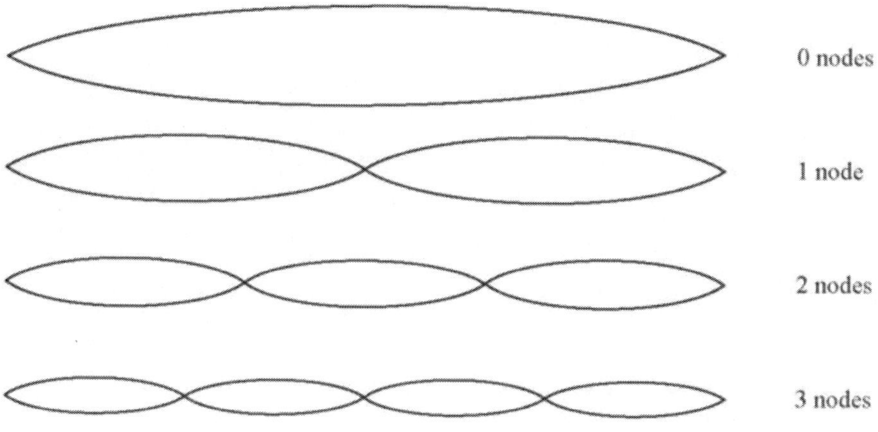

Figure 1. A swinging strings node points. This system is developed so the swings corresponds to 1,2,3, etc nodes. Each node has its own swing speed, corresponding to its frequencies, equivalent to the original strings frequency multiplied by the amount of nodes. By this, a number of tones called the number of overtones are formed (see also Figure 2).

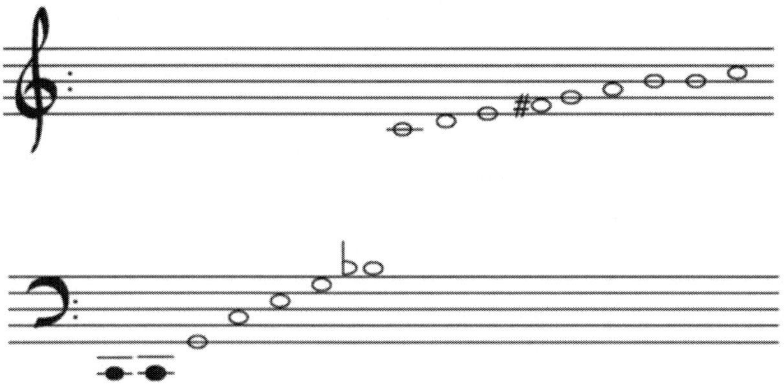

Figure 2. The number of overtones (nodes gives only an approximate description).

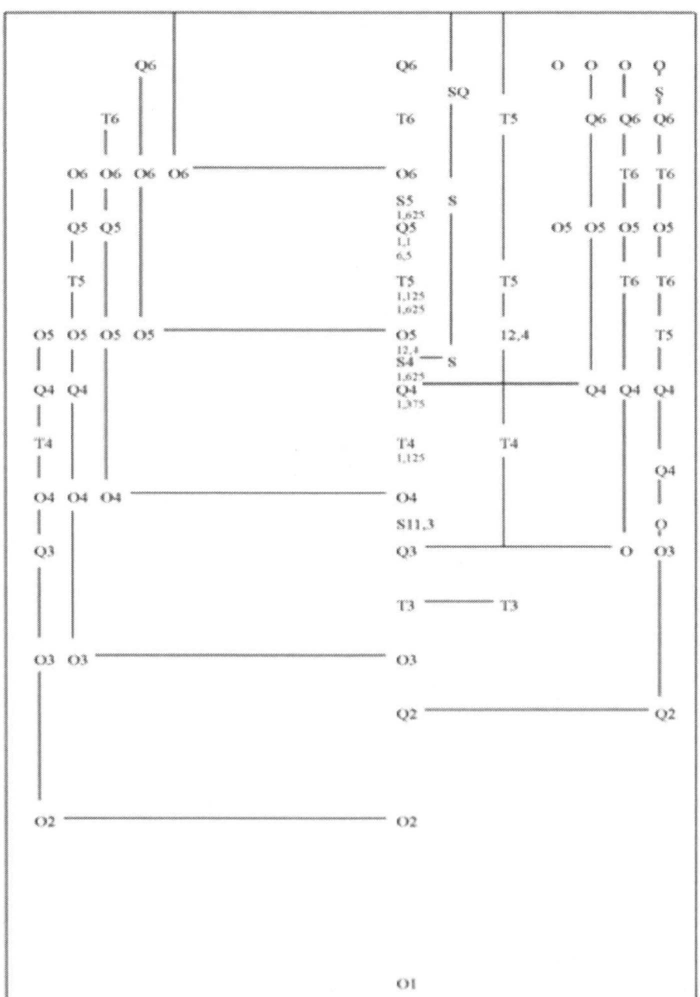

Figure 3. The tree of overtones. The image is shown, as its pattern would look on a millimetre paper.

This makes every individual instrument having its own sound – just think of the Stradivarius. The sound is closely related to the overtones from the materials, and both sound and tone are depending on the particular instruments form (6).

When a string swings, different tones can be distinguished. Almost all can hear the clean fifth that appears with the swing created by two nodes and a portioning in three parts of the string. Many people can also hear the huge clean third, while it demands some exercise to separate the highest overtones from the sound. Therefore the sound is particularly formed by the seventh and higher overtones (from the seventh or higher nodes) (Professor Yngve Trade, personally communication). Seven nodes are equivalent to the node B (approx. 30 cent too low, see later) with the C as the ground tone.

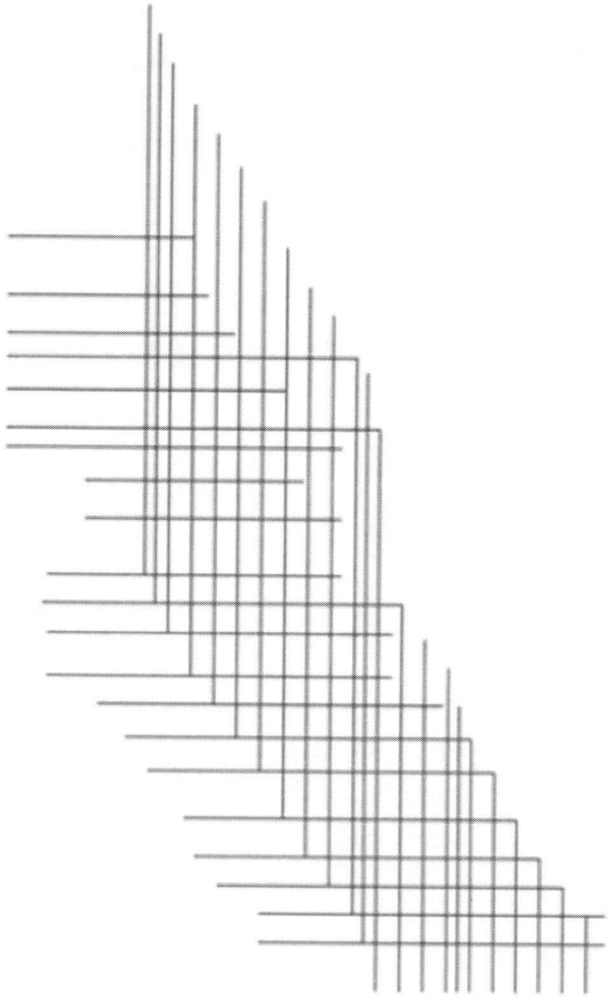

Figure 4. The tree of overtones in a simpler version. This figure illustrates a part of the pattern, as it would look on a millimetre paper.

Through the second transition from quantity to quality, tone is transferred to sound. This time, multiple tones are transferred and interpreted as sound because the brain has not capacity to interpret too many tones otherwise. Because it is limited how much complexity the consciousness is able to deal with simultaneously, this phenomenon can be defined as tolerance broadness. The brain may distinguish between rhythm, tone and sound (7).

The relation between wavelength, frequency and the cent value of the sound

A specific length of a string captures a specific swing speed (corresponding to a specific wavelength). This swing speed is dependent on the string material, thickness, restringing, temperature and other physically relations. The wavelength is reversed proportionally with the frequency. This means that if the wavelength is halved then the frequency is doubled, and vice versa. The closest overtone is found by doubling the frequency, corresponding to an octave and sensed as the keynote. This makes a sound, sounding as if the difference represented the distance between the octaves. Seemingly, the tone system is grounded on such repetition because the keynote and the octave are considered as the same tone (5).

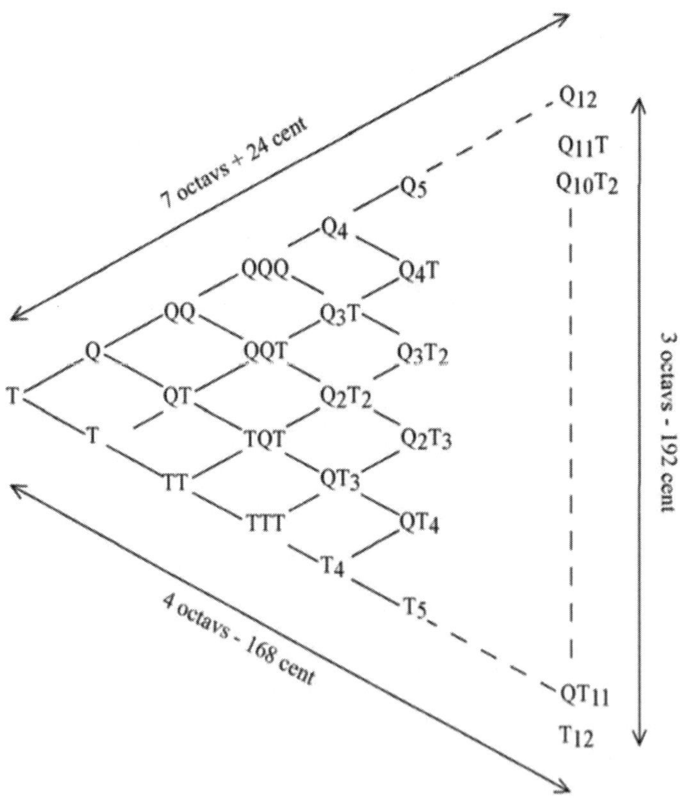

Figure 5. An equational representation of the truncated tree of overtones including information about the deviations from the simple 12-tone system. The pattern is converted from a millimetre paper.

A frequency of octaves divided in 12 equally peaces of 100 cent give the equally floating temperate chromatic scale (8,9). Such division of octaves is necessary to compare the heights of two tones, because both the wavelengths and the frequencies are misleading. This is the case, because the same difference in frequency or wavelength makes a different tone distance if the tones are located differently in the system of octaves. The ear is build to grasp the height of the tone as a logarithmic function of the frequency. This means that when the frequency is doubled multiply times, this is only experienced as a doubling of the tone height.

The intervals and the extended system of overtones

The system of overtones is more complicated than illustrated in Figure 2, because every overtone in an ideally elastic material, in itself is the keynote of its own overtones (5). In this way a complete, highly branched, tree of overtones is formed like a fractal system, where a specific pattern is repeated in itself on all levels (see Figure 3).

This simple tree can be represented in a simple way by focusing on the fifth and third and theirs overtones (see Figure 4). This figure shows all tones involved in a specific tone-height and compared to the keynote (each line represents a triad). The figure is illustrated through the clean fifths and thirds, where the most intervals (the distance between the horizontal lines) are formed around quarter-tones and therefore do not form a nice tonal system.

Concerning the trimmed truncated overtone trees we search for a scale (as we know from the C-major scale), or intervals enough to form a scale. The truncated tree of overtones can be adapted to the system (see Figure 5). By this adaptation the row in which fifths and thirds goes up has no connotation with the reached tone-height. By this the fifths third is the same tone as the thirds fifth, and the third's third fifth equal with the fifth's third third or the third's fifth third. Because we work in clean fifths and thirds we will never find the same tone more then once through the system that includes 1 + 2 + 3 + 4 + ... + 13 (=82) tones totally.

Such a system is extremely complex, because all included tones are different. However, it is possible to make such a system simpler if the lowest corner is drawn by T (12) (as if the system was a plate of rubber) so the (T3), T(6), T(9) and T(12) are going to be the octaves above the keynote, and Q(12) at the same time is pressed with the top side a bit up. In this way also Q(12) forms an octave, and a system with only 12 tones is created. This illustrates our 12 tone system. By this the third and fifth get a size called temperate in the equally floating temperate scale– the same as Johann Sebastian Bach (1685-1750) used because of its huge degree of freedom. This freedom is exactly illustrated by the freedom to move around in the system as illustrated in Figure 5.

If all the fifths and thirds by now are plotted onto a millimetre paper (illustrated by drawing in Figure 4), a simple system compared to the naturally overtones system are formed (see Figure 6).

To define the tones for an ordinary C-major is a difficult work because e.g. the tone F is defined as QQQTT, a very complicated composition. However, we have not directly admission to other intervals than fifths and (huge) thirds under such circumstances, and it is only these and the octave that are conceived as harmony between the overtones (caused by the truncation of the sound).

The idea is to release the whole language of musicality from the overtones, demanding that we discover the tones of the scale in a reasonably way. However, this has not been possible yet indicating the exclusion of an important principle. What is found is all the triads and the naturally ways to follow between these.

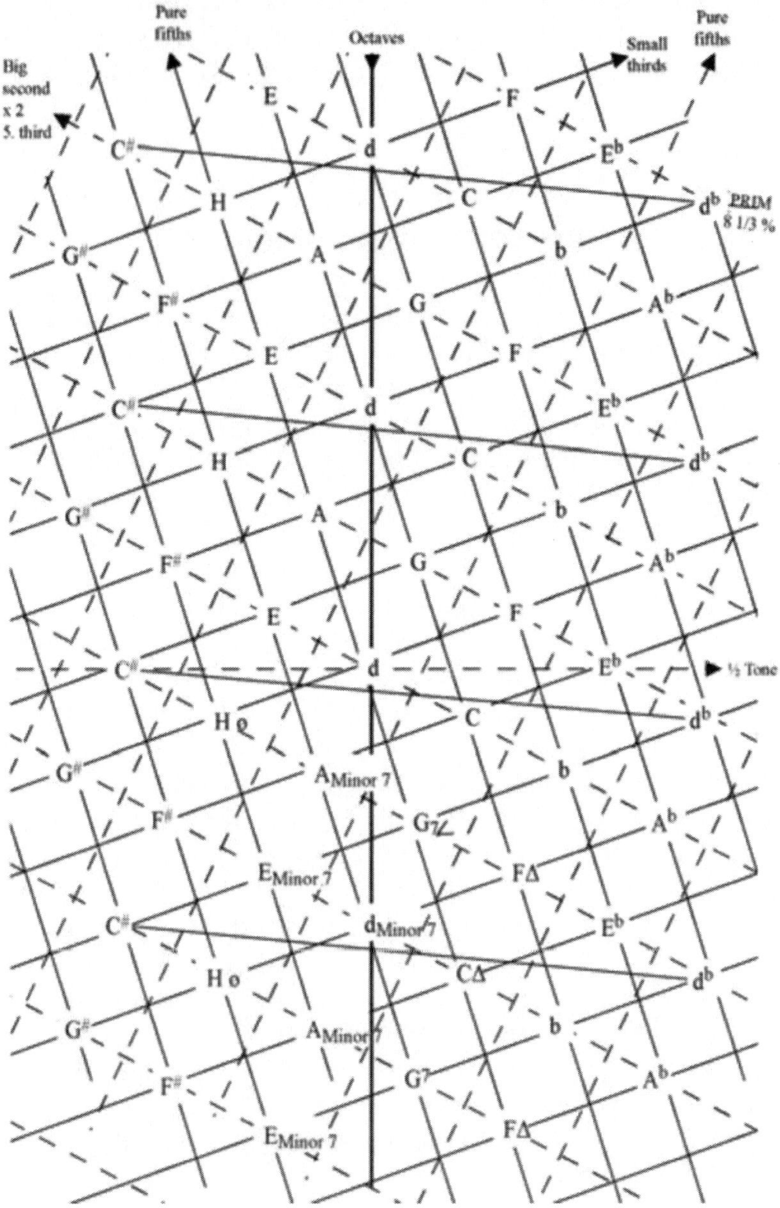

Figure 6. An illustration of the pattern of a classical tone tissue with all its intervals.

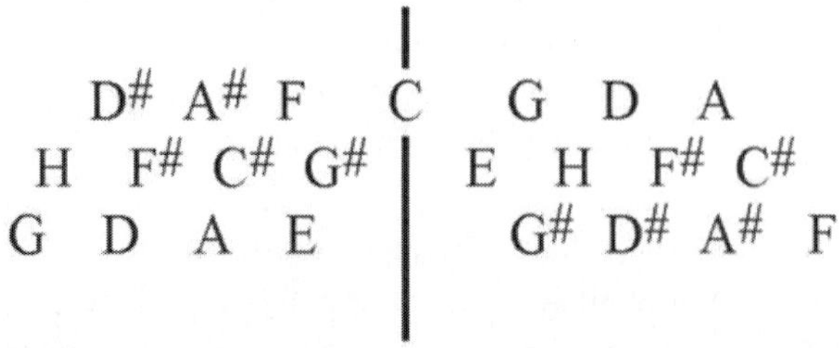

Figure 7. A reflection of the system around the tone C.

Namely the above described jumps of thirds and fifths. But, the scales and the complete melodic aspects are missing. Below we will expand the system to include these concepts.

The reflection around tonic

In our system we miss the foundation for minor, the fourth (F) and the small third. Leading the fifth downwards gives immediately the fourth. Though, this is not allowed following our rule saying that the overtones should be understood as being over the tonic. However, the brain could also have the capability of forming overtones downwards. So, we will try to make a functionally reflection making the downward fifth and third equally with the upward equivalents. Figure 7 shows that this gives the tone F as the missing downwards fifth of the tonic's (C's). F (the sub-dominant) is a really important tone ranging exactly under the fifth. If the third, however, is reflected, the result will be the tone G, not necessary to form the C-major. The problematic tone G interferes with the idea that a simple reflection can eliminate our problems, that two huge thirds sound awfully together. Or expressed in another way, two huge thirds give a very tense sound only enjoyable when the tension brakes. The question then, is which sound is the most beautiful? The simplest, or the most complex, tensioned and disharmonic? We think the answer is that the disharmonic sound can be separated in harmony (discussed later).

The complex overtone-multi-dimensionality

The simple reflection could not solve our problem and we still need to find a useable expansion to our system. An intimate investigation of the number of overtones shows that the 1st octave (sounds boring and normally overlooked) in the reality is stretched out over two complete tone tissues, analogue to those illustrated in Figure 5 (see also Figures 8A and 8B). Together these two-tone tissues may cover more then the fifth-third tone tissue, because the octave is more fundamental and of higher priority than the fifth and third. Furthermore, seventh (the 6st overtone) is stretched out entirely over 3 tone tissues, namely the tone tissues of the octave, the fifth and the third (Figures 8CDE). All the following overtones, consisting of a number of nodes equivalent to the prime numbers, similarly, stretch out an expanded number of tone tissues. In this way, the theoretical infinite number of overtones is stretching out as still a larger number of tone tissues all being tied in a huge multi dimensional structure sourcing all musical experiences. We call this the complex overtone-multi-dimensionality.

Because all nodes represent a number of overtones, the complex overtone-multi-dimensionality can be simplified as a roomy latticework. Professor Yngve Trade (personal communication) has explained to us that overtones normally are organized with seven or more nodes in the sound universe instead of in the harmonic world. Such reduction of the complex overtone-multi-dimensionality makes it exactly 3-dimentional and related to our well-known 3-dimentionally visually world. Therefore, the harmonic universe is exactly experienced as a roomy development contrary to the rhythms development in time. This makes a 3-dimensional model of the reduced complex overtone-multi-dimensionality in with the number of overtones are presented as nodes (see Figure 9).

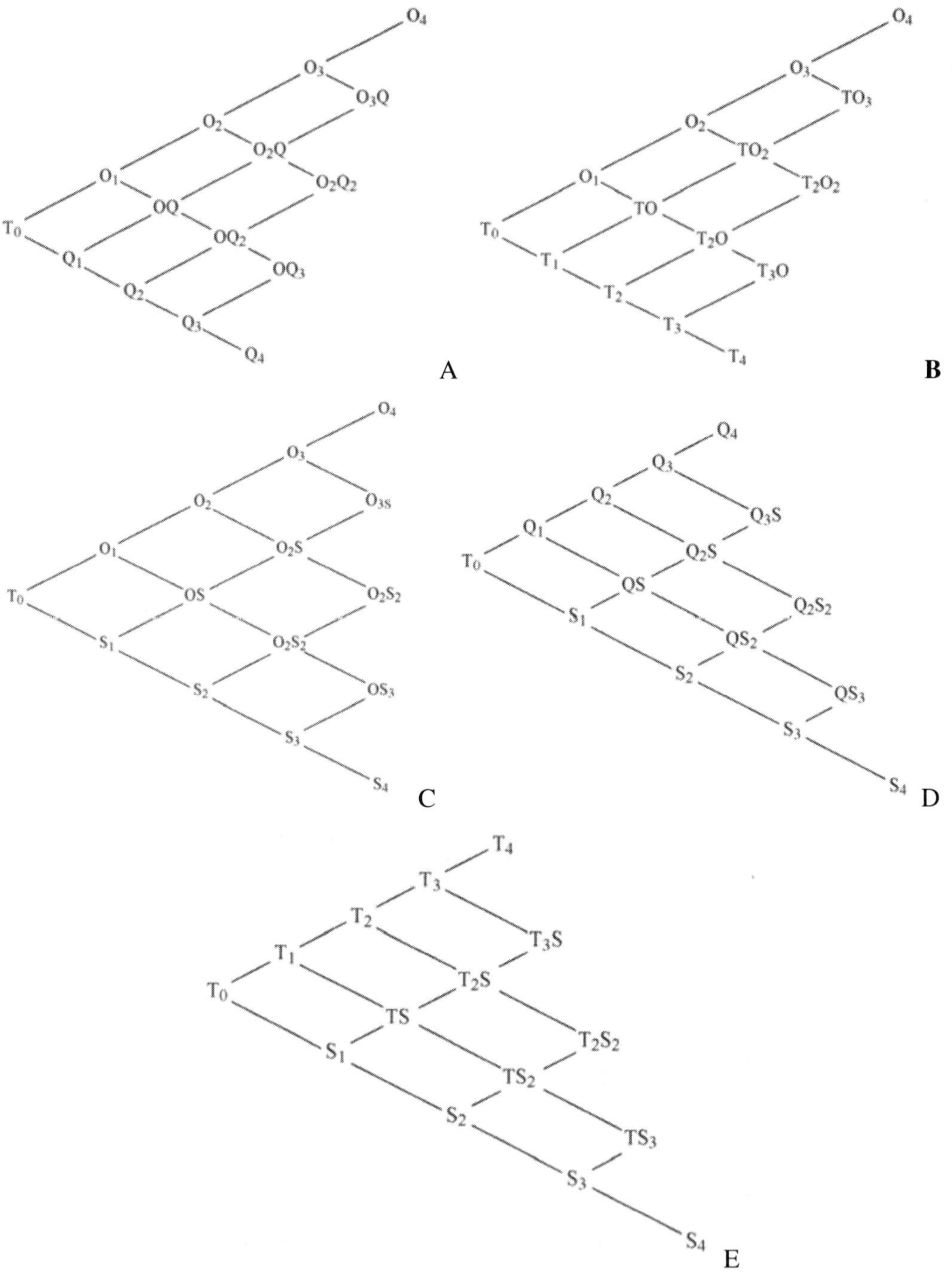

Figure 8. ABCDE. A. Illustrates the octave-fifth tone tissue. B. Illustrates the octave-third tone tissue. CDE. Illustrated the seventh's tone tissues.

In this model the reduced complex overtone-multi-dimensionality is expanded from a reflection of the fifth and third, also to include the octave and (in the principle) all overtone intervals of the tissue. It should now be possible in the principle, musically to move in all directions from a selected tone.

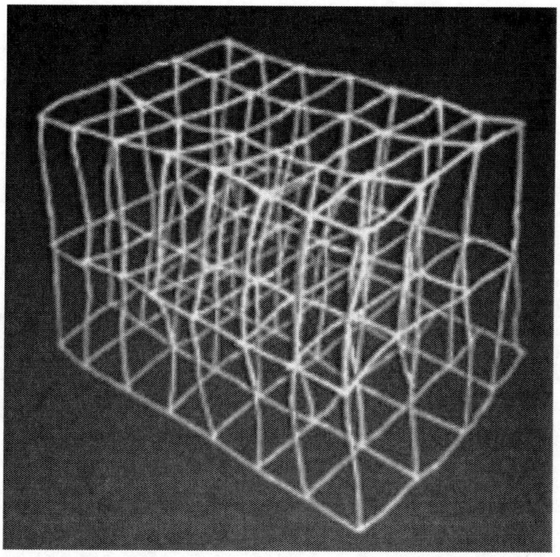

Figure 9. The reduced complex overtone-multi dimensionality as a 3-dimentional roomy representation of the harmonic sound universe of a human being.

Practically, we prefer to move the tone height upwards from the musically starting point. Such action looks to be attached to the functionally dynamics operating around the complex overtone-multi-dimensionality that makes the music. The reduced complex overtone-multi-dimensionality makes by this only the frame around the music, while the dynamics and the fundamental functions make the content. In a philosophic way the understanding of the complex overtone-multi-dimensionality can be interpreted as the structure in the human brain the emotions use to create the music.

The 3-dimentional model of the sound reducing complex overtone-multi-dimensionality gives a complete new and clearly understanding of the concept, to temperate the music, possible. Already the fifth-third tone tissue in Figure 5 shows that it would be possible to temperate the system especially by stretching the fifths 82 tones are reduced to 12. But this is no argument for postulating that the brain should prefer the equally floating temperate scale. However, from the model we can see that the number of octaves forms the most fundamental structure in the complex overtone-multi-dimensionality. If we project the fifths and the thirds on this structure that already is contained by the complex overtone-multi-dimensionality (this may be reasonable because it would simplify the system and make it harmonic) it will result in an equally floating temperate chromatic scale. This is illustrated by folding our complex overtone-multi-dimensionally model (see Figure 10).

This interpretation is the condition of all scales and the scale is the condition for the melodic tone world (9). By this the melodic concept can be understand as a freedom within the borders of our musically tolerance, when we choose not to use the roomy expression of the harmonies. Through a melody we can get a compact harmonic serial of tones presented in a high speed.

In this way, the harmonic series of tones are reduced to weak memories about harmony. Thereby, the human being gets the joy of being able to experience huge sequences of tones. Concerning the conditions between melody and harmony, this explains how the melody gets its huge degree of freedom compared to the harmony.

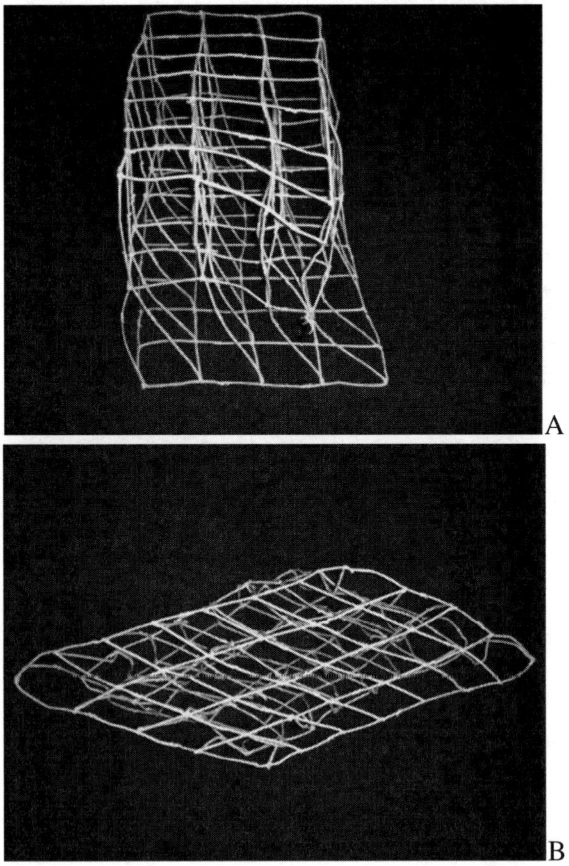

Figure 10. A folding of the complex overtone-multi-dimensionality that projects all the tones on the octave axis and by this immediately forms an equally floating temperate scale. A. A folding of a corner of the structure is transmitted throughout the structure. B. A complete stretch of the system from the two horizontal corners.

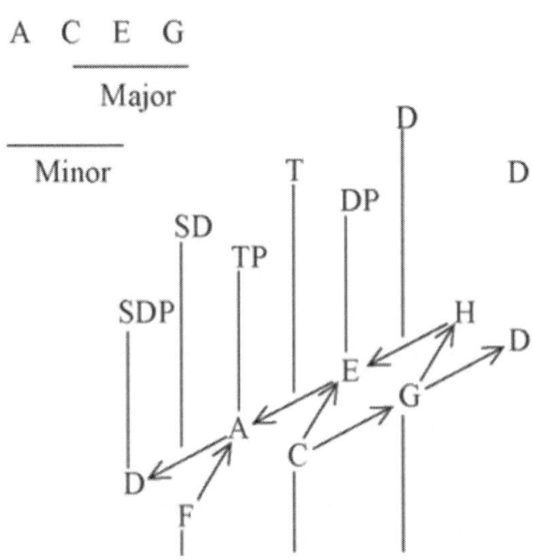

Figure 11. This figure shows the strategy for the development of the major and minor parallelism.

The release of the scale of major and minor from the model of the complex overtone-multi dimensionality

The up-folding of the 3-dimentionally complex overtone-multi-dimensionality around the octave axis gives a chromatic scale without any structures of major and minor. Therefore, this will not be able to source these tones. The major is, in a way, formed out of the fifth and third, because it is defined by a huge third as beginning point in the same way as the minor is defined with a little third as beginning point (8,9). The major and the huge third are involved in our terms, but the little third is not. The little third exists only as the difference between the fifth and the huge third. This difference indicates that one of the intervals should go up and the other down. This again indicates that, exactly going a third up and a fifth down is leading to the three parallel kinds of minor-tones. If we only go one fifth down will we, together with the tonic (C), get the three kinds of major tones (see Figure 11). It is exactly the mentioned jump that makes the "well known" tones. Here, the structure of the complex overtone-multi-dimensionality gives the possibility of a number of combinations of theoretically equally fifth-third intervals (see Figure 12).

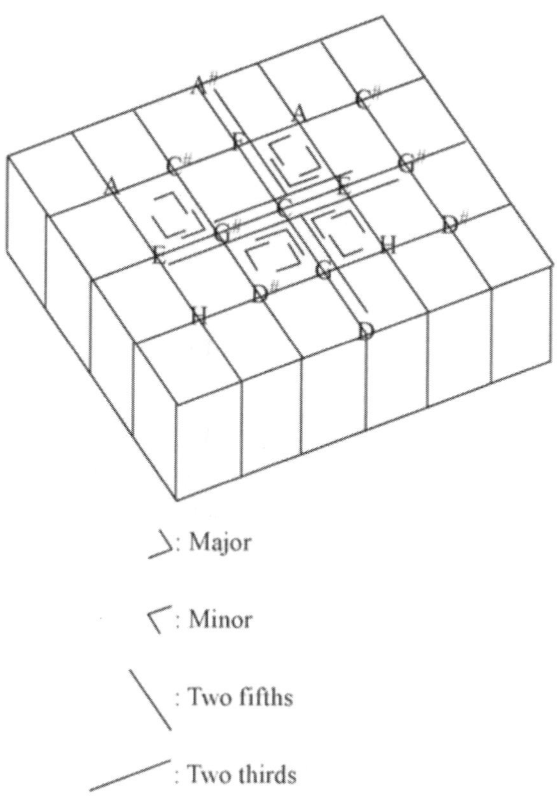

Figure 12. This figure shows the theoretical possibilities for fifth-third interval jumps in the reduced complex overtone-multi dimensionality.

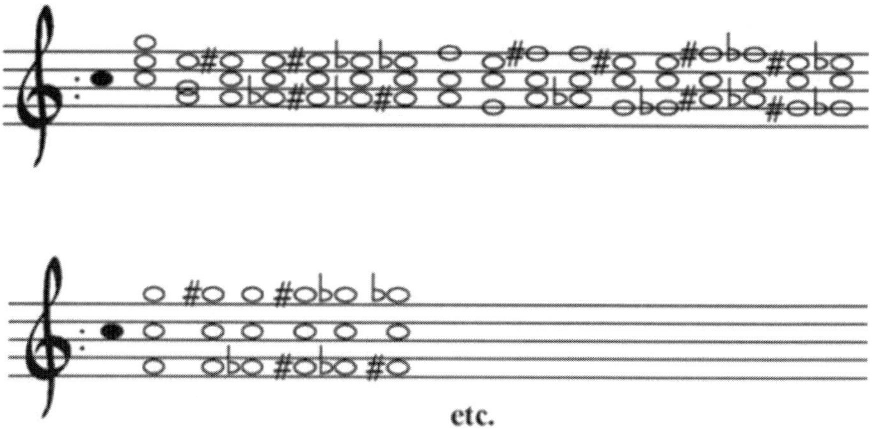

Figure 13. In this figure is shown the theoretical possibilities for triads that contain the tone C.

Interval-complexity as fundament of the choices of harmony-intervals

Are the intervals the final factor of the functional chose of intervals? This work has especially been based on computer analysis of the mutually conditions of the tones interference. This has lead to the conclusion that, for isolated intervals, it counts that simplicity is beautifulness and complexity is ugliness. Furthermore it is important to point out that this looks very different from a harmonic connection where the complexity is the same as musically tension, without which the music would be dull and boring. This shows that isolated intervals make the beautiful (resting) music, while the ugly is made of the compact (tense) intervals.

Interval complexity as fundament for the harmonic intervals

Our work (not published) concerning the conditions of interference in music has, as mentioned above, led to the conclusion that for isolated intervals simplicity is beauty and the complexity is ugliness. However, in a harmonic connection where the complexity is equivalent with musically tension, without which the music would be colourless and boring, these relations are a kind of reversed. The real beauty of the music seems to come from transformation of disharmony into harmony through time. When we forget the dimension of time and dynamic developments, the simple and harmonic becomes the beautiful. There seems to be an objective, static, beauty of simple intervals, but in addition an objective dynamic beautifulness coming from the gradual approximation to more simple intervals through the time.

Below we will try to show that the simple intervals are the beautiful ones. Simplicity results from a pattern of intervals happening through addition of the difference in the sinus curves between two different tones, because these contains only a few simple vibrations in a pattern that very fast repeats themselves. Therefore, beautifulness is represented by a periodic function and the periods are only brief when an interval is harmonic.

The possibilities for interval composition in the new model

Figure 13 shows the triads that could form the tone C. Theoretical there are following possibilities for combination of the fifth-third tone tissue. Two fifths on line, two thirds on line, and the ones that forms an angle, an upwards fifth and an upwards third, a downwards fifth and a downwards third, and the two combinations of an upwards and a downwards fifth and third.

First we analyse the two fifths and the two thirds. No one of these intervals is experienced as beautiful, when standing alone. The difference between the two fifth-intervals is called a "none" (9 tones), and the difference between the two third-intervals is a fifth. The straight lines in Figure 13 therefore, form a great amount of disharmony. When the third and the fifth of the third or fifth and the third of the fifth are gathered, the interval of half tones are also gathered. This interval looks to be very complex. A tension so huge demands a great musical talent to dissolve it. Therefore this interval is very seldom used, and standing alone, it is one of the ugliest intervals.

A fifth and an upwards third form the well-known and harmonic triad, the fundament of the major. This can also be formed by the combination, a downwards fifth and an upwards third. But this combination has another keynote. The combination with the upwards fifth and the downwards third forms the small third, the fundament for the minor triads.

Discussion

Our ability to understand perception of music from simple mathematic analysis indicates that the human brain uses a simple mathematical language to interpret the world. We have not yet identified this language, but from what we learned in this paper the brain seems to use a fractal model of reality, like a three with many branches, but going not just up like a normal tree, but also in a similar way going down from the point of focus. We have seen quantities turning into qualities, which seems to be a universal feature of how the nervous system creates meaning out of chaos in the world.

Our analysis of music seems to confirm our understanding of the brain as presented in an earlier paper (10). Seemingly the functioning brain behaves like one highly structured, but extremely coherent ball of energy, making the idea of a quantum field carrying the consciousness very likely (11-21). If that is the case, the structure of the experience of music might be a quite unique window into a hidden quantum reality of brain and life. The structure of music is then a mirror of a much more complex, but similar, structure of the energetic field of the working brain.

The minor and the major intervals are the fundaments of the most beautiful scales of music. The minor intervals are more complex then the major intervals and it can be discussed if their well-known saddening sound is grounded in this complexity (a disharmony) or in the fact that they are constructed by a downwards third that could symbolize a downward aim, or movement towards the earth.

When we move upwards in the system of tones (common major), we feel energized and happy, and this upward movement adds speed and movements in the music. The picture of a downward movement is often experienced as a loss of energy, and a slowing down. If the minor was sad because it contains the small disharmony, maybe this could be dissolved to major by going from the small to the huge third. This on the other hand, happens extremely seldom.

But if the minor is saddening, because it represents an energetic downward movement, this demands a fundamental reinterpretation of how the brain creates music; we suggest the existence of a multi-dimensional perceptional space, very similar to the complex three of overtones.

We know that our analysis of music is not yet thoroughly enough to answer the fundamental questions of music satisfactory. We have found proportionality between the complexity of the intervals and their subjective tension and beauty. A harmonic tension – a musical ugliness - that isolated could be experienced as painfully bad sound, can in a peace of music be transformed into a wonderful, harmonic, dynamic, and moving energy.

We learn, yet again, that the interesting human experience is created at the border between chaos and order; and we always love when we go from chaos to order. But without darkness and disharmony life is just a doll and indifferent affair.

References

[1] Vander AJ, Sherman JH, Luciano DS. Human physiology: The mechanisms of body function. New York: McGraw Hill, 2001.

[2] Gray H. Gray's anatomy: The classic collector's edition, 15th ed. New York: Gramercy, 1988.

[3] Arnold B. Rhythm primer: Music sight reading exercises. New York: Muse Eek Publ, 1998.

[4] Marieb EN. Human anatomy and physiology, Sixth ed. San Francisco: Benjamin-Cummings, 2003.

[5] Ardsley G. Temple of tone. Music book index. Temecula, CA: Reprint Services Corporation, 1925.

[6] Miller DC. (1916) Science of musical sounds. Music book index. Temecula, CA: Reprint Services Corporation, 1916.

[7] Jourdain R. Music, the brain, and ecstasy: How music captures our imagination. New York: Harper, 1998.

[8] Smith RL. Fingering for the accordion. Livermore, CA: Robert L Smith, 2005.

[9] Slonimsky M. Thesaurus of scales and melodic patterns. Music book index. New York: Schirmer Books, 1947.

[10] Ventegodt S, Hermansen TD, Kandel I, Merrick J. Human development XII: a theory for the structure and function of the human brain. ScientificWorldJournal 2008;8:621-42.

[11] Hermansen TD, Ventegodt S, Rald E, Clausen B, Nielsen ML, Merrick J. Human development I. Twenty fundamental problems of biology, medicine and neuropsychology related to biological information. ScientificWorldJournal 2006;6:747-59.

[12] Ventegodt S, Hermansen TD, Nielsen ML, Clausen B, Merrick J. Human development II. We need an integrated theory for matter, life and consciousness to understand life and healing. ScientificWorldJournal 2006;6:760-6.

[13] Ventegodt S, Hermansen TD, Rald E, Flensborg-Madsen T, Nielsen ML, Clausen B, Merrick J. Human development III. Bridging brain-mind and body-mind. Itroduction to "deep" (fractal, poly-ray) cosmology. ScientificWorldJournal 2006;6:767-76.

[14] Ventegodt S, Hermansen TD, Rald E, Flensborg-Madsen T, Nielsen ML, Clausen B, Merrick J. Human development IV. The living cell has information-directed self-organization. ScientificWorldJournal 2006;6:1132-8.

[15] Ventegodt S, Hermansen TD, Flensborg-Madsen T, Nielsen ML, Clausen B, Merrick J. Human development V: Biochemistry unable to explain the emergence of biological form (morphogenesis) and therefore a new principle as source of biological information is needed. ScientificWorldJournal 2006;6:1359-67.

[16] Ventegodt S, Hermansen TD, Flensborg-Madsen T, Nielsen M, Merrick J. Human development VI: Supracellular morphogenesis. The origin of biological and cellular order. ScientificWorldJournal 2006;6:1424-33.

[17] Ventegodt S, Hermansen TD, Flensborg-Madsen T, Rald E, Nielsen ML, Merrick J. Human development VII: A spiral fractal model of fine structure of physical energy could explain central aspects of biological information, biological organization and biological creativity. ScientificWorldJournal 2006;6:1434-40.

[18] Ventegodt S, Hermansen TD, Flensborg-Madsen T, Nielsen ML, Merrick J. Human development VIII: A theory of "deep" quantum chemistry and cell consciousness: Quantum chemistry controls genes and biochemistry to give cells and higher organisms consciousness and complex behavior. ScientificWorldJournal 2006;6:1441-53.

[19] Ventegodt S, Hermansen TD, Flensborg-Madsen T, Rald E, Nielsen ML, Merrick J. Human development IX: A model of the wholeness of man, his consciousness and collective consciousness. ScientificWorldJournal 2006;6:1454-9.

[20] Hermansen TD, Ventegodt S, Merrick J. Human development X: Explanation of macroevolution — top-down evolution materializes consciousness. The origin of metamorphosis. ScientificWorldJournal 2006;6:1656-66.

[21] Hermansen TD, Ventegodt S, Kandel I. Human development XI: The structure of the cerebral cortex. Are there really modules in the brain? ScientificWorldJournal 2007;7:1922-9.

Section V: Understanding pathogenesis and salutogenesis

Chapter XIV

Degeneration and regeneration of existence

In this chapter we see the living organism as an information-directed, self-organizing, complex, dynamic entity, closely integrated with the outer world. From the informational field of the organism, intent emerges and intent also direct the informational field, being a mysterious force in life. We discuss two different types of conscious learning of higher vertebraes: conditioning and recognition and relates these to the degeneration and regeneration of intent. We analyse the psychophysical problem and the problem of the free will. We also discuss the nature of conditioning as information storage in the complex dynamics at organism level instead of brain level. Conditioning can be seen as a consequence of a power struggle with suppression and submission, especially obvious between parents and their children. We analyse the mind-body problem, and discuss some factors interacting with and influencing the human development and the individual ability to heal spontaneiously by use of the human adult metamorphosis.

Introduction

Degeneration and regeneration of intend are concepts that are important to understand in the development from child to the independent life of an adult. Small children live in the emotional interpretation of reality and as they grow up to learn how to handle the outer world they gradually acquire the rational, mechanistic interpretation of reality. Under ideal conditions of growth the child unfolds larger and larger areas of its biological program and becomes a healthy individual living in harmony with itself and the world, a person with a spontaneous insight into reality.

However, the ideal conditions almost never prevail. Parents are not fully unfolded persons themselves, and therefore they force their children to adapt to their own reality by means of punishment and reward. This is what Skinner (1) called operant conditioning. In both cases, chronically ungratified needs appear. Learning is connected with behaviour through cessation or production of suffering. Fore example, the children are subjected to conditioned learning that shows the conditional nature of mental diseases (2-4).

In his later work Freund (5) assumed that man from nature possessed a sexual drive, the libido, together with a death drive, the destrudo. While libido is the urge to create, destrudo is the urge to destroy both yourself and everything around you. Together these two antagonistic drives were supposed to explain man's often contradictory actions. Freud supposed that the inherited intentions were only primitive instincts that had to be cultivated through conditioned learning (5,6). In contrast to this we believe that conditioned learning encourage to create diseases and evilness (7), while intentions of constructive activity in the world as expressed in love life, work life and family life, as well as all understanding, stems directly from the biological nature of man (4,7). Conditioning is a repression of the natural biological unfolding and growth of the child. Conditionings operate by undermining the original, constructive intention; the result is destructive intentions such as revenge and unconstructive assertiveness, and often a total loss of the intention of love or work (4,7). Conditioning often alienates man from the emotional interpretation and reifies him; man becomes more unloving and unconstructive (4,7,8). The purpose by this paper is to discuss these concepts and consider degeneration and regeneration of human intent in the view of our holistic biological ideas.

Degeneration and regeneration of intent

The nature of conditioning: information storage in the complex dynamics at organism level instead of brain level. Reality can be represented in two ways: directly through the wholeness of the organism, and indirectly through the brain (7). Information can generally be stored in the complex dynamics through information transmitting reciprocal interactions in the same way as retrieved (7).

The patterns at the deepest levels may be assumed to correspond to the biological purposes, and in the complex dynamics between the deep, fine levels and the superior level of the whole, patterns can be stored, when they are represented in the wholeness of the organism with sufficient strength. In the same way as the brain is able to save its representation of reality, the whole body is able to save the representation too. The nature of these representations differs: the brain administrates incidents in time and space, while the whole administrates the highest purposes of the organism (7). Due to the interactions between the wholeness and the lowest space-time groupings in the brain, we assume that data stored in the wholeness at the same time is represented in the lower levels of the brain.

Conditioning as a consequence of power struggle, suppression and subjection between people, especially between parents and children. Two people confronting each other (see Figure 1) interact in three ways: through speech, through action, and through their will and awareness. When their purposes conflict, the two people persons conflict, and they may behave aggressively towards one another, try to persuade each other through words, or to subject the other to their own purpose directly through use of will. In confronting each other people will most often use one of these three possibilities. If one of the two persons in the conflict is weaker than the other, he succumbs.

Parent-child relations are examples of the uttermost unequal distribution of power, because children have not developed their model of reality and therefore are not able to defend themselves with words.

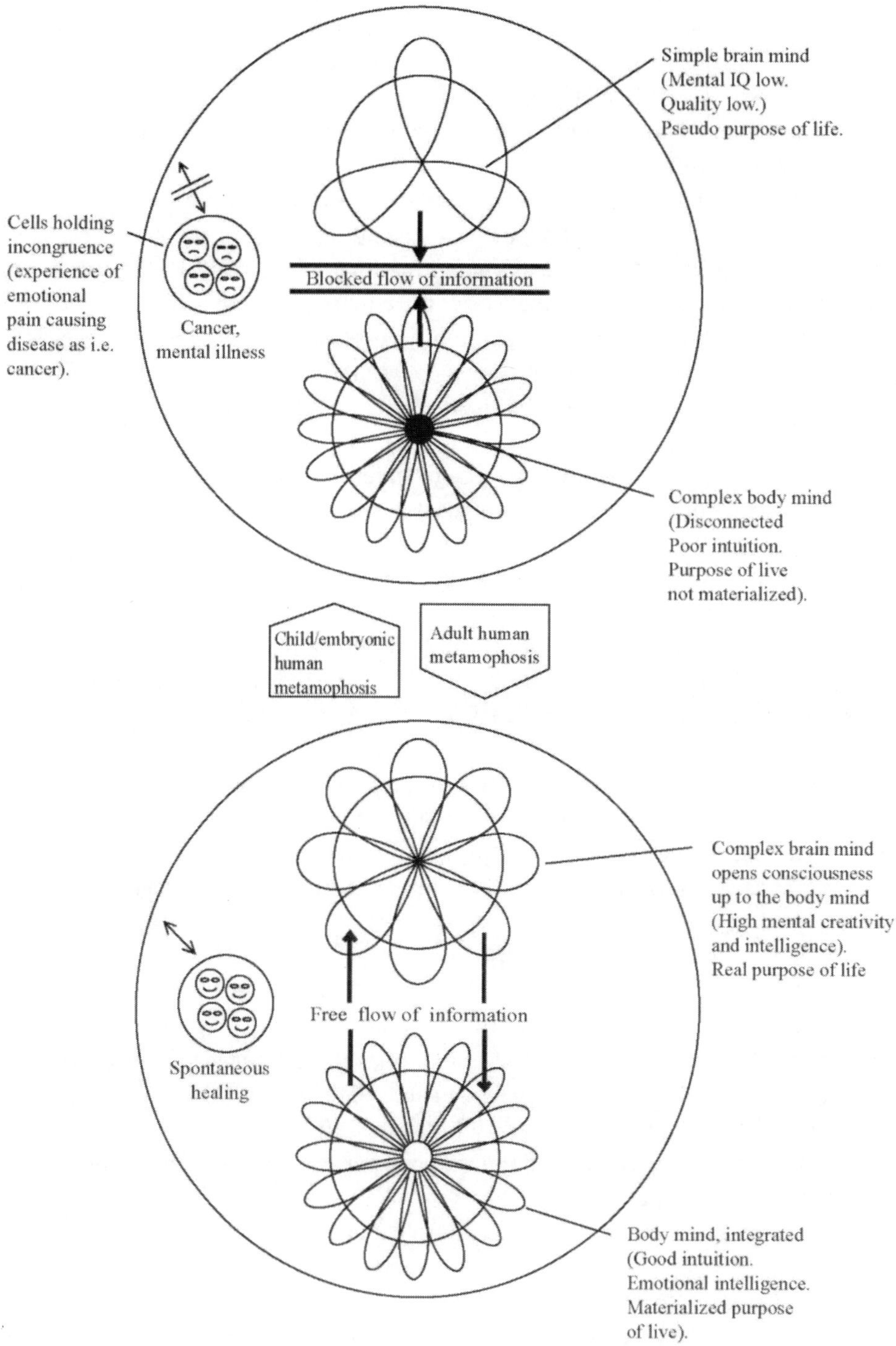

Figure 1. This figure illustrates what happens when two people are in confrontation.

They are physically weak and thus not able to defend themselves through action. Their will is undeveloped and they cannot resist the will of their parents. Not being able to escape, they will always and inevitably become subjected to the purpose of the parents. If the intentions of the parents are not in accordance with the biology of man, because they themselves have become conditioned by their own parents, the children in their turn are forced to take upon themselves these non-ideal, degenerate purposes and make a compromise of striving for lesser goals than the ideals defined by man's biological nature itself. We believe that it is a central problem of mankind that these broken purposes repeat themselves from generation to generation.

The children are subjected to the parents with the greatest possible irresistible force: "If you don't adapt, you'll die." Throughout their childhood the children become filled with poor pictures of love, because of their parents' imperfect love relation. When the parents cannot unfold their inner man and woman in the world, the children become filled with broken man-woman pictures. The children have no opportunity to sense what is going on, because their model of reality within the brain is too undeveloped. In the worst cases this kind of behaviour from the parents can lead to mal-functions and disease.

The life situation of the conditioned person. When the child is ready to leave its parents in order to live on its own, their wholeness and their complex dynamics is loaded with distorted pictures that correspond to undermined purposes and hinder the unbroken purposes to manifest themselves (8).

The original life purpose of the child has been broken. The model of reality in its brain has been imperfectly developed, because the broken purposes have served as an overall frame of its interpretation. Add to this the number of untruths about reality that serve as a defence of the individual, especially for its self-esteem (9). The nature of the defence is to pretend to one-self and others, that your intentions are better than the ones you have in your biological totality. The consequences of defence are, that the conditioned person does not acknowledge his own degenerate purposes (8), and has a great risk of getting into trouble" and "accidentally happens to do everything wrong" and believes, that "it is human to fail" (it is commonly seen, that the very conditioned person allows "emotions to run away with him", which means that he shifts to a condition, where he can freely act out his degenerate purposes, like the abuse of alcohol).

The degenerate purposes can very seldom be drawn from what people say, but almost only from the result of their actions – and as a result of their actions summed up throughout their life, that is: the life situation they are situated in.

The founders of the typical conditioned family have degenerated purposes according to both love and career. This gives man a fundamental division into love and power. Love is distributed from the mother, and power is distributed from the father. The father often becomes a tyrant towards the children in order to compensate for his shortcomings as regards his wife and the world at large (8).

When the children grow up and try to unfold beyond the level of their parents, the complete defeat of the life of their parents is revealed. In order to avoid this situation the parents prevent the unfolding of the children (8). The mother-son relation and the father-daughter relation often seem supportive instead of oppressive (10,11), but when mother and father could not manage love together, these parent-child relations compensate the mutual love between the parents. The spouses betray each other and cling to the children of the opposite sex. This in turn prevents the children in unfolding love together with the "right" third person, because such a person will be seen as a rival by the father or the mother and will be repelled. Degenerate purposes create degenerate family patterns, which in turn create degenerate purposes in the children and so on and so forth (8).

The insane and the sound patterns

Depression

A very conditioned person only has a few human qualities at his disposal and is stiff and non-dynamic (12). Whenever in life the off-conditioned qualities are needed, such as feelings and love, this person will experience an insurmountable wall of suppression, when the conditioned suppression of the biological unfolding surfaces. The emotional content of suppression is a lethal weight, a feeling of worthlessness and giving up. This depressive pattern colours the existence of the person. There are two possible strategies for the lifting of the depression. If possible you can normalize the situation by returning to the level of liveliness you had before the depression, or you can go forward and integrate the conditionings in your knowledge and obtain a new degree of freedom in your life. This last possibility is of course better than the first; which is most often what takes place in psychotherapy.

Neurosis

When the child reaches the age of puberty and is ready to separate himself from his parents, this can happen in either of two extreme ways or halfway in between these. One of the extremes is when the child succeeds in realizing exactly the life the parents have promised the child. This leads the child to idealize its parents and broadly live like them. Often the parents are not ideal, but have great faults, and in this situation the child uncritically adopts its parents' faults. This is the so-called neurotic person, who spends much of his energy in defending himself and his parents. The faults are passed on to the children as conditioned learning. The degenerate intentions and defects have been built into the model of reality in the brain of the child.

Schizophrenia and psychosis

The other extreme is reached, when it comes to a split between the parents and the young person. This can happen either because the parents deny the child or because they commit obvious mistakes against it, or because the child does not succeed in living his life at the conditions of his parents. The split between the child and his parents may have immense consequences, when it results in a pronounced denial of one, or both, of the parents. Because of the child's previously great adaptation to the parents he denies himself and the greater part of his strong qualities. The result of a weakly founded and immature person's violent rupture and violent denial is a condition of powerlessness and lack of motivation, which constitutes the schizophrenic pattern.

The young person's weakness is due to the fact that there are very great discrepancies between the child's emotional interpretation of reality and the child's rational interpretation of reality. A child like that can only function when it is supported very strongly. As long as the child lives at home with the parents this is almost never a problem, because the parents, that have conditioned the child, have adapted it to the family in order for this to function as a stable unit.

But when the child as a young person is going to leave home and manage on his own, his continued function is threatened, and only in situations where he learn by his experiences, or gets other people to help him, he will cope with leaving home.

It often happens, that the young person in his emotional interpretation of reality realizes that the parents with their conditionings have been destructive and "evil", which the parents of course deny. The child, still being adapted to the parents, in his rational interpretation of reality also denies the suppression by the parents. This child contains a difficult conflict: his opinions of his parents according to his emotions and according to his reason are at variance. If the emotions win, and the child will be living in the emotional interpretation of reality, the outcome will be both a denial of the parents and of his own reason. This is the psychotic, schizophrenic pattern, where the child either appears as giving up and totally loosing zest for life, or where mental activity is channelled in an unconstructive way, a state of mind that is called productive psychosis. When reason wins, and the child will be living in the rational interpretation of reality, we have a condition, where the emotional interpretation of reality manifests itself as "evil" voices and visual hallucinations. Often, in a spontaneous attempt at integration, a severe mingling of the rational and the emotional interpretation of reality occur. This is called disorganization of logical thought, and often neologisms and an elaborate private reality is found.

The sound pattern

Between these two extremes of idealization and denial of the parents lies the sane and sound pattern, where one take the good, constructive points and dissociate oneself from the bad qualities of one's parents. This pattern presupposes a determined search to find out which of the qualities of the parents are good and sound, and which are not.

Solutions of the problems

Knowledge as a way out of conditionings

Concerning the human process of knowledge is precisely the opposite of the conditioning process. In the former the human aims his purposes towards the deep truths of his biological nature. The key to the process of knowledge is integration of incidents into the model of reality in the brain (13), so the emotional and the rational interpretation of reality meet in a common interpretation. Information stored at the concrete and the abstract levels, corresponding to lower and higher space-time groupings, is interwoven into a continuous model of reality, with all levels of grouping placed in continuation of each other without discontinuities. The integration process forerunner as an elaborate examination of the, often, rough events of life. The rationale of the examination is to extract the knowledge content of each event: what can be learned from this? This is a process of systematic reorganization of the information stored in the organism. The totality is cleansed for bad pictures, when the bridge from the higher levels to the lower levels with considerable power forces the lower levels to shift representation. The lower levels intensively interact with the totality of the organism, and the bad pictures are wiped away from the totality and are hereafter only stored deeply within the complex dynamics of the brain, tagged as past. When eventually the bad pictures and intentions are removed, intent recovers, and better pictures can be created as goals.

Through the knowledge process the human adapts to his new wider frame after having obtained greater freedom in a new life situation. The first steps of the knowledge process does not presuppose insight into the mechanisms behind it, but demands that the incidents of life are confronted again, this time with a more developed model of reality. Many practical techniques have been developed aimed at supporting a beginning knowledge: psychoanalysis (14), psychotherapy (15) or meditation techniques (16). However we believe that all these techniques have their limits. As far as we can see a full insight into reality and as a consequence a full regeneration of purpose is not developed through the use of techniques, but only through a hard and determined fight for truth encompassing the whole life of the individual.

Solution of the mind-body problem

It can be difficult to propose a solution of the mind-body problem, but we will try to do it by looking at the origin of the psychophysical problem. In the Western tradition the human is seen as a spiritual being (17). The communication between these two is usually thought of as being accomplished in terms of the brain, as in the classical dualistic explanation of Descartes (16). The classical paradox concerning the discrepancy between thoughts and action, and feeling and reason, respectively, has the conditioned learning as their source. The mind-body problem (the psycho-physical problem) describes the problems that can arise in the relation between the body and psyche. Traditionally, this has been interpreted as a conflict between the structural and the functional part of the brain. This is understood as a split described as if a "ghost being" - our real self - was able to communicate with the body through the brain; this is a kind of dualism (18), in which the body is understood as a robot. Another kind of dualism is the possibility that we should consist of a physical and a psychical body able to integrate with each other. Both of these interpretations is an expression of that the biological reality has been confused by the experienced reality. The psychophysical problem seems also to have its roots in the fundamental split of the individual between body (symbolizing the wholeness) and head (that follows the conditioned learning). Following the co-ordinations the brain learns an over-simplified interpretation that cleaves off "body" (as the physical) and "spirit" (as the psychical) of the wholeness of the man. The transition between body and psyche (spirit and matter) is completely fluid following the description of the complex dynamics – in the reality it is possible to separate them. The system has to be understood as a wholeness where body and psyche are separated through a memorised construction of two seemingly different functional domains. The description of biological systems as complex dynamics then, gives a monistic solution of the psycho-physical problem. This is the case because when the information transforming interactions are binding all parts and levels together in a structural, functional unit it has really no meaning to talk about the body and psyche separately. The experienced reality of the psychophysical problem could be a result of conditioned learning, and acknowledgement is the way to solve the problem and a complete experience of the human reality.

Solution of the problem of free will versus determinism

A problem closely related to the psycho-physical problem is the concept of the free will. We all have the experience of being able to choose by ourselves, free and independently, and no one, not even ourselves, is able to predict, what we are going to choose.

If we have a free will, which means an ability to choose freely according to our circumstances (inheritance and environment), we also have a true responsibility for everything that happens to us. Without the free will we are determined slaves simply having to follow our destiny, and we don't have any possibility of forming our destiny ourselves. None of the two possibilities seem immediately reasonable.

It has to be noted, that we are not conscious of the sources to the data we use for the evaluation of the value of the different possibilities, and we are not aware of the fundamental source of the choices we are presented for. We often create these situations of choice without our knowing and often as a direct consequence of conditioning.

We suggest a complex determinism as a solution of the problem of the free will. Man is living in a causal world, obeying to the laws of nature. However, the data sources are immeasurably complex and the biological information system as a whole shows a large degree of instability, which makes man's life completely unpredictable.

Apparently the instability shows up as insignificant influences – for instance the pushes of realization that we as people give each other – may lead to conscious choices of either seeking knowledge or resign to knowledge. Such choices have decisive influence on the life of the individual in the future.

The very conditioned person functions like a robot programmed by conditionings, but such a person may undergo a change to full autonomy through knowledge that gives access to the unfolding of the biological nature of one's self. The "free" man is never freer than his overriding genetic programs permanently determine him. But man has his freedom if he uses his genetic programs in the "perfect" way, and therefore has the possibility of unfolding himself in a constructive way towards a more perfect family life and career.

Knowledge about the nature of the human reality will be able to get man to search the realization that will improve, not only his own life, but also improve the conditions for people on earth in general.

Good intension, adult human metamorphosis and self-healing

When a human child is influenced by impulses from the outer world as e.g. the parental influence from both parents and this influence is conflicting, because the parents give different impulses to the child, this will interfere with the child's understandings of the outer world and lead to a psychological crises, because its own interpretation of the world does not match with the inputs from the parents. Such problematic disturbances of the child's way to understand the world can follow the child through its development and thereby be a main problem for the resulting adult human being, because such disturbances interfere with the concerned human's intension (19). The intension interacts with the fractal levels of the human body (read wholeness) (20) and therefore such a human being is in danger for, through its intension that has become bad, to generate disturbances in the positional information that organizes the body structure through the ontogenesis (21,22). By help of a therapist, it will be possible for such a human being to recover from psychical illness by reversing the bad intension to a good intension.

By such activity the individual human being can, by help from the therapist influence the positional information of his body's fractal levels, and through this activate the adult human metamorphosis (23). In this way the degenerated human being can by his own consciousness cure himself in a process of human self-healing, as it is often seen in clinical holistic medicin and holistic sexology (24-33).

Conclusion

We think the experienced reality of the psychophysical problem is a result of conditioned learning, and the acknowledgement is the way to solve this problem and gain a complete experience of the human reality. Another problem, closely related to the psychophysical problem is the concept of the free will. We suggest a complex determinism as a solution of this problem. Man has his freedom if he uses his genetic programs in the "perfect" way, and therefore has the possibility of unfolding himself in a constructive way towards a more perfect family life and career.

Knowledge about the nature of human reality will enable man to search the realization that will improve his life and the conditions for people on earth in general. The process of knowledge is therefore the opposite of the conditioning process, a way out of "disorder". Through the knowledge process man adapts to a wider frame after having obtained greater freedom in a new life situation.

Parent-child relations are examples of the uttermost unequal distribution of power, because children have not developed their model of reality and therefore are not able to defend themselves through words. Children will always and inevitably become subjected to the purpose of the parents. If the intentions of the parents are not in accordance with the biology of man, because they themselves have become conditioned by their own parents, the children in their turn are forced to adapt themselves to these non-ideal, degenerate purposes and make a compromise of striving for lesser goals than the ideals (defined by man's biological nature itself?).

We believe that it is a central problem of mankind that these degenerate purposes repeat themselves from generation to generation. Degenerate purposes create degenerated family patterns, which in turn create degenerate purposes in the children etc. Between the existing extremes of idealization and denial of an individual's parent lies the sane and sound pattern, where man takes the constructive points and dissociates himself from the bad qualities of his parents.

This pattern presupposes a determined search to find out which of the parents qualities that are good and sound, leading to a life without failure and diseases.

References

[1] Skinner BF. The behaviour of organisms: An experimental analysis. London: Skinner Foundation, 1938.
[2] Beck A. Cognitive therapy and the emotional disorders. New York: Int Univ Press, 1976.
[3] Beck A. Cognitive therapy of depression. New York: Guildford, 1979.
[4] 4. Ventegodt S, Hermansen TD, Kandel I, Merrick J. Human development XV: The biochemical hypothesis for the etiology of the mental diseases is not substantiated. J Altern Med Res 2009;1(4):419-32.
[5] Freud S. New introductory lectures on psycho-analysis. London: Hogarth Press, 1933.
[6] Freud S. The ego and the Id. The standard edition of the complete psychological works of Sigmund Freud. New York: WW Norton, 1962.
[7] Ventegodt S, Andersen NJ, Merrick J. The life mission theory V. A theory of the anti-self and explaining the evil side of man. ScientificWorldJournal 2003;3:1302-13.
[8] Bateson G. Steps to an ecology of mind: Collected essays in anthropology, psychiatry, evolution, and epistemology. Chicago: Univ Chicago Press, 2000.
[9] Clark AJ. Defense mechanisms in the counseling process. Thousand Oaks, CA: Sage, 1998.
[10] Canfield J, Hansen MW, Aubrey P, Autio NM, Thieman L. Chicken soup for the father and daughter soul: Stories to celebrate the love between dads and daughters throughout the years. Deerfield Beach, FL: Health Communications (HCI), 2005.
[11] Bassoff ES. Between mothers and sons: The making of vital and loving men. New York: Dutton, 1994.
[12] Costello CG. Symptoms of depression. New York: Wiley, 1993.
[13] Ventegodt S, Hermansen TD, Kandel I, Merrick J. Human development XII: a theory for the structure and function of the human brain. ScientificWorldJournal 2008;8:621-42.
[14] Freud S. The origin and development of psychoanalysis. New York: Gateway, 1910.
[15] Fuglsang AK, Moergeli H, Hepp-Beg S, Schnyder U. Who develops acute stress disorder after accidental injuries? Psychother Psychosomat 2002;71(4):214-22.
[16] Curley E. Analysis in the meditations: The quest for clear and distinct ideas. In: Rorty AO, ed. Essays on Descartes' meditations. Berkeley, CA: Univ California Press, 1986:153-76.
[17] Giles C. The nature of spirit, and of man as a spiritual being. North Ryde, NSW: Swedenborg, 1997.
[18] Popper KR, Eccles JC. The self and its brain. London: Routledge Kegan Paul, 1984.
[19] Ventegodt S, Andersen NJ, Merrick J. Quality of life philosophy V. Seizing the meaning of life and becoming well again. ScientificWorldJournal 2003;3:1210-29.
[20] Ventegodt S, Hermansen TD, Flensborg-Madsen T, Nielsen ML, Merrick J. Human development VIII: A theory of "deep" quantum chemistry and cell consciousness: Quantum chemistry controls genes and biochemistry to give cells and higher organisms consciousness and complex behavior. ScientificWorldJournal 2006;6:1441-53.
[21] Ventegodt S, Hermansen TD, Flensborg-Madsen T, Nielsen ML, Clausen B, Merrick J.Human development IV: the living cell has information-directed self-organisation. ScientificWorldJournal 2006;6:1132-8.
[22] Ventegodt S, Hermansen TD, Flensborg-Madsen T, Nielsen M, Merrick J. Human development VI: Supracellular morphogenesis. The origin of biological and cellular order. ScientificWorldJournal 2006;6:1424-33.
[23] Hermansen TD, Ventegodt S, Merrick J. Human development X: Explanation of macroevolution — top-down evolution materializes consciousness. The origin of metamorphosis. ScientificWorldJournal 2006;6:1656-66.
[24] Ventegodt S, Kandel I, Merrick J. Clinical holistic medicine (mindful short-term psychodynamic psychotherapy complimented with bodywork) in the treatment of schizophrenia (ICD10-F20/DSM-IV Code 295) and other psychotic mental diseases. ScientificWorldJournal. 2007;7:1987-2008.
[25] Ventegodt S, Kandel I, Merrick J. Clinical holistic medicine: how to recover memory without "implanting" memories in your patient. ScientificWorldJournal 2007;7:1579-89.
[26] Ventegodt S, Clausen B, Omar HA, Merrick J. Clinical holistic medicine: holistic sexology and acupressure through the vagina (Hippocratic pelvic massage). ScientificWorldJournal 2006;6:2066-79.

[27] Ventegodt S, Clausen B, Merrick J. Clinical holistic medicine: pilot study on the effect of vaginal acupressure (Hippocratic pelvic massage). ScientificWorldJournal 2006;6:2100-16.
[28] Ventegodt S, Clausen B, Merrick J. Clinical holistic medicine: the case story of Anna. III. Rehabilitation of philosophy of life during holistic existential therapy for childhood sexual abuse. ScientificWorldJournal 2006;6:2080-91.
[29] Ventegodt S, Thegler S, Andreasen T, Struve F, Enevoldsen L, Bassaine L, Torp M, Merrick J. Clinical holistic medicine (mindful, short-term psychodynamic psychotherapy complemented with bodywork) in the treatment of experienced impaired sexual functioning. ScientificWorldJournal 2007;7:324-9.
[30] Ventegodt S, Kandel I, Neikrug S, Merric J. Clinical holistic medicine: holistic treatment of rape and incest trauma. ScientificWorldJournal 2005;5:288-97.
[31] Ventegodt S, Morad M, Hyam E, Merrick J. Clinical holistic medicine: holistic sexology and treatment of vulvodynia through existential therapy and acceptance through touch. ScientificWorldJournal 2004;4:571-80.
[32] Ventegodt S, Morad M, Kandel I, Merrick J. Clinical holistic medicine: problems in sex and living together. ScientificWorldJournal 2004;4:562-70.
[33] Ventegodt S, Morad M, Merrick J. Clinical holistic medicine: holistic pelvic examination and holistic treatment of infertility. ScientificWorldJournal 2004;4:148-58.

Section VI:
Implications for medical disciplines

Chapter XV

Biochemistry and mental diseases

We review the understanding of the etiology of the mental diseases, which has changed considerably during the last two decades. We consider the results from psycho-neuro-pharmacology and the derived historical, biochemical hypotheses for the mental diseases, and find that they have not been substantiated. We analyse the biochemical hypothesis of the diseases depression and schizophrenia and find that the standard biochemical hypotheses are not substantiated. We suggest a mechanism for the pasifying effect of the antipsychotic drugs. When it comes to the etiolology of the two most common mental diseases both of these seem to be caused by disturbed psychosexual development, causing a degenerated intent, in accordance with Bleuler's classical description and understanding of schizophrenia. We present a psychobiological and more holistic models of the mental diseases and their etiology, which we find more plausible than the historical, biochemical and genetic models.

Introduction

All transmitter systems known so far are present in both rats and humans; they must therefore be considered to be very stable in an evolutionary perspective (1). But here are significant differences also in the vertebrate brains; it looks as if man to a much higher degree than animals is in control of his fundamental mental functions such as attention, sleep, and motor activity. In man all these functions are under control of the will and are assumed to be ruled from the cerebral cortex. This is in accordance with the known development of the cerebral neocortex and above all the fronto-orbital lobes. Because mental diseases such as depression and schizophrenia are hardly found in animals, it seems reasonable to believe that mental diseases are caused by regulation from the cerebral cortex rather than deregulation in one or several of the ascending systems. The rational-mechanical interpretation of reality favours the hypothesis of a defect in the ascending systems, while the energetic-informational interpretation of reality favours a complex cerebral deregulation. Research has concentrated on the ascending systems, partly because of the documented effect of the antidepressants and the neuroleptics. However little is known about the descending, neural systems.

Seen from a rational interpretation of reality we think it is natural to assume, that mental diseases are caused by an inherited, mechanical defect in the brain such as a defect in an enzyme or a receptor. When psychopharmacological drugs were invented, they wereseemingl seeamed toy be able to reduce many of the symptoms of mental illnesses, i.e. the hallucinatory behaviour of schizophrenic patients. The mentally ill patients became much more calm and easier to care for, and members of the patients family often appreciated the results of the treatment, in spite of the rate of cure was rarely found to be significantly improved compared to the traditional methods of psychotherapy and holistic therapy. Especially the obedience of the patients was often radically improved, and the lack of resistance and the cooperation was seen as important signs of improvement.

This led to a number of hypotheses, as it was assumed that the effect of the drugs was a simple compensation for specific defects behind the mental diseases. This interpretation has, however, in spite of many years of research and thousands of published papers not lead to the expected full understanding of the aetiology of mental diseases; quite on the contrary mental illness and human brain function seems more mysterious and hard to understand from a chemical perspective than ever.

This paper provides arguments to reject the hitherto proposed simple hypotheses about a link betweenbiochemical, molecular defects and mental disease. For instance, in spite of a great scientific effort neither schizophrenia nor depression has been linked to such defect genes or other specific biochemical defects; and this fact was clear to researchers already in 1989 (2).

Psychoneuropharmacology and the biochemical hypotheses of the etiology of mental diseases

Almost every drug that affects the brain does so in terms of influences on synaptic transmissions (3). Antidepressants and neuroleptics as well as several other types of psychotropic drugs often work on one or more ascending mono-aminergic transmitter system (with serotonine (5HT), norepinephrine (NE) or dopamine (DA) as transmitter), whose nuclei are present in the brain stem or the diencephalon. Pharmacological interaction with systems with acetylcholine (Ach) or gamma-aminobutyrate (GABA) can, however, also give a similar effect. Reserpin was the first antipsychotic drug that were shown to empty the monoamines from their storage vesicles, making them accessible to degradation by mitochondria-tied monoamine oxidase. The discovery of such compounds lead to the formation of the hypothesis of genetically specified dysfunction in these systems as a possible cause of schizophrenia, and serotonine and norepinephrine of depression.

The monoamine systems (5HT, NE, DA) are supposed to be the primary site of action of many psychotropic drugs and have therefore been the subject of intensive research activity. But in spite of this the precise nature of the exact relation between the monoamine systems, the mental effect of the drugs, physiology, and behaviour is still not understood in details. Because of their great diversity the monoaminergic systems can be assumed to attend to general regulatory functions. The function of the serotonergic systems is the least understood. They seem to function through a neural inhibiting tonus especially to the limbic system, correlated to muscle tonus. Dopaminergic systems possibly regulate motor activity and the activity of thoughts. Regulation of outwardly directed attention is possibly connected to the noradrenergic systems.

Model examples of mental disorders

Depression and schizophrenia can be seen as examples of mental illness, which is the heavy workload in any psychiatric centre or practice. Depression is characterized by constantly bad mood, self-disapproval, a low self-confidence, and a negative self-image (4). Schizophrenia most frequently occurs at the age 15-35 years and is characterized by lack of zest for life and disorganization of logical thought most often together with auditory hallucinations and paranoid delusions (5), together with emotional flattening and social withdrawal (comp. Bleuler 1911). Depressed and schizophrenic patients constitute far the greatest part of the mentally ill patients. These two mental diseases will be the subjects of the further discussion in this paper, because they represent the best known and investigated mental disorders existing today.

Depression

The inheritance of affective diseases

Studies of monozygote twins showed that if twins grow up together, the concordance as regards depression is higher for monozygote twins (33-79%) than for dizygote twins (54%) (6). A concordance of 79% in the heaviest cases showed that even in these cases monozygotic twins have a considerable freedom to develop in their own way despite having the same genes and living in the same environment.

Studies of adoptation

A study of 29 adopted bipolar depressive patients showed that 31% of the patients had adoptive parents who suffered from affective diseases, in contrast to only 9% of the biological parents. In the case of non-depressive adopted children 2% of their adoptive parents suffered from affective diseases. These data, reviewed by (7) suggested that the environmental factors dominate in the ratio 2:1 against the "early factors", at least in this case. A simple mendelian inheritance seems unlikely. We hypothesize that intrauterine information transmitting interactions between the mother and child are responsible for the remaining cases not explained by genetics and environment.

Most affective diseases have a cyclic nature, most notably bipolar depression that hardly tallies with a genetic defect or a descending cerebral dysfunction. In the well-known "winter-depression" the light factor seems to play a key role.

It seems that environmental factors as well as genetic factors and other early factors, of which we favour intrauterine information transmitting interactions, play a role in the development of depressive diseases. As there is a very close contact between mother and child in utero, and adopted children of course spend their foetal lives in their mother's womb, twin studies are not able to distinguish between the two kinds of factors.

The effectiveness of antidepressants compared to placebo

For many years it was thought that antidepressants were more active than placebo, but around year 2000 the understanding of the active placebo effect led to re-investigation and comparison of the most efficient antidepressants (the tricyclic antidepressant) with active placebo, and quite chocking it was found that these drugs, being the most potent antidepressant drugs know, was not at all better that placebo (8). Before that it was generally believed that app. 65% of non-psychotic depressive patients respond to imipramin, while app. 30% responds to placebo; all heterocyclic antidepressantshad the response figures of 65-75% after one month treatment compared to 20-40% response to placebo. Generally speaking, the worse a depression is, the better the antidepressant fares compared to the passive placebo (7), but this difference is obviously annihilated when it comes to active placebo.

The above-mentioned old studies were conducted in a way that opened up to criticism. Even a 50% reduction in the Hamilton rating score (9) were clinically called a response, but in fact this was only a mild relief of symptoms. More importantly most researchers only included patients that completed the experiment, but did not inform about how many patients that did not complete the experiment – and they did not count these as non-responders, what most of them perhaps were. Moreover it was rarely told how many patients the subjects are chosen from, and there were serious problems with the criteria for election for the experiment.Only patientsthat do respond positively to the drugs were included in most of the studies, creating a tremendous bias: "Most of the controlled clinical studies exclude patients who have not responded to antidepressants in the past" (10). In most cases the studies report no results of the duration of the recovery or of the frequency ofrelapse, and long-term follow-ups are extremely rare. So it is easy to see to day how the inefficient, antidepressive drugs were artificially turned into active and valuable drugs by the research that was almost always paid for by the pharmaceutical industry.

The hypothesis of depression

Around 1990 research of the effect of antidepressants led to a fundamental revision of the earlier hypotheses. Most researchers seem to agree that the simple hypothesis of depression as caused by lack of monoamines was not verified. This hypothesis was partly based on the assumption that the drugs worked through an increased supply of transmitter substances (NE and 5HT) in the synapses in a clinical time variance of the effect of between 7- 45 days (7), a much longer time than the quickly induced biochemical effect. Moreover drugs e.g. iprindol and mianserin, that are not reuptake- or MAO-inhibitors in a significant way were found to be as efficient as the classical drugs. In addition amphetamine, which is a reuptake-inhibitor, cannot effectively be used in the treatment of depressions.

Studies of monoamine-turnover in laboratory animals, long term treated with antidepressants or electroshock, showed no significant deviation from normal. The influence of the reuptake-blocking seems to be submitted to a feedback-regulated mechanism exerting its effect through auto-receptors.

None of these findings are sufficient in themselves in order to reject the hypothesis, "but together they provide a powerful argument for its re-examination and suggest that antidepressants act in a more complex manner than that envisaged by the monoamine deficiency hypothesis of depression" concluded Elliot and Stephenson in 1989 (4).

Psychotherapy, holistic therapy and depression

Since 1975 great methodological and technical improvements have been made, including improvement of methods of evaluating 1) the condition of the patient, 2) the qualifications of the therapist, 3) the contents of the therapy and 4) the improved situation of the patient.

Central coordination of therapist training and evaluation programs has apparently resulted in a minor revolution in this area. The results are mixed, but some researchers have concluded that there are large differences in effectiveness between the different therapies (9).

Resent research has documented that depressed patients were helped better by psychotherapy than by psychiatric standard treatment (11-13). One of the best predictors for response to interpersonal therapy is a pathological picture indicating endogen depression!

It seems reasonable to conclude, that mental factors are of tremendous importance concerning affective diseases (14,15). We interpret the existing data in the following way: Mental illnesses are caused primarily by psychological factors, not by genes, as genes cannot be changed by psychotherapy. More efficient that psychotherapy alone is the combination of psychotherapy and bodywork, and holistic therapy like clinical holistic medicine, also including work with philosophy of life and sexuality (16-25).

Schizophrenia

The inheritance of schizophrenia

Evidently schizophrenia is not randomly distributed within a population, but is more frequent within exposed families than in the population as a whole ($p<0.001$) (26). In recent twin studies the concordance between pairs of twins growing up together is larger between monozygote (31-78%) than betweendizygote twins (6- 28%). These studies also show that 22-69% of monozygote twins growing up together do not both develop schizophrenia (ibid.). This points towards a dominant influence of environmental factors. Since several investigations have shown that monozygote twins to a far higher degree than dizygote twins share the same environment, friends, and attitudes of their parents and teachers (ibid.), the greater concordance might as well be attributed to these environmental likenesses.

Studies have found a significantly higher occurrence of schizophrenia among children adopted by schizophrenic parents and among adopted children, whose biological parents were schizophrenic, compared to the average population. In some studies no significant difference was found, however. Studies of monozygote twins not growing up together were of much greater value than studies of monozygote twins growing up together. The adopted children have, however, often spent a smaller or larger time together with the mother, and in all circumstances they had had the opportunity to adapt to the mother within the womb. The information transmitting interactions have had the time to work.

What is transferred from one generation to the next is not a simple tendency towards the development of schizophrenia, not even a non-specific tendency towards the development of psychiatric diseases, but a tendency towards bad psychosocial functioning, "These findings provide an increasingly complex, but informative, picture of the nature of transmitted liability to schizophrenia" (26).

Since 1916 it has been known, that schizophrenia does not follow a classic Mendelian inheritance pattern, thus, it is obvious to imply a polygenic inheritance. Models for polygenic inheritance, however, are flexible, because they are very difficult to falsify.

From all this, it seems that the studies of adopted monozygote children could suggest the importance of " early factors", but it is not known, whether these are of a genetic or an intrauterine nature. In addition, it is evident, that environmental factors play a great role.

Hypotheses for schizophrenia

Several hypotheses for schizophrenia have been proposed, but the dopamine hypothesis seemed for many years to be the only transmitter hypothesis, that could not be definitively falsified (see historical review in (5)). The dopamine hypothesis was founded in the effect on dopaminergic systems of manyof the original neuroleptics. The hypothesis says that schizophrenia is caused by a (genetically inherited) hyper-activity of the dopaminergic system.

The hypothesis came in different versions each considering one of several possibilities with regard to the function of DA in the psychotic brain. Either there was too fast a DA-metabolism, or a too large a receptor sensitivity (5). Unfortunately for the believer in the dopamine hypothesis, post mortem studies or other studies did not provide evidence for an increased DA-turnover in the brain of people with schizophrenia. An up regulation of D2-receptors as a consequence of the administration of neuroleptics in the brain of schizophrenics was not found. There is still no positive evidence for occurrence of changed D2-frequency in untreated schizophrenics. Thus there is no evidence for any molecular hypothesis for schizophrenia, not even for the dopamine hypothesis. This is in agreement with the fact that schizophrenia occurs in episodes - a fact that is very difficult to explain in terms of a defect in a transmitter system. The earlier mentioned simple hypotheses about schizophrenia as caused by simple biochemical defects or disturbances were around 1990 abandoned in favour of more complex explanations of the brain function.

From 1990 to 2008 came a large series of Cochrane metaanalysis analysing the effect of all kinds and types of antipsychotic drugs on a number of different illnesses and mental states (27-103). Quite surprisingly it was found that every time an antipsychotic drug was tested against placebo, the patients' mental state was not found to be significantly improved. Behaviour was still found to be modified, but the effect of the behaviour was just pacification, not an improvement. Hallucinations and other symptoms of mental illness was not at all relieved by the drugs, which does not deserve its name "antipsychotic medicine" anymore, as the drugs are not at all antipsychotic, only tranquillising the patients. It was also documented in these studies that the adverse effects of the drugs were very severe; they basically took the patient's energy and autonomy away, thus giving the obedient and more socially acceptable picture of an improved patient, that from an existential perspective was actually loosing his quality of life as a sad consequence of the treatment.

The Cochrane analyses finally made it impossible to believe in the biochemical hypothesis of schizophrenia and the other psychotic mental illness; they were simply not substantiated.

Negative and positive symptomatology

The subdivision of schizophrenics into a positive and a negative symptomatology has a long history and seems to be supported by morphological studies (104). The positive symptoms are hallucinations, delusions and some types of thought disturbances as derailment, neologisms and incoherency. The negative symptoms are lack of function in a number of areas, such as social withdrawal, weakened affect, reduced motivation, psycho motor retardation, and poverty of speech.

It was around 1990 commonly understood, that the negative symptoms are not disappearing by use of neuroleptics, but only with the Cochrane studies systematic exploration of the effects of the drugs on the positive effects was it documented that the positive symptoms were not improved either.

There is clinical evidence showing that negative symptoms may be connected with too low a dopamine activity. It is known that Parkinson disease is often associated with social withdrawal and deflated affect. Large doses of neuroleptics may trigger the negative symptoms, besides motor inhibition, while chronic l-dopa administration, which counteracts neuroleptics, sometimes is able to alleviate deflated affect, withdrawn emotions, and apathy.

This means that schizophrenic patients can be divided into a "hyper-dopaminergic" group with positive symptoms, and a "hypo-dopaminergic" group with negative symptoms. The words "hyper" and "hypo" refer to the pharmaceutical compensation that seems to remove the symptoms, and not necessarily to the DA-activity of the patients.The variance in neuropsychological state cannot in itself support a biochemical hypothesis for schizophrenia.

Taken all together it is clear today that the biochemical hypothesis for schizophrenia is in no way substantiated.

Discussion

The passive placebo effect seems to be the same for anti-schizophrenic (antipsychotic drugs) as for antidepressants (5); the active placebo effect is only known for the antidepressant drugs, as nobody yet has investigated this with the antipsychotic drugs. The psychological and sexual factors seems to be dominant in schizophrenia as well as in depression. In studies of neuroleptics, the fact that 2 of three or more were non-responders showed that the brain has a great adaptive capacity to compensate thesedating influences ofthe neuroleptics.

It has been known for a long time that the side effects of neuroleptics closely resemble Parkinson's disease, which is known to be associated by the decay of dopaminergic neurons; the strongest evidence for the dopaminergic effect is the fact that the clinical efficiency of many neuroleptics is closely correlated to their displacement of 3H-spiroperidol and 3H-haloperidol from D2-receptors (5,104). When it was shown, that there was a good correlation between the clinical efficiency of neuroleptics and D2-binding, it seemed reasonable to assume that neuroleptics worked through the D2-receptor. Today a whole new generation of neuroleptics with quite different affinity profiles (27-103,105) have been created. Among the newly identified neuroleptics are compounds that by thorough clinical testing has been shown to be as effective as the old ones, while they on the whole have no affinity to DA-receptors (e.g. clozapin; less tested is flulerlapin, and BW 234 U). These compounds are all found to be "effective" in animal models. It has been shown, that they in general are clinically effective, since it is no longer possible to associate neuroleptic activity with D2-binding. Hence there is no pharmacological evidence, that psychosis is associated with theDA-sytems (5). Webster and Jordan concluded in 1989: "The controversy over neuroleptic treatment and the state of D2-receptors remain unsolved." To day this is finally solved: The illness called schizophrenia is not at all connected to the D2-receptors.

The considerable time-elapse, before the effect of the neuroleptics occurs, points to a complex interaction between drug and brain. As the discontinuation of the drug rarely leads to an immediate aggravation of symptoms, it is evident, that the effects of the drugs cannot be explained by a simple interaction between a drug and a transmitter system.

Neuroleptics have not improved during the past 50 years (28) and while patients' mental health according to the many new Cochrane metaanalysis stays totally unaffected their bodies suffers. A vast fraction of the patients get serious side effects, such as tardive dyskinesia and tardive psychosis, the consequences of which are still uncertain. In spite of intensive studies the patients that selectively respond to these drugs have not been characterized (106).

Finally there is no clinical evidence that neuroleptics should be more active against schizophrenic psychoses than against any other kind of psychosis (107). Therefore there is no reason to limit the DA-hypothesis to schizophrenia; it should comprise all kinds of psychoses. Discontinuation of neuroleptics rarely seems to result in acute aggravation of the schizophrenic symptoms. The schizophrenic symptoms seem to arrive in episodes, a detail that proposes a very complex mechanism.

One of the strangest arguments of the 80'ies, interesting for its historical value, is that the pharmacological effect is due to adaptation to the drug. This hypothesisare of cause not plausible, because adaptation should lower the effect of the neuroleptics, not increase them, but in the 80'ies researches in antipsychotic drugs often suspended all reason to prove what they believed was be true. But this is not so rare in science.

A suggestion of the mechanism of psychopharmacological drugs

The key problem in understanding the mechanism of antidepressants and neuroleptics seems to be the great time delay of their effect. The pharmacological effect takes a few hours, the central nervous system adaptation to this effect presumably takes a few days, but the clinical effect often takes a month or more. The hypothesis of adaptation at receptor level as a mechanism behind the clinical effect does not seem plausible given the time discrepancy. The pharmacological effect of antidepressants generally seems to be an argumentation of synaptic activity, where neuroleptics (e.g. reserpine) may induce depression.

About 1990 it seemed reasonable to assume, that antidepressants respectively neuroleptics compensated a hypo-activity respectively a hyper-activity in the brain as a whole, not at any specific site of action for a specific drug. This compensation could give the complete neural system a "push" in the right direction towards normal function and normal interpretation of reality. According to this interpretation the time delay of the clinical effect was seen as inertia in the adaptation at the higher (mental) levels of the brain.

We suggest that the cognitive content of mental disease corresponds to a large number of considerations and decisions that take a long time to accumulate in one's model of reality in the brain. This inertia in the change of perception of reality leads to the time delay in any treatment of depression and schizophrenia, whether it is done by pharmacological means or by electro chock (ECT), psychotherapy and holistic therapy.

But the most obvious hypothesis for the function of the drugs is much simpler: Poisoning. As time goes by, and the patients loose energy due to severe poisoning, the behaviour becomes more and more obedient, passive and without the initiative and rebellion that characterizes autonomous beings. The psychopharmacological drugs are simple socializing the patients by depraving them of their life energy. This interpretation seems to be in almost perfect accordance with the findings of the Cochrane studies of antipsychotic drugs.

The etiology of depression and schizophrenia

In 1990 it was found that depression could be counteracted through interaction with many different transmitter systems. This pointed towards a complex mechanism and not a simple one tied to a single transmitter system. Reuptake in itself could also be excluded as a mechanism, because cocaine and amphetamine did not act as antidepressants. Compensatory up regulation of beta-receptors was often seen, but not always (107), thus this could not be the general regulation mechanism. Adaptation to a drug, including receptor adaptation through increased sensitivity, was suggested as a mechanism. This did not seem likely, because such an adaptation should eliminate the disturbance and thus decrease rather than increase. In this way reduce instead of increase the effect of the antidepressants. It seems absurd to suppose that such an adaptation should give a whole new effect as for example to alleviate a depression. An adaptation to a psychotropic drug normally takes about four days (108), whereas the effect of antidepressants often does not assert its effect before about six weeks (7).

The long interval before the effect shows up indicated a very complex mechanism instead of a simple molecular mechanism. The same conclusion was indicated by the fact that about one third of the patients did not respond to antidepressants at all. The placebo effect – known today to account for the full effect (8) of the antidepressant drugs – caused by the expectations to a treatment, indicated an important mental factor. Inheritance studies suggested that a certain amount of genetic transmission could not be excluded. Spontaneous remission was well known in patients with depression, but would not be likely in the case of a genetic programmed biochemical error. The periodical nature of manic depression (bipolar depression) was also difficult to connect to a genetic deficiency.

Today we know that psychotherapy is superior to drugs, and we know that the psychopharmacological drugs themselves are only giving positive effects though psychological mechanisms – the placebo effect.

The conclusion therefore is, that environmental factors are more important for the etiology of mental illness than genetic defects. As defect genes causing mental illness has never been found, the “early factor” seemingly important in the etiology of schizophrenia is more likely to be information-transmitting interactions between mother and child in and outside the womb. Inheritance studies showed that environmental factors played a decisive role in the etiology of schizophrenia. Early factors, such as genetic and/or intrauterine factors, were of minor importance. We hypothesize that information-transmitting interactions in utero and in early childhood were more important than genetic factors.

Studies of neuroleptics have shown a considerable placebo effect and a substantial group of non-responders, as is also the case in antidepressants. New generations of neuroleptics forced researchers around 1990 to reject the earlier assumption of schizophrenia was tied to the dopaminergic transmitter system. The long lapse of time before the effect manifest itself (7-30 days) corresponded badly with the time for the chemical effect (2 hours) or the time of adaptation at receptor level (a few days). Moreover, the episodic occurrence of schizophrenia makes it hard to maintain simple, molecular hypotheses for schizophrenia.

The positive and negative symptomatology seems to show, that schizophrenia covers a broad spectrum from “hypo“ to “hyper“ dopaminergic activity. Finally, neuroleptics assert their effect non-specifically against all psychoses, not only against schizophrenia.

All in all no evidence for any molecular hypothesis seems to have been found. On the contrary, there is clear evidence for the importance of environmentaland psychological factors.

Conclusions

We have the following final remarks after our review of the literature:

Neuroleptics

Almost hundred Cochrane metanalysis seems to have documented that psychotic mental illness in general are not causally associated with the DA-systems, or any other transmitter system. The time-elapse from ingestion of neuroleptics to the effect occurs indicates that simple interaction between drug and transmitter system is not a plausible explanation. According to the receptor hypothesis concerning the effect of neuroleptics the pharmacological effect is due to adaptation to drugs, but we believe this is wrong, because adaptation should lower the effect of the neuroleptics, not increase them. It is most likely that the effect of antipsychotic drugs on behaviour – in our analysis seen as the reduction of the autonomy, libido and life energy from the patient, thus pacifying him or her and depraving the patient his basic life, is due to simple poisoning by the drugs. The Cochrane studies have systematically documented that compared to placebo; none of the existing not a single type of antipsychotic drugs had a positive effect on did improve the mental state, neither for the schizophrenics nor for other patients.

Antidepressants

In 1990 the pharmacological effect of antidepressants was used as argumentation of a biochemical hypothesis for depression linked to serotonine. This interpretationcame into difficulties as the time delay of the clinical effect was seen as inertia in the adaptationto the drugs at the mental levels of the brain. A Cochrane study showed in 2004 that the most effective of the antidepressant drugs are not more effective than active placebo (8). Any hypothesis based on antidepressant drugs is therefore not substantiated.

Etiology of the mental diseases in general

Neither in the case of depression nor schizophrenia there seems to be evidence of genetic or molecular defects from the literature. Inheritance studies show a very important environmental factor, and a less important “early” factor, which may be due to genetic defects, but in our view far more likely, may be due to information transmitting interactions in the uterus.

The placebo effect seems to be what gives the effect of the psychopharmacological drugs; therefore it is hardly surprising that psychotherapy and holistic therapy has been found to be more effective than psychopharmacological drugs. We conclude that the mental diseases have an etiology based on psychosexual developmental factors, not genetics.

A psychosexual etiology of the mental diseases is in accordance with recent research and opens up for psychodynamic (11-13) and scientific holistic therapy as the rational cure for mental diseases (16-25,109-130) and psychoform pain (131-133). It might also explain the most interesting connection between the sense of coherence and disease (134-140).

References

[1] McGeer PL, Eccles JC, McGeer EG. Molecular neurobiology of the mammalian brain. New York: Plenum Press, 1987.

[2] Meltzer HY. Psychopharmacology. The third generation of progress. New York: Raven, 1987.

[3] Snyder SH. Molecular strategies in neuropsychopharmacologic research. In: Meltzer HY, ed. Psychopharmacology. The third generation of progress. New York: Raven, 1987.

[4] Elliot JM, Stephenson JD. Depression. In: Webster RA, Jordan CC, eds. Neurotransmitters, drugs and disease. Oxford: Blackwell, 1989.

[5] Webster RA, Jordan CC. Neurotransmitters, drugs and disease. Oxford: Blackwell, 1989.

[6] Gershon ES, Berrettini W, Nurnberger J, Goldin LR. Genetics and affective illness. In: Meltzer HY, ed. Psychopharmacology. The third generation of progress. New York: Raven, 1987.

[7] Brotman AW, Falk WE, Gelenberg AJ. Pharmacologic treatment of depressive subtypes. Psychiatr Med 1988;6(3):92-113.

[8] Moncrieff J, Wessely S, Hardy R. Active placebos versus antidepressants for depression. Cochrane Database Syst Rev. 2004;(1):CD003012.

[9] Williams JBW. A structured interview guide for the Hamilton Depression Rating Scale. Arch Gen Psychiatry 1988;45(8):742-7.

[10] Weissman MM, Jarrett RB, Rush AJ. Psychotherapy and its relevance to the pharmacotherapy of major depression: a decade later (1976-1985). In: Meltzer HY, ed. Psychopharmacology. The third generation of progress. New York: Raven, 1987.

[11] Leichsenring F, Rabung S, Leibing E. The efficacy of short-term psychodynamic psychotherapy in specific psychiatric disorders: a meta-analysis. Arch Gen Psychiatry 2004;61(12):1208-16

[12] Leichsenring F. Are psychodynamic and psychoanalytic therapies effective? A review of empirical data. Int J Psychoanal 2005;86(Pt 3):841-68

[13] Leichsenring F, Leibing E. Psychodynamic psychotherapy: a systematic review of techniques, indications and empirical evidence. Psychol Psychother 2007;80(Pt 2):217-28.

[14] Beck A. Cognitive therapy and the emotional disorders. New York: Int Univ Press, 1976.

[15] Beck A. Cognitive therapy of depression. London: Guildford, 1979.

[16] Ventegodt S, Kandel I, Merrick J. Clinical holistic medicine (mindful short-term psychodynamic psychotherapy complimented with bodywork) in the treatment of schizophrenia (ICD10-F20/DSM-IV Code 295) and other psychotic mental diseases. ScientificWorldJournal 2007;7:1987-2008.

[17] Ventegodt S, Kandel I, Merrick J. Clinical holistic medicine: how to recover memory without "implanting" memories in your patient. ScientificWorldJournal 2007;7:1579-89.

[18] Ventegodt S, Clausen B, Omar HA, Merrick J. Clinical holistic medicine: holistic sexology and acupressure through the vagina (Hippocratic pelvic massage). ScientificWorldJournal 2006;6:2066-79.

[19] Ventegodt S, Clausen B, Merrick J. Clinical holistic medicine: pilot study on the effect of vaginal acupressure (Hippocratic pelvic massage). ScientificWorldJournal 2006;6:2100-16.

[20] Ventegodt S, Clausen B, Merrick J. Clinical holistic medicine: the case story of Anna. III. Rehabilitation of philosophy of life during holistic existential therapy for childhood sexual abuse. ScientificWorldJournal 2006;6:2080-91.

[21] Ventegodt S, Thegler S, Andreasen T, Struve F, Enevoldsen L, Bassaine L, Torp M, Merrick J. Clinical holistic medicine (mindful, short-term psychodynamic psychotherapy complemented with bodywork) in the treatment of experienced impaired sexual functioning. ScientificWorldJournal 2007;7:324-9.

[22] Ventegodt S, Kandel I, Neikrug S, Merric J. Clinical holistic medicine: holistic treatment of rape and incest trauma. ScientificWorldJournal 2005;5:288-97.

[23] Ventegodt S, Morad M, Hyam E, Merrick J. Clinical holistic medicine: holistic sexology and treatment of vulvodynia through existential therapy and acceptance through touch. ScientificWorldJournal 2004;4:571-80.

[24] Ventegodt S, Morad M, Kandel I, Merrick J. Clinical holistic medicine: problems in sex and living together. ScientificWorldJournal 2004;4:562-70.

[25] Ventegodt S, Morad M, Merrick J. Clinical holistic medicine: holistic pelvic examination and holistic treatment of infertility. ScientificWorldJournal 2004;4:148-58.

[26] Kendler KS. The genetics of schizophrenia: A current perspective. In: Meltzer HY, ed. Psychopharmacology. The third generation of progress. New York: Raven, 1987.

[27] Abhijnhan A, Adams CE, David A, Ozbilen M. Depot fluspirilene for schizophrenia. Cochrane Database Syst Rev 2007;(1):CD001718.

[28] Adams CE, Awad G, Rathbone J, Thornley B. Chlorpromazine versus placebo for schizophrenia. Cochrane Database Syst Rev 2007;(2):CD000284.

[29] Amato L, Minozzi S, Pani PP, Davoli M. Antipsychotic medications for cocaine dependence. Cochrane Database Syst Rev 2007;(3):CD006306.

[30] Arunpongpaisal S, Ahmed I, Aqeel N, Suchat P. Antipsychotic drug treatment for elderly people with late-onset schizophrenia. Cochrane Database Syst Rev 2003;(2):CD004162.

[31] Bagnall A, Lewis RA, Leitner ML. Ziprasidone for schizophrenia and severe mental illness. Cochrane Database Syst Rev 2000;(4):CD001945.

[32] Bagnall A, Fenton M, Kleijnen J, Lewis R. Molindone for schizophrenia and severe mental illness. Cochrane Database Syst Rev 2007;(1):CD002083.

[33] Ballard C, Waite J. The effectiveness of atypical antipsychotics for the treatment of aggression and psychosis in Alzheimer's disease. Cochrane Database Syst Rev 2006;(1):CD003476.

[34] Basan A, Leucht S. Valproate for schizophrenia. Cochrane Database Syst Rev 2004;(1):CD004028.

[35] Belgamwar RB, Fenton M. Olanzapine IM or velotab for acutely disturbed/agitated people with suspected serious mental illnesses. Cochrane Database Syst Rev 2005;(2):CD003729.

[36] Binks CA, Fenton M, McCarthy L, Lee T, Adams CE, Duggan C. Pharmacological interventions for people with borderline personality disorder. Cochrane Database Syst Rev 2006;(1):CD005653.

[37] Carpenter S, Berk M, Rathbone J. Clotiapine for acute psychotic illnesses. Cochrane Database Syst Rev 2004;(4):CD002304.

[38] Chakrabarti A, Bagnall A, Chue P, Fenton M, Palaniswamy V, Wong W, Xia J. Loxapine for schizophrenia. Cochrane Database Syst Rev 2007;(4):CD001943.

[39] Chua WL, de Izquierdo SA, Kulkarni J, Mortimer A. Estrogen for schizophrenia. Cochrane Database Syst Rev 2005;(4):CD004719.

[40] Coutinho E, Fenton M, Quraishi S. Zuclopenthixol decanoate for schizophrenia and other serious mental illnesses. Cochrane Database Syst Rev 2000;(2):CD001164.

[41] Cure S, Rathbone J, Carpenter S. Droperidol for acute psychosis. Cochrane Database Syst Rev 2004;(4):CD002830.

[42] David A, Adams CE, Eisenbruch M, Quraishi S, Rathbone J. Depot fluphenazine decanoate and enanthate for schizophrenia. Cochrane Database Syst Rev 2005;(1):CD000307.

[43] David A, Quraishi S, Rathbone J. Depot perphenazine decanoate and enanthate for schizophrenia. Cochrane Database Syst Rev 2005;(3):CD001717.

[44] DeSilva P, Fenton M, Rathbone J. Zotepine for schizophrenia. Cochrane Database Syst Rev 2006;(4):CD001948.

[45] Dinesh M, David A, Quraishi SN. Depot pipotiazine palmitate and undecylenate for schizophrenia. Cochrane Database Syst Rev 2004;(4):CD001720.
[46] Duggan L, Brylewski J. Antipsychotic medication versus placebo for people with both schizophrenia and learning disability. Cochrane Database Syst Rev 2004;(4):CD000030.
[47] Duggan L, Fenton M, Rathbone J, Dardennes R, El-Dosoky A, Indran S. Olanzapine for schizophrenia. Cochrane Database Syst Rev 2005;(2):CD001359.
[48] El-Sayeh HG, Morganti C. Aripiprazole for schizophrenia. Cochrane Database Syst Rev 2006;(2):CD004578.
[49] Elias A, Kumar A. Testosterone for schizophrenia. Cochrane Database Syst Rev 2007;(3):CD006197.
[50] Fenton M, Rathbone J, Reilly J, Sultana A. Thioridazine for schizophrenia. Cochrane Database Syst Rev 2007;(3):CD001944.
[51] Gibson RC, Fenton M, Coutinho ES, Campbell C. Zuclopenthixol acetate for acute schizophrenia and similar serious mental illnesses. Cochrane Database Syst Rev 2004;(3):CD000525.
[52] Gilbody SM, Bagnall AM, Duggan L, Tuunainen A. Risperidone versus other atypical antipsychotic medication for schizophrenia. Cochrane Database Syst Rev 2000;(3):CD002306.
[53] Gillies D, Beck A, McCloud A, Rathbone J, Gillies D. Benzodiazepines alone or in combination with antipsychotic drugs for acute psychosis. Cochrane Database Syst Rev 2005;(4):CD003079.
[54] Hartung B, Wada M, Laux G, Leucht S. Perphenazine for schizophrenia. Cochrane Database Syst Rev 2005;(1):CD003443.
[55] Hosalli P, Davis JM. Depot risperidone for schizophrenia. Cochrane Database Syst Rev 2003;(4):CD004161.
[56] Huf G, Alexander J, Allen MH. Haloperidol plus promethazine for psychosis induced aggression. Cochrane Database Syst Rev 2005;(1):CD005146.
[57] Hunter RH, Joy CB, Kennedy E, Gilbody SM, Song F. Risperidone versus typical antipsychotic medication for schizophrenia. Cochrane Database Syst Rev 2003;(2):CD000440.
[58] Jayaram MB, Hosalli P, Stroup S. Risperidone versus olanzapine for schizophrenia. Cochrane Database Syst Rev 2006;(2):CD005237.
[59] Jesner OS, Aref-Adib M, Coren E. Risperidone for autism spectrum disorder. Cochrane Database Syst Rev 2007;(1):CD005040.
[60] Joy CB, Adams CE, Rice K. Crisis intervention for people with severe mental illnesses. Cochrane Database Syst Rev 2006;(4):CD001087.
[61] Joy CB, Adams CE, Lawrie SM. Haloperidol versus placebo for schizophrenia. Cochrane Database Syst Rev 2006;(4):CD003082.
[62] Kennedy E, Kumar A, Datta SS. Antipsychotic medication for childhoodonset schizophrenia. Cochrane Database Syst Rev 2007;(3):CD004027.
[63] Kumar A, Strech D. Zuclopenthixol dihydrochloride for schizophrenia. Cochrane Database Syst Rev 2005;(4):CD005474.
[64] Leucht S, Hartung B. Benperidol for schizophrenia. Cochrane Database Syst Rev 2005;(2):CD003083.
[65] Leucht S, Hartung B. Perazine for schizophrenia. Cochrane Database Syst Rev 2006;(2):CD002832.
[66] Leucht S, Kissling W, McGrath J, White P. Carbamazepine for schizophrenia. Cochrane Database Syst Rev 2007;(3):CD001258.
[67] Leucht S, Kissling W, McGrath J. Lithium for schizophrenia. Cochrane Database Syst Rev 2007;(3):CD003834.
[68] Lewis R, Bagnall AM, Leitner M. Sertindole for schizophrenia. Cochrane Database Syst Rev 2005;(3):CD001715.
[69] Macritchie KA, Geddes JR, Scott J, Haslam DR, Goodwin GM. Valproic acid, valproate and divalproex in the maintenance treatment of bipolar disorder. Cochrane Database Syst Rev 2001;(3):CD003196.
[70] Macritchie K, Geddes JR, Scott J, Haslam D, de Lima M, Goodwin G. Valproate for acute mood episodes in bipolar disorder. Cochrane Database Syst Rev 2003;(1):CD004052.
[71] Marques LO, Lima MS, Soares BG. Trifluoperazine for schizophrenia. Cochrane Database Syst Rev 2004;(1):CD003545.

[72] Marriott RG, Neil W, Waddingham S. Antipsychotic medication for elderly people with schizophrenia. Cochrane Database Syst Rev 2006;(1):CD005580.
[73] Marshall M, Rathbone J. Early intervention for psychosis. Cochrane Database Syst Rev 2006;(4):CD004718.
[74] Matar HE, Almerie MQ. Oral fluphenazine versus placebo for schizophrenia. Cochrane Database Syst Rev 2007;(1):CD006352.
[75] Mota NE, Lima MS, Soares BG. Amisulpride for schizophrenia. Cochrane Database Syst Rev 2002;(2):CD001357.
[76] Nolte S, Wong D, Lachford G. Amphetamines for schizophrenia. Cochrane Database Syst Rev 2004;(4):CD004964.
[77] Pekkala E, Merinder L. Psychoeducation for schizophrenia. Cochrane Database Syst Rev 2002;(2):CD002831.
[78] Premkumar TS, Pick J. Lamotrigine for schizophrenia. Cochrane Database Syst Rev 2006;(4):CD005962.
[79] Punnoose S, Belgamwar MR. Nicotine for schizophrenia. Cochrane Database Syst Rev 2006;(1):CD004838.
[80] Quraishi S, David A. Depot flupenthixol decanoate for schizophrenia or other similar psychotic disorders. Cochrane Database Syst Rev 2000;(2):CD001470.
[81] Quraishi S, David A. Depot haloperidol decanoate for schizophrenia. Cochrane Database Syst Rev 2000;(2):CD001361.
[82] Rathbone J, McMonagle T. Pimozide for schizophrenia or related psychoses. Cochrane Database Syst Rev 2007;(3):CD001949.
[83] Rendell JM, Gijsman HJ, Keck P, Goodwin GM, Geddes JR. Olanzapine alone or in combination for acute mania. Cochrane Database Syst Rev 2003;(3):CD004040.
[84] Rendell JM, Gijsman HJ, Bauer MS, Goodwin GM, Geddes GR. Risperidone alone or in combination for acute mania. Cochrane Database Syst Rev 2006;(1):CD004043.
[85] Rendell JM, Geddes JR. Risperidone in long-term treatment for bipolar disorder. Cochrane Database Syst Rev 2006;(4):CD004999.
[86] Rummel C, Hamann J, Kissling W, Leucht S. New generation antipsychotics for first episode schizophrenia. Cochrane Database Syst Rev 2003;(4):CD004410.
[87] Rummel C, Kissling W, Leucht S. Antidepressants for the negative symptoms of schizophrenia. Cochrane Database Syst Rev 2006;3:CD005581.
[88] Soares BG, Fenton M, Chue P. Sulpiride for schizophrenia. Cochrane Database Syst Rev 2000;(2):CD001162.
[89] Soares BG, Lima MS. Penfluridol for schizophrenia. Cochrane Database Syst Rev 2006;(2):CD002923.
[90] Srisurapanont M, Kittiratanapaiboon P, Jarusuraisin N. Treatment for amphetamine psychosis. Cochrane Database Syst Rev 2001;(4):CD003026.
[91] Srisurapanont M, Maneeton B, Maneeton N. Quetiapine for schizophrenia. Cochrane Database Syst Rev 2004;(2):CD000967.
[92] Tharyan P, Adams CE. Electroconvulsive therapy for schizophrenia. Cochrane Database Syst Rev 2005;(2):CD000076.
[93] Trevisani VF, Castro AA, Neves Neto JF, Atallah AN. Cyclophosphamide versus methylprednisolone for treating neuropsychiatric involvement in systemic lupus erythematosus. Cochrane Database Syst Rev 2006;(2):CD002265.
[94] Tuominen HJ, Tiihonen J, Wahlbeck K. Glutamatergic drugs for schizophrenia. Cochrane Database Syst Rev 2006;(2):CD003730.
[95] Tuunainen A, Wahlbeck K, Gilbody SM. Newer atypical antipsychotic medication versus clozapine for schizophrenia. Cochrane Database Syst Rev 2000;(2):CD000966.
[96] Volz A, Khorsand V, Gillies D, Leucht S. Benzodiazepines for schizophrenia. Cochrane Database Syst Rev 2007;(1):CD006391.
[97] Wahlbeck K, Cheine M, Essali MA. Clozapine versus typical neuroleptic medication for schizophrenia. Cochrane Database Syst Rev 2000;(2):CD000059.

[98] Waraich PS, Adams CE, Roque M, Hamill KM, Marti J. Haloperidol dose for the acute phase of schizophrenia. Cochrane Database Syst Rev 2002;(3):CD001951.

[99] Webb RT, Howard L, Abel KM. Antipsychotic drugs for non-affective psychosis during pregnancy and postpartum. Cochrane Database Syst Rev 2004;(2):CD004411.

[100] Whitehead C, Moss S, Cardno A, Lewis G. Antidepressants for people with both schizophrenia and depression. Cochrane Database Syst Rev 2002;(2):CD002305.

[101] Wijkstra J, Lijmer J, Balk F, Geddes J, Nolen WA. Pharmacological treatment for psychotic depression. Cochrane Database Syst Rev 2005;(4):CD004044.

[102] Wong D, Adams CE, David A, Quraishi SN. Depot bromperidol decanoate for schizophrenia. Cochrane Database Syst Rev 2004;(3):CD001719.

[103] Young AH, Geddes JR, Macritchie K, Rao SN, Vasudev A. Tiagabine in the maintenance treatment of bipolar disorders. Cochrane Database Syst Rev 2006;3:CD005173.

[104] Losonczy MF, Davidson M, Davis KL. The dopamine hypothesis of schizophrenia. In: Meltzer HY, ed. Psychopharmacology. The third generation of progress. New York: Raven, 1987.

[105] Tamminga CA, Gerlach J. New neuroleptics and experimental antipsychotics in schizophrenia. In: Meltzer HY, ed. Psychopharmacology. The third generation of progress. New York: Raven, 1987.

[106] Hollister LE. Novel drug treatments for schizophrenia. Psychopharmacol Bull 1987;23(1):82-4.

[107] Blackwell B. Newer antidepressant drugs. In: Meltzer HY, ed. Psychopharmacology. The third generation of progress. New York: Raven, 1987.

[108] Falk JL, Feingold DA. Environmental and cultural factors in the behavioural action of drugs. In: Meltzer HY, ed. Psychopharmacology. The third generation of progress. New York: Raven, 1987.

[109] Ventegodt S, Flensborg-Madsen T, Andersen NJ, Nielsen M, Mohammed M, Merrick J. Global quality of life (QOL), health and ability are primarily determined by our consciousness. Research findings from Denmark 1991-2004. Soc Indicator Res 2005;71:87-122.

[110] Ventegodt S, Merrick J. Clinical holistic medicine: Applied consciousness-based medicine. ScientificWorldJournal 2004;4:96-9.

[111] Ventegodt S, Morad M, Andersen NJ, Merrick J. Clinical holistic medicine Tools for a medical science based on consciousness. ScientificWorldJournal 2004;4:347-61.

[112] Ventegodt S, Morad M, Kandel I, Merrick J. Clinical holistic medicine: a psychological theory of dependency to improve quality of life. ScientificWorldJournal 2004;4:638-48.

[113] Ventegodt S, Kandel I, Neikrug S, Merrick J. Clinical holistic medicine: Holistic treatment of rape and incest traumas. ScientificWorldJournal 2005;5:288-97.

[114] Ventegodt S, Kandel I, Neikrug S, Merrick J. Clinical holistic medicine: The existential crisis – life crisis, stress and burnout. ScientificWorldJournal 2005;5:300-12.

[115] Ventegodt S, Gringols G, Merrick J. Clinical holistic medicine: Holistic rehabilitation. ScientificWorldJournal 2005;5:280-7.

[116] Ventegodt S, Andersen NJ, Neikrug S, Kandel I, Merrick J. Clinical holistic medicine: Mental disorders in a holistic perspective. ScientificWorldJournal 2005;5:313-23.

[117] Ventegodt S, Andersen NJ, Neikrug S, Kandel I, Merrick J. Clinical holistic medicine: Holistic treatment of mental disorders. ScientificWorldJournal 2005;5:427-45.

[118] Ventegodt S, Merrick J. Clinical holistic medicine: The patient with multiple diseases ScientificWorldJournal 2005;5:324-39.

[119] Ventegodt S, Clausen B, Nielsen ML, Merrick J. Clinical holistic medicine: Advanced tools for holistic medicine. ScientificWorldJournal 2006;6:2048-65.

[120] Ventegodt S, Clausen B, Merrick J. Clinical holistic medicine: The case story of Anna: I. Long term effect of child sexual abuse and incest with a treatment approach. ScientificWorldJournal 2006;6:1965-76.

[121] Ventegodt S, Clausen B, Merrick J. Clinical holistic medicine: the case story of Anna. II. Patient diary as a tool in treatment. ScientificWorldJournal 2006;6:2006-34.

[122] Ventegodt S, Clausen B, Merrick J. Clinical holistic medicine: The case story of Anna. III. Rehabilitation of philosophy of life during holistic existential therapy for childhood sexual abuse. ScientificWorldJournal 2006;6:2080-91.

[123] Ventegodt S, Merrick J. Suicide from a holistic point of view. ScientificWorldJournal 2005;5:759-66.

[124] Ventegodt S, Clausen B, Omar HA, Merrick J. Clinical holistic medicine: Holistic sexology and acupressure through the vagina (Hippocratic pelvic massage). ScientificWorldJournal 2006;6:2066-79.
[125] Ventegodt S, Clausen B, Merrick J. Clinical holistic medicine: Pilot study on the effect of vaginal acupressure (Hippocratic pelvic massage). ScientificWorldJournal 2006;6:2100-16.
[126] Ventegodt S, Thegler S, Andreasen T, Struve F, Enevoldsen L, et al. Clinical holistic medicine: Psychodynamic short-time therapy complemented with bodywork. A clinical follow-up study of 109 patients. ScientificWorldJournal 2006;6:2220-38.
[127] Ventegodt S, Thegler S, Andreasen T, Struve F, Enevoldsen L, et al. Clinical holistic medicine (mindful, short-term psychodynamic psychotherapy complemented with bodywork) in the treatment of experienced impaired sexual functioning. ScientificWorldJournal 2007;7: 324-9.
[128] Ventegodt S, Thegler S, Andreasen T, Struve F, Enevoldsen L, et al. Clinical holistic medicine (mindful, short-term psychodynamic psychotherapy complemented with bodywork) improves quality of life, health, and ability by induction of Antonovsky-salutogenesis. ScientificWorldJournal 2007;7:317-23.
[129] Ventegodt S, Thegler S, Andreasen T, Struve F, Enevoldsen L, et al. Clinical holistic medicine (mindful, short-term psychodynamic psychotherapy complemented with bodywork) in the treatment of experienced mental illness. ScientificWorldJournal 2007;7:306-9.
[130] Ventegodt S, Thegler S, Andreasen T, Struve F, Enevoldsen L, et al. Self-reported low self-esteem. Intervention and follow-up in a clinical setting. ScientificWorldJournal 2007;7:299-305.
[131] Ventegodt S, Merrick J. Clinical holistic medicine: Chronic pain in the locomotor system. ScientificWorldJournal 2005;5:165-72.
[132] Ventegodt S, Merrick J. Clinical holistic medicine: Chronic pain in internal organs. ScientificWorldJournal 2005;5:205-10.
[133] Ventegodt S, Thegler S, Andreasen T, Struve F, Enevoldsen L, et al. Clinical holistic medicine (mindful, short-term psychodynamic psychotherapy complemented with bodywork) in the treatment of experienced physical illness and chronic pain. ScientificWorldJournal 2007;7:310-6.
[134] Flensborg-Madsen T, Ventegodt S, Merrick J. Sense of coherence and physical health. A review of previous findings. ScientificWorldJournal 2005;5:665-73.
[135] Flensborg-Madsen T, Ventegodt S, Merrick J. Why is Antonovsky's sense of coherence not correlated to physical health? Analysing Antonovsky's 29-item sense of coherence scale (SOCS). ScientificWorldJournal 2005;5:767-76.
[136] Flensborg-Madsen T, Ventegodt S, Merrick J. Sense of coherence and health. The construction of an amendment to Antonovsky's sense of coherence scale (SOC II). ScientificWorldJournal 2006;6: 2133-9.
[137] Flensborg-Madsen T, Ventegodt S, Merrick J. Sense of coherence and physical health. A cross-sectional study using a new SOC scale (SOC II). ScientificWorldJournal 2006;6:2200-11.
[138] Flensborg-Madsen T, Ventegodt S, Merrick J. Sense of coherence and physical health. Testing Antonovsky's theory. ScientificWorldJournal 2006;6:2212-9.
[139] Flensborg-Madsen T, Ventegodt S, Merrick J. Sense of coherence and health. The emotional sense of coherence (SOC-E) was found to be the best-known predictor of physical health. ScientificWorldJournal 2006;6: 2147-57.
[140] Ventegodt S, Flensborg-Madsen T, Andersen NJ, Merrick J. Life Mission Theory VII: Theory of existential (Antonovsky) coherence: a theory of quality of life, health and ability for use in holistic medicine. ScientificWorldJournal 2005;5:377-89.

Chapter XVI

The role of adult human metamorphosis in spontaneous recovery

Mentally ill patients can enter the state of adult human metamorphosis, to re-do the juvenile metamorphosis into non-human forms that was necessary for their survival in childhood. The adult human metamorphosis looks at first glance like a normal psychotic crisis of a schizophrenic patient, but when you study it closer it becomes obvious that it is actually a healing crisis, an unusual state of accelerated healing process of the type "Antonovsky-salutogenesis". The healing crisis and the process of adult human metamorphosis seems to be the pathogenesis reversed: an inverted event of a juvenile metamorphosis, where the patient originally turned him- or herself into a non-human (often quite alien) form to survive a hostile and unfriendly invironment. This often leads to severe developmental disturbances and thus to physical or mental health problems later in life. We have observed several cases of spontaneous remission of a number of different mental diseases induced by holistic existential therapy: schizophrenia, borderline, anxiety, and bulimia. We propose that the general etiology of mental diseases is juvenile metamorphosis intended for survival, not defect genes disturbing the patient's brain chemistry. This understanding empowers us to induce salutogenesis and spontaneous recovery in also the most ill, mental patients with clinical holistic medicine. The healing crisis could be wrongly diagnosed as a brief reactive psychosis, if the psychiatrist is without practical knowledge of the phenomenon of Antonovsky-salutogensis; this could explain why intensive psychotherapy has been considered dangerous by some psychiatrists.

Introduction

As we have discussed in earlier papers in this series on human development, shape and function is closely related to the organism's consciousness. Intent seems to be able to change and modify shape and function. The modifying force of intent is active though life from its very beginning, and the individual seems to be able to read what is going on in its world though the exchange of information on many levels in the individual and collective informational systems.

This opens up for radical self-modifications early in life. We suggest that the individual is actually able to transform itself into a being more fit for survival and the family it is going to soon be a part of. These transformations can be caused by dramatic events in the family, or by the environment being so rough that a normal child could not survive in it. Many dysfunctional families could qualify for this description. The most dramatic change we know of is the metamorphosis; the radical and complete transformation of an individual through an interaction between the level of intent and consciousness to the level of form and function.

We suggest that early adaptation to severely dysfunctional parents, in the womb or in early childhood, will render the individual so changed that it can present itself almost non-human; both consciousness and functions will be severely disturbed, and the person can even sometimes look not only ugly, but "alien" in bodily appearance and energy.

Adult human metamorphosis is re-metamorphosis of childhood adaptations

Some of the authors have examined and treated a number of severely mentally ill patients and we have observed that they often appear very unappealing physically and very alien mentally, as if they have turned themselves into some kind of "monsters" early in life. In surprisingly many cases they have adapted to extreme environments dominated by violent or sexual abuse, or severe neglect and fail. Spontaneously in the holistic existential therapy (1-6) they have entered the state of what we call adult human metamorphosis and they have spent weeks to reconnect to the human collective consciousness, as if they have lived in a system of their own non-connection to the human universe but to some strange parallel reality. It is like they are coming back from a monstrous world into the world of human beings.

Often they have felt alien; felt that they were like aliens waling the earth. So we suggest that they have metamorphosed themselves into beings with non-human intent (3,4,6-10) in some cases even using biological information from alien life forms, in fantasy or for real, if biological information can ride even the energy of the galaxies. Changing intent and the patterns of thought and perception early in life away from the typical human patterns into something else, giving the individual severe problems with relating to other people and the human world at large, is from this perspective what causes mental diseases of the skizotypical types. Presumably this also goes for the affective disorders, as the root of affective aberration likewise seem to be cognitive disturbances. So we suggest that mental diseases are caused by the individual entering metamorphosis early in life to transform into a being with non-human consciousness better able to survive an inhuman and severely dysfunctional environment. This give some meaning to the extremely scary effects of aliens in movies: that they are here already, walking amongst us, disguised as humans. And we think they are.

All aspects of a human being except its fundamental purpose of life can be changed by human juvenile metamorphosis; the purpose of life cannot be changed, but repressed and forgotten. The person going through the process of metamorphosis seems to change functioning purpose of life and fundamental intent, and both mind and body seem to follow. Love, consciousness and sexuality is often severely affected, and the ability to love, understand and enjoy is often damaged. The gender is often annulated or even inverted energetically; the feelings are often completely withdrawn, and mental and intellectual capacity is changed into focusing on non-living issues, very much like the loss of a "I-thou" relationship described by Martin Buber (1878-1965) (11).

The healing of mental diseases through re-metamorphosis

The metamorphosis of the human mind during the psychotic crises seems to be the most radical and drastic of the processes of healing. It happens spontaneously, when development has been radically arrested early in life, normally because of a trauma forcing the patient to let go of his or her purpose of life and substitute it with another purpose. The patients have seemingly had a psychotic episode in connection to this, often in early childhood. Often many aspects of psychosexual development has been arrested, and the people likely to go into human metamorphosis are in many ways like the butterfly's larvae, which grow and grow in size but not in any other way, until the day of metamorphosis, where the information linked to being an adult is finally accessed at once, giving a complete transformation of character, personality and consciousness.

Entering the metamorphosis is helped by intense holistic existential therapy combined with intense holding and combined with the patient's extreme intent to heal here and now and today! Healing existence must be more important that anything else, and the process of metamorphosis is taking the patient to a place of continuous healing for days. Recreation for a week or two after the introvert, psychotic crises seems mandatory, and the patient cannot work or look after kids or have other normal obligations in this period, and sometimes for an extended period of time.

Once the process of personal transformation hasve started it must run to its natural end. If the metamorphosis is disturbed, when the patient is in the most vulnerable state, severe harm can be done. It is very important that spouses and other family members are informed about the natural course of this kind of spontaneous healing, and they must be carefully informed about the urgent needs for tranquillity and loving care and support for days or weeks. The danger of poor living for months or years if the healing process is disrupted and the patient is sedated or drugged must be severely stressed.

An important ethical problem is if the physician or the relatives should judge what is in the best interest of the patient, when the patient enters the introvert, "psychotic" state similar to the butterfly's pupae. Often the spouse of family members not completely trusting the natures ways will react with fear and want the patient back in the old condition, while the patient him- or herself is doing everything possible to transform into a new and better version of self. Coming from fear and anger the family can cause severe damage to the patient, by interrupting the metamorphosis and arresting it at a transformational state where the patients mind is not very functional. Another dangerous aspect of this is that the person needing the transformation is often victimised early in life, and therefore of a week character allowing other people to exploit them, when they evolve into a more responsible and whole version of themselves they will often rebel against the dominance and "ownership", to become free and autonomous. It is important that the physician notice patterns of abuse in the relationship and helps the patient to understand the consequences of submitting to dominance and being owned by i.e. a spouse. If the behaviour of relatives is obviously threatening the patient's health, the physician should inform the social authorities and other relevant authorities for them to take the appropriate action.

Most interestingly human metamorphosis seems to be able to explain many religious experiences, like the 40 days in the desert where Jesus meets his creator, or Gautama Buddha's famous enlightenment, where he reaches Nirvana, the cosmic emptiness creating the world. A simple way of understanding human metamorphosis is as re-establishing the coherence with the world, getting direct access to the "web", "the nest of the world" (in prep. for publication), or the deep level of the universe we call "the matrix of energy and information"(in prep. for publication), feeding all organisms with qualities like intent and talent.

Case stories

At many occasions during our 14 years of research in quality of life and holistic medicine at the Quality of Life Research Clinic in Copenhagen we have seen patients with the diagnoses of schizophrenia or borderline enter an extremely intense, accelerated process of healing we have called "adult human metamorphosis", because of its remarkable resemblance of a butterfly's larva entering the pupae and metamorphosing into the butterfly. The patients even sometimes look like larvae, with plump poorly demarcated body contours, immature, clumsy movement patterns, and they are characterised by having poor reality testing, as if they lived in their own world, in a parallel universe. Sometimes they have been students of remarkable intelligence, studying medicine or psychology at the university, sometimes they have been very intelligent, but not able to study, sometimes they have chosen to study nursing or occupational therapy. The patients that have come to our research clinic for holistic medicine come to our private clinic by own choice to enter our research protocol on healing the mentally ill (12,13); they normally get a grant from the Quality of Life Research Center so they are able to participate; they typically pay 25% of the therapy themselves.

Female university student of medicine, 24 years, borderline

At the beginning of the treatment she was 30 kg overweight, a poor reality testing, no close friends; she was still a virgin with no interest in men, and a strange non-human uni-sex appearance. All body movements was impaired, she was slow, clumsy, and seemingly depressed, but with no emotional problems, except feeling like an alien. She was not able to look into other people's eyes. In holistic existential therapy she confronted that her 10-year-older big brother had raped her when she was five years old. The energy was that of war-rape: she was raped in the intent of repressing her. Strangely her parents wanted her eliminated and her brother was in alliance with the parents against her and he was not punished. She entered a psychotic crisis that lasted for 14 days during which she was hospitalised at the clinic. She experienced that she melted down and re-entered the human stream of consciousness – the collective human consciousness. She entered a visionary state of mind and for days she received thousands of pictures of human life from an inner source of wisdom and knowledge. When she re-appeared to the surface she was completely transformed into a wise young human remembering her true human nature. She changes her life completely, started dating boys, doing exercise, changing diet and loosing weight. In art-therapy she painted hundreds of paintings of the scenarios she has visualised during the metamorphosis.

Female university student of psychology, 22 years, bulimia

Severe problems with self-esteem, looks, sexuality – not able to enjoy sex or intercourse – self-confidence, presenting severely disturbed eating patterns of overeating and vomiting. She believed she must weigh 50.00 kg. If not, she found her body disgusting. Sometimes she dressed extremely feminine in skirt and appears as a beautiful young woman, this interchanging with a much more male appearance where she varies men's pants. She also suffers from anxiety, and her sexual borders are 2 meters from her body; if men get closer she often feels intimidated. In holistic existential therapy the therapist (SV) could not get close to her for many sessions. When trust was won she finally allowed the therapist to get close emotionally and she melted down in a psychotic crises, where all her problems of sexuality of low bodily self-esteem exploded. She entered a mental state of feeling totally unreal and stayed like this for days. She was hospitalised at our clinic for four days, before she can integrate the painful childhood events of failure and neglect that she has confronted; obviously she has then as a child metamorphosed into a person living in her own world not needing any contact to her parents, to avoid the pain of feeling not loved and not cared for. After the psychotic crises her condition slowly normalised and she was able to function again as a human being.

We have earlier described this kind of spontaneous healing of two schizophrenic patients (14), and we have often seen holistic sexological therapy and holistic gynaecology lead to accelerated existential healing (Antonovsky salutogenesis) (15-24).

Discussion

In a number of cases in our clinical work we have observed a radical process of spontaneous healing that seems to be adult human metamorphosis, parallel to the metamorphosis of many insects and some vertebrates. In its most radical version the person's mind is melting down and all behavioural and mental patterns are disintegrated for up to 10 days. The function of the human metamorphosis is to allow a person to catch up after many years of arrested psychosexual development. It happens spontaneously or provoked by holistic existential therapy, and is most likely to happen with people who have been violently or sexually abused in early childhood. The process is often initiated by a catharsis and a break through into old trauma of extremely intense emotional pain; rape in early childhood seems to be the paradigm. Other traumas of sufficient intensity to cause the developmental arrest which is set free by the metamorphosis is: surgery, social isolation, violence, and other events giving the child a near-death or psychic death experience. During the metamorphosis the person will re-live the extreme neural arousal and temporary psychosis of the traumatic childhood event.

It is most important to stress that the metamorphosis is not a mental disease - but a state of spontaneous healing - and it should not be treated as such. In this spontaneous healing event the patient needs loving care and tranquillity, while being "in the pupae": introvert, mentally disintegrated, seemingly psychotic and not in present time, sinking into the sea of biological information of being human, to finally reappear the way nature originally meant this person to be.

We suggest that human metamorphosis is really a biological process, and we believe that there is a complete set of genes and chemical mediators (hormones, neurotransmitters, or neuropeptides) to handle the biological side of the human metamorphosis. We believe that human metamorphosis is the most efficient healing process known to this day, and that a successful metamorphosis can save the patient from many years of therapy and sufferings. Human metamorphosis cannot only change the mind. We believe that spontaneous healing of cancer and other diseases is also often caused by human metamorphosis.

Conclusions

Interestingly, mentally ill patients can enter the state of adult human metamorphosis, to re-do the juvenile metamorphosis into non-human forms that was beneficial for their childhood survival. The adult human metamorphosis looks at first glance like a normal psychotic crisis of a schizophrenic patient, but when you study it closer it becomes obvious that it is an unusual state of accelerated healing, a healing crisis. It is important to understand that the adult human metamorphosis is the inverted event of a juvenile metamorphosis where the patient turned him- or her self into some non-human alien form to survive; the self-transportation of human existence into a parallel dimensionof existence often gives severe developmental problems and both physical and mental problems to the person later in life but was necessary for survival from an emotional point of view.

We have now carefully observed several cases of spontaneous remission of a number of different mental diseases: schizophrenia, borderline, anxiety, and bulimia. We understand the adult human metamorphosis as a confirmation of our hypothesis of aetiology of mental diseases: that they are in general caused by juvenile metamorphosis (12,13), not by defective genes and disturbed brain chemistry.

We believe that we in cracking this "secret code" have found a general way to understand and heal mental diseases, by taking them into adult human metamorphosis with clinical holistic medicine.

References

[1] Ventegodt S, Andersen NJ, Merrick J. Holistic medicine III: The holistic process theory of healing. ScientificWorldJournal 2003;3:1138-46.

[2] Ventegodt S, Andersen NJ, Merrick J. Holistic medicine IV: Principles of existential holistic group therapy and the holistic process of healing in a group setting. ScientificWorldJournal 2003;3:1388-1400.

[3] Ventegodt S, Andersen NJ, Merrick J. The life mission theory V. A theory of the anti-self and explaining the evil side of man. ScientificWorldJournal 2003;3:1302-13.

[4] Ventegodt S, Andersen NJ, Merrick J. The life mission theory VI: A theory for the human character. ScientificWorldJournal 2004:4:859-80.
[5] Ventegodt S, Morad M, Andersen NJ, Merrick J. Clinical holistic medicine Tools for a medical science based on consciousness. ScientificWorldJournal 2004;4:347-61.
[6] Ventegodt S, Flensborg-Madsen T, Andersen NJ, Merrick J. Life mission theory VII: Theory of existential (Antonovsky) coherence: a theory of quality of life, health and ability for use in holistic medicine. ScientificWorldJournal 2005;5:377-89.
[7] Ventegodt S. The life mission theory: A theory for a consciousness-based medicine. Int J Adolesc Med Health 2003;15(1):89-91.
[8] Ventegodt S, Andersen NJ, Merrick J. The life mission theory II: The structure of the life purpose and the ego. ScientificWorldJournal 2003;3:1277-85.
[9] Ventegodt S, Andersen NJ, Merrick J. The life mission theory III: Theory of talent. ScientificWorldJournal 2003;3:1286-93.
[10] Ventegodt S, Merrick J. The life mission theory IV. A theory of child development. ScientificWorldJournal 2003;3:1294-1301.
[11] Buber M. I and thou. London: Free Press, 1971.
[12] Ventegodt S, Andersen NJ, Neikrug S, Kandel I, Merrick J. Clinical Holistic medicine: Holistic treatment of mental disorders. ScientificWorldJournal 2005;5:427-45.
[13] Ventegodt S, Andersen NJ, Neikrug S, Kandel I, Merrick J. Clinical holistic medicine: Mental disorders in a holistic perspective. ScientificWorldJournal 2005;5:313-23.
[14] Ventegodt S, Kandel I, Merrick J. Clinical holistic medicine (mindful short-term psychodynamic psychotherapy complimented with bodywork) in the treatment of schizophrenia (ICD10-F20/DSM-IV Code 295) and other psychotic mental diseases. ScientificWorldJournal. 2007;7:1987-2008.
[15] Ventegodt S, Kandel I, Merrick J. Clinical holistic medicine: how to recover memory without "implanting" memories in your patient. ScientificWorldJournal. 2007 Sep 17;7:1579-89.
[16] Flensborg-Madsen T, Ventegodt S, Merrick J. Sense of coherence and physical health. The emotional sense of coherence (SOC-E) was found to be the best-known predictor of physical health. ScientificWorldJournal. 2006;6:2147-57.
[17] Ventegodt S, Clausen B, Omar HA, Merrick J. Clinical holistic medicine: holistic sexology and acupressure through the vagina (Hippocratic pelvic massage). ScientificWorldJournal 2006;6:2066-79.
[18] Ventegodt S, Clausen B, Merrick J. Clinical holistic medicine: pilot study on the effect of vaginal acupressure (Hippocratic pelvic massage). ScientificWorldJournal 2006;6:2100-16.
[19] Ventegodt S, Clausen B, Merrick J. Clinical holistic medicine: the case story of Anna. III. Rehabilitation of philosophy of life during holistic existential therapy for childhood sexual abuse. ScientificWorldJournal 2006;6:2080-91.
[20] Ventegodt S, Thegler S, Andreasen T, Struve F, Enevoldsen L, Bassaine L, Torp M, Merrick J. Clinical holistic medicine (mindful, short-term psychodynamic psychotherapy complemented with bodywork) in the treatment of experienced impaired sexual functioning. ScientificWorldJournal 2007;7:324-9.
[21] Ventegodt S, Kandel I, Neikrug S, Merric J. Clinical holistic medicine: holistic treatment of rape and incest trauma. ScientificWorldJournal 2005;5:288-97.
[22] Ventegodt S, Morad M, Hyam E, Merrick J. Clinical holistic medicine: holistic sexology and treatment of vulvodynia through existential therapy and acceptance through touch. ScientificWorldJournal 2004;4:571-80.
[23] Ventegodt S, Morad M, Kandel I, Merrick J. Clinical holistic medicine: problems in sex and living together. ScientificWorldJournal 2004;4:562-70.
[24] Ventegodt S, Morad M, Merrick J. Clinical holistic medicine: holistic pelvic examination and holistic treatment of infertility. ScientificWorldJournal 2004;4:148-58.

Chapter XVII

Jerne's anti-idiotypic network theory

In this chapter we propose that self-nonself discrimination takes place at a supra-cellular systemic level and involves selective activation of immune cell precursors. We discuss this activation, the further differentiation to active lymphocytes, and the following immune response induced by the system permitting all relevant information to be involved in the "decision" process in a dynamic way. We discuss Jerne's immunological network theory based on the capability of self-nonself discrimination and we discuss if his immunological network theory is able to carry out such self-nonself discrimination. We argue that this is not the case. We discuss the immune tolerance that seems not to be transferred with the immune cells, what Jerne's network model involves as a necessity. We discuss Jerne's idiotypical network's capability of being expanded to include T-cells. Also here we argue that this is not the case. Furthermore, We discuss Jerne's own arguments for his immunological network theory and discuss different experiments capability of supporting or rejecting Jerne's immunological network theory. We found that not enough evidence and proof for his model have yet been presented in the forum of discussion or in the scientific literature to support the existence of Jernes network theory. Finally we argue that his model in principle can not explain the immunological self-nonself discrimination.

Introduction

In this paper we discuss self-nonself discrimination, mainly in the view of the Danish immunologist Niels Kaj Jerne (1911-1994) immunological network theory (1,2). In 1974, Niels Jerne came up with his network theory for the immune system (1). It was an a priori hypothesis that had no experimental foundation. It was what Jerne himself later called "a preconceived idea" (2). Jerne did take an important step in 1974, because the investigation of systemic information was and is essential. Because the scientific community at that time had no clue about the function of the immune system as a system, Jerne's hypothesis was very well received. Because a network theory was expected at that time, many immunologists did not question the nature of Jerne's theory. The hypothesis was in its basic elements simple and easy to understand, but in its totality impossible to get an overview of, even incomprehensible. Nobody had expected the regulation of the immune system to be simple, and fundamentally Jerne's theory did not deter many immunologists. Today opponents against Jerne's theory reject his network theory as a formal absurdity (3-5), and its spokesmen admit that self-nonself discrimination also in the view of Jerne's theory lack the capability of controlling the network (6). Jerne's network theory only takes care of the regulation of the immune response after activation (when the "choices" has been taken), and a problem concerning the spokesmen is that Jerne rarely cites the experimental results that contradict his network theory (7,8).

According to Klein (9) a network is a possibility, and he discusses the existence of idiotype specific cells (Th-cells). There are as many experiments that confirm the existence of these cells as there are experiments that contradict their existence. We think the existing proof for confirmation of Jerne's network theory has not been sufficient. In this paper we argue that Jerne's network theory is not able to explain the mechanism behind self-nonself discrimination, what our holistic paradigm is (10).

Jerne's network model

Jerne (1) at first defined a few concepts: the antigen determinant he called the epiotope, and the antibody combining site he called the paratope. Each individual antibody molecule carries a specific paratope, but also a number of its own specific determinants that under specific circumstances are able to give an immune response. These are situated at the variable part and are called idiotopes (idios = self). Jerne assumes that the existence of identical paratopes not necessarily means that the two antibodies have the same idiotopes. He furthermore assumes that each antibody recognizes through its paratope and is recognized through all its idiotopes.

Jerne emphasized, that the resulting network implies both free and cell bound Ig, why the B-cells are involved. He mentioned that Myeloma Ig gives idiotypespecific tolerance, when it is administered untreated in a mouse, while the polymerised or modified Myeloma Ig gives anti-idiotypic antibody response. From this he concludes, that the B-lymphocyte can react with response or with tolerance at the Ig (this is not a logical consequence but seems very reasonable). He further mentions, that most of the experiments suggest that B-cells become suppressed, when a paratope of an alien Ig finds an idiotope at the B-cell (but the evidence seems weak). Finally, Jerne (1) emphasized the importance of suppression: "I have become increasingly convinced of its lymphocytes", he wrote.

Experiments with tolerance induction show that a cell also can become suppressed when it meets the specific epiotope for which it has a paratope, wherefore the inhibiting forces in the network seem to outdo the activating forces. Jerne imagined a dynamic equilibrium, where the immune cells through inhibition are prevented from working, except when they are needed. Jerne mentioned the problem of low-dose tolerance and wrote that the concentration of Ig-idiotopes maybe exceed the threshold of low-dose tolerance. The low-dose interval, however, seemed to be 106-1012 epitopes pr. ml., while the average of Ig-idiotopes taken as an average of some millions, is about 1010. This value should furthermore be adjusted in a downward direction for the peaks of specific Ig's formed in connection with an immune response. Thereby the idiotopes seem to fall within the low-dose tolerance interval. From this we conclude that the immune system has an eigen-behaviour and constantly reacts towards itself.

In Figure 1 for each Ig plotted a paratope (P) and an idiotope (I) are also. P(Ig1) (also referred to as AB1 below) recognizes the antigen (E) (left). But, at the same time, the paratope P(Ig1) fits to an idiotope I(Ig2) on the Ig2. P(Ig1) have its own idiotope I(Ig1) that is recognized on P(Ig3). To a stranger epitope E, a set of Ig's (P-I) exists with different affinity. These affinities can be produced through an immune response. They recognize a great amount of idiotopes said to be the E's "inner picture". These Ig's have a huge amount of idiotopes (I's) that are recognized by an extended set of paratopes (P's).

Jerne argued that the inner picture may have the same effect as the presentation of an outer antigen, and therefore may stimulate the B-lymphocytes. The extended set that recognizes the Ig-idiotopes, on the other hand, should be inhibiting. However, the immune system is after this in a stage of extended suppression that has to be overcome through an outside stimulus, the antigen. This may be the first step in an immune response and may correspond to a suddenly miss of AB1, because this then, is tied to the antigen. This leads to absence of inhibition on a part of the inside picture, that has to be activated to give an increased response. Simultaneously, the Ig's that were stimulated by inhibiting the AB1 should be further inhibited by now. Following Jerne, this should stimulate the immune response against the foreign antigen. Another set that inhibited this set, should be inhibited itself, etc. etc.. The activated inner picture, have an inner picture itself that have to be activated, etc. etc..

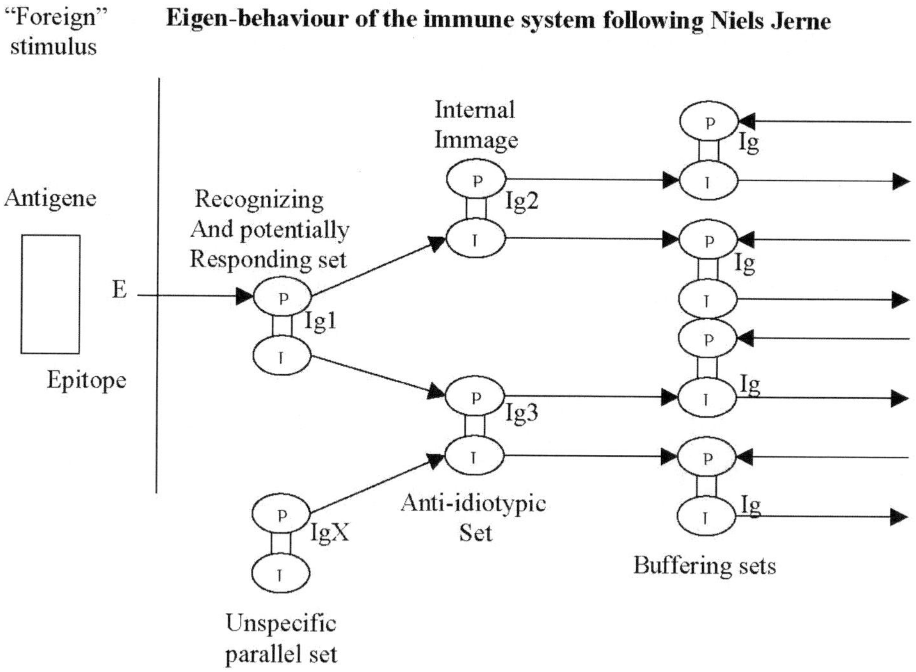

Figure 1. A model of Niels Jerne's immunological network modified from ([1)]. For each Ig a paratope (P) and an idiotope (I) are plotted. P(Ig1) recognizes the antigen (E). The paratope P(Ig1) fits to an idiotope I(Ig2) on the Ig2. P(Ig1) have its own idiotope I(Ig1) that is recognized on P(Ig3). To a foreign epitope E, a set of Ig's (P-I) exists with different affinity produced through an immune response. A modification of Jerne ([1)] (Figure 7).

After this the strengthened inner picture uninhibited provoke an activation (and propagation?) of AB1-cells that may soon exist in huge amounts. Jerne thought that this reinforcement would tend against counteracting the mentioned avalanche in such way that the network attempts to regain its balance. But we ask how this should be able to happen. The enforcement is possible as a consequence of the unbalance in the system. So, will it be able to bring it in balance again? We do not think so. Analogically, this could be compared to a man that, to find the balance again, grabs the stone he caused to unbalance and just has lost. However, Jerne criticizes his own hypothesis for lack of precision, and admitted that it is not clear what elements that are activated, and which that are inhibited, and why.

Discussion

Does a regulatory, idiotypical network as Jerne's exist? According to our theoretical discussion, it is clear that such a network even if it regulates the ordinary immune response, not is able to make the choice of self-nonself discrimination because it is not perfect since it is able to exclude parts of itself. We think that such network can hardly be expanded to include T-cells. It may be considered which expansions that are theoretically possible. Can the idiotypical network be expanded also to include the lymphocytes with T-cell receptors? The common opinion is yes, but we predict that such expansion would lead to big problems caused by the difference between Ig- (among others) and the T-cell receptor, because the T-cell receptor has a much smaller affinity. But also specially grounded the MHC-restriction. Furthermore, the problem that all information is distributed between cells could be an obstruction for the expansion with T-cells. Because Th- and Tc-cells seem not to communicate directly, the network may happen inside the Th-cell population alone. On the other hand, these cells operate as if they were primed and not as if they were functionally suppressed and only wait for escaping their neighbours to be able to break out in full activity. This means that systemic information cannot be excluded because this seems to be a condition for priming.

The T-cell receptor does not seem to function as if it recognizes idiotypes on other T-cell receptors but as if it read an antigen-determinant in a MHC-site. If this is the case, an idiotypic network can be excluded unless also the T-cell receptors are represented in these sites. But we think this is unlikely. However, a problem is that even if the network on the T-cell level is a system that functions and makes choices, it does not seem to be an idiotypical specific network.

Evidence for the existence of Jerne's network does not seem overwhelming. Holmberg et al (11) thought to have the first direct indication of the existence of a formal idiotypical network. But, everything they have shown is as far we can see that antibodies can react mutually, in a broad way of understanding. This is also the condition of the existence of such a network, but if this network really existed, this would not be enough to confirm if the network has the immunological function as supposed by Jerne. His network is not yet supported by existing data. On the other hand, evidence against Jerne's network exists. As mentioned in the section on the self-Ig's theoretical status of immune response, these are induced by tolerance concerning dose, size, self-nonself status, administration route, and untreated Myeloma-Ig gives a specific idiotype tolerance. Other experiments have shown that antibodies first are able to give an immune response if they are given together with an adjuvant, or if they are tied to an immunogenic molecule (9). This could be the best evidence against Jerne's network theory – it is unlikely that the immune system at all will react on the idiotypes through normal conditions.

In experiments using ongoing (un-physiologic) immunizing, anti-idiotypic antibodies have been formed, but the anti-AB is first created after culmination of the AB's one month later (12). Also idiotype specific antibodies without previous immunizing are identified in vivo. Especially in connection with auto-immunity as the idiotypical antibodies are proposed to regulate (13,14). On the other hand, it can also be the ongoing presence of Ig's that gradually triggers its own immune response analogue to the experiment described below.

In germ free environments the Ig concentration is found to be extremely low (9). This alone, seems to exclude that an idiotypic network should direct the immune system, because such a direction presupposes that the net is present before the release of antigens.

When animals are repeatedly immunized with antibodies the AB1's are first created, then AB2 against AB1, so AB3 against AB2, and AB4 against AB3 etc., then AB3 imitates AB1, and AB2 imitate AB4 (9). This is in contradiction with the idiotypic network that predicts a branch that is able to realize the network. The experiments indicate that the paratope is the most preferred goal for an immune response. If this is the case, it makes the future antibodies insignificant concerning the immune response, simply because AB2 is a copy of the original antigen epitope.

Experiments indicate that everything happening concerning the formation and activation of the immune cells is antigen dependent. According to the network theory this should be net-regulated (9). Many examples exist where anti-idiotypic antibodies alone, have stimulating or inhibiting effects (9). But these effects can easily be understood if it is assumed that they have exactly the same configuration in the paratope as the original antigenic epitope – in this way the ordinary possibilities for antigenic-response and tolerance-induction appear.

The idiotypical network, apparently, is without experimental evidence and cannot explain the self- nonself discrimination. The only possibility left is to keep going with the remnants and modify the network theory so profoundly that it no longer has any similarities with its beginning point. This has been tried (9,15), but apparently, not with any further luck. As said by Schrödinger "he, that seeks the truth, has to guard himself from ad hoc modifications".

Other previous attempts to make immune regulation happen on the level of organisms (16,17) seem not very successful. The biggest problem concerning Jerne's network theory is that any useable alternatives do not exist in the literature because people reluctantly reject the existing fundament for the theory if no better has arisen.

Conclusion

Jerne criticizes his own hypothesis for lack of precision and admits that it is not clear what elements are activated, and which are inhibited. It seems to be a very bad procedure to change a construction shown not to be in agreement with the experimental facts in fundamental and central fields because it is not obvious that such a network exists. We conclude that no experimental evidence exist for the network theory of Jerne and think that his theory is not able to explain self-nonself discrimination because the immune system apparently is not regulated through an idiotypical network. We also conclude that a network like Jerne's is not able to make the choice of self-nonself discrimination since it is not complete, and we think that it is not plausible to expand it to include the T-cells. We predict that such an expansion would lead to big problems caused by the difference between Ig- and the T-cell receptors.

Some experiments show that antibodies only are able to give an immune reaction together with adjuvants or are tied to an immunogenic molecule. This could be the best evidence against Jernes network theory – it is unlikely that the immune system will react at all to the idiotypes under normal conditions. Because idiotopes seem to fall within the interval of low-dose tolerance we conclude that the immune system has an eigen-behaviour and constantly reacts towards itself.

References

[1] Jerne NK. Towards a network theory of the immune system. Ann. Immunel. (inst. Pasteur) 1974;125C:373-89.

[2] Jerne NK. Idiotypic network and other preconceived ideas. Immunol Rev 1984;79:5-24.

[3] Langman RE, Cohn C. The "complete" idiotype network is an absurd immune system. Immunol Today 1986;7(4):100-1.

[4] Cohn M. Foreword: clippings from one immunologist's journal. In: Langman, RE, ed. The immune system. San Diego, CA: Academic Press, 1989.

[5] Podolsky S, Tauber A. The generation of diversity. Harvard: Harvard Univ Press, 1997.

[6] Paul WE. Fundamental immunelogy. New York: Raven, 1984.

[7] Slaoui M, Urbain-Vansanten G, Demeur C, Leo O, Marvel J, et al. Idiotypic games within the Immune network. Immunol Rev 1986;90:73-91.

[8] Kohler H. Internal antigen and immune network. Proc Soc Exp Biol Med 1985;178(2):189-95.

[9] Klein J. Natural history of the major histocompability complex. New York: John Wiley, 1986.

[10] Ventegodt S, Hermansen TD, Flensborg-Madsen T, Rald E, Nielsen, ML, Merrick J. Human development VII: A spiral fractal model of fine structure of physical energy could explain central aspects of biological information, biological organisation, and biological creativity. ScientificWorldJournal 2006;6:1434-40.

[11] Holmberg D, Holmberg D, Forsgren S, Ivars F, Coutinho A. Reactions among IgM antibodies derived from normal, neonatal mice. Eur J Immunol 1984;14:435-41.

[12] Roitt I, Brostoff J, Male D. Immunology. London: Gower Med Publ, 1985.

[13] Abdou N. The idiotype-antiidiotype network in human autoimmunity. J Clin Immunol 1985;5(6):365-9.

[14] Zanetti M. Idiotypic regulation of autoantibody production. Crit Rev Immunol 1986;6(2):751-3.

[15] Urbain J, Urbain J, Brait M, Bruyns C, Demeur C, et al. The idiotypic network: Order from the beginning or order out of chaos? Curr Top Microbiol Immunol 1985;119:127-42.

[16] Vas NM, Varela FJ. Self and non-sense: An organism¬-centered approach to immunelogy. Med Hypotheses 1978;4:231-67.

[17] Coutinho A, Forni L, Holmberg D, Ivars F, Vas N. From an antigen-centered, clonal perspective of immune responses to an organism-centered, network perspec¬tive of autonomous activity in a self-referential immun system. Immunol Rev 1984;79:151-68.

Chapter XVIII

A theory for immune system regulation (The "self-nonself discrimination")

The immune system is the most important defence against dangerous attack from the outside world and a strong immune response is necessary for a human to stay healthy. In this chapter the immunological self-nonself recognition is discussed from the view of the holistic paradigm. The biological immunological self-nonself discrimination is discussed in context to the mental, consious self-nonself discrimination and the possible informational connections between these two levels are discussed. Immunological self-nonself discrimination and the non-genetic somatic diseases is discussed and we hypothesize that these diseases could be a consequence of incorrectly adjusted immunological self-nonself discrimination due to unsolved emotional, sexual and existential problems on the consious, organismic level.

Introduction

A key problem in immuno-regulation today is the self-nonself discrimination, characterized by the ability of the organism to react toward the nonself-antigens and to avoid to react against self-antigens. By the 1970s it had become increasingly clear, that the immuno regulation was taking place at a systemic level. Then, however, there was no general conceptual handling of biological systems at a systemic level. The existing cybernetic analyses were summarized as a network of interactions. A number of network theories appeared to describe biological systems, e.g. network theories of gene expression in eukaryotes (1), for the function of the cerebellum (2), and later also for the function of the cerebrum (3,4).

However, the Danish immunologist Niels Kaj Jerne (1911-1994) (5) developed an innunological network model trying to explain the capability of a body to discriminate between organismic self and nonself through its immune system (5). We have discussed Jernes immunological network model (6) and came to the conclusion that Jerne's network was not able to explain the crucial immunological self-nonself discrimination essential for a body to stay fit and survive, and for people with non-genetic, somatic diseases, to be able to overcome their illnesses and stay fit. Therefore we, in this paper, give our explanation of how the immune system can be able to discriminate between self and nonself in the human body. We discuss the immunological self-nonself discrimination in the view of the holistic paradigm (7-13).

Because of the very complicated mechanism behind the immunological self-nonself discrimination, there have been some attempts to explain this mechanism. But no model proposed in the past has been sufficient enough to be able to explain such mechanisms in a satisfactory way. On the other hand, we think the holistic paradigm is able to give scientists a tool to explain the immunological self-nonself discrimination in a more satisfactory way then the past models, because it predicts an interaction between the psychical level (through the consciousness and the intent), and the immunological-systemic level. In this paper we will discuss the immunological self-nonself discrimination and its interaction with the mental self-nonself discrimination that following the holistic paradigm are a consequence of the interaction between the intent and the systemic fractal levels of a human body, and thereby, are under tight control from the super orbital of the individual human being (8).

Self-nonself discrimination of the human wholeness

Morphogenetic analyses show that biological system function like sensitive control systems, which naturally have the ability to evaluate the placement of the cells and correct this, if it is wrong. This ability of biological systems is the foundation of the self-nonself discrimination.

Psycho-neuro-immunology is a rather new discipline. It aims to expounding the control of the psyche through the immune system and brain. Self-nonself discrimination is a crucial point of this control. Besides the morphological self-nonself discrimination, this new science deals with immunological-systemic self-nonself discrimination and mental self-nonself discrimination. The psychoneuroimmunologists believe the immune defence of the body can be conditioned in a pavlovian way, because the mental self-nonself discrimination is able to influence the immunological-systemic self-nonself discrimination. These two kinds of self-nonself discrimination are only seen as the specific expressions of the self-nonself discrimination of the organism pertaining to the brain and the immune system respectively, explaining the close affinity between these two different ways of discrimination.

The immunological-systemic self-nonself discrimination is governed by the biological information carried by the quantum field of the organism (the "super orbital")

The super orbital is united by all quantum chemical energies of a body and following the holistic paradigm it is the source of the consciousness and thereby of the intention (8). The super orbital are following the holistic paradigm governing everything in a biological body that demands an energetic source, including all biochemical and biophysical functions (8). Following the holistic paradigm this means that the immunological-systemic self-nonself discrimination is tuned by the super orbital /read consciousness (8).

The intention that is sourced from the super orbital interacts with the energies of positional information of the fractal levels that exists in a human body (8). These fractal levels (7) represent the different levels of the building plane of the body as e.g. the cells, the organs (build by the cells), the tissues (gathered by the organs) and the complete organic body (gathered by the tissues). The imunnological-systemic self-nonself discrimination is directed through the interaction between the intention and the systemic fractal levels (meaning all the fractal levels represented in the body).

The consciousness that sources the intention is also the source of the mind and the psyche. The psyche is representing the mental self-nonself discrimination. Because of the psychological relation between the consciousness, the intension, the mind and the psyche, and thereby the clear connection between the intension and the positional information of the systemic fractal levels of the human body, there is also a connection between the mental self-nonself discrimination and the immunological-systemic self-nonself discrimination that interacts in a united co-operation to fine tune the body to be fit and capable of survival.

The interaction between the psychical- and the immunological self-nonself discrimination

Following the holistic paradigm the interaction between the physical body and the consciousness is very tight coupled. This gives the individual human being the possibility through his consciousness and his intention (9) to govern the positional information (10,11) of the systemic fractal levels (7) in a way so that he is able to control the coordination between the mental self-nonself discrimination and the immunological-systemic self-nonself discrimination. This means that these two kinds of self-nonself discrimination perform a very close interaction between these two different ways of the human body and mind to discriminate between itself and the outside world. How effective this discrimination is depends on the interaction between the intention and the systemic fractal levels (8). This interaction makes the holistic paradigm very powerful, because it makes it possible for the individual human being to adjust his immune system after his immediate psychological state. Following the holistic paradigm this means that the interaction between the intention and the positional information of the systemic fractal levels also are directly dependent of the individual human being's immediate psychically state. Thereby, the human self-nonself discrimination becomes an important part of the united system that controls the most important mechanisms leading to fitness and survival, namely human self-nonself discrimination and adult human metamorphosis (12,13), but the self-nonself discrimination can also lead to decay, self destruction and dead, because a sick mind can direct a human being's immune system to attack his own body. Therefore, it is of an immensely importance for a human being to be able to fine tune his immunological self-nonself discrimination to be as optimal as possible, because only in this way he will be able to maintain the appropriate fitness necessary for survival, the main function of the evolution (12).

Immunological self-nonself discrimination and evolution

Without the capability of being able to fine tune the immunological self-nonself discrimination the human being would probably never have been created. We think this is the case, because by this capability the immunological-systemic self-nonself discrimination, in a way, side by side with the naturally selection proposed by Charles Darwin (14) and the "metamorphous top down" evolution described by Hermansen et al (12) is responsible for the evolutionary mechanisms that leads towards fitness and survival and probably towards further evolutionary development of the human being (12). By this, the immunological-systemic self-nonself discrimination becomes a key function of human fitness and survival, but also a key function of the evolution.

The "metamorphous top down" evolution (12), is coupled to adult human metamorphosis (12,13), because the last one is a reminiscent of the first (12). Following the holistic paradigm, this means that the interaction between the immunological-systemic self-nonself discrimination and the adult human metamorphosis, in the human body, both have the mission of getting the individual human being capable of staying fit for fight and capable of survival, and to cooperate to insure the farther evolution of the human being.

Table 1. Development of immunological ideas according to Jerne (5)

Aims	Periods	Pioneers /Types of theory	Notions
Application	1870-1890	Pasteur Metchnikoff	Immunization Phagocytosis
Description	1890-1910	Bering Ehrlich	Antibodies Cell receptors
	1910-1930	Bordet Landsteiner	Specificity Haptens
Mechanisms	1930-1950	Sub-cellular	Antibody synthesis Antigen template
	1950-1970	Cellular	Selection Clones
System analysis	1970-1990	Multi-cellular	Network Cooperation and suppression

Table 2. Results of imbalance in the self-nonself discrimination

Origin of stimulus	Too low a discrimination	Too high a discrimination
Outer stimulus	Infection	Hyperreactivity
Inner stimulus	Neoplasm	Autoimmunity

The connection between self-nonself discrimination and somatic diseases

Immunology has been a scientific discipline for about 120 years. According to Niels Jerne it has been developed through a number of phases, each based on specific ideas (see table 1). It is a plausible hypothesis that choices of self-nonself discrimination throughout life are the cause of such diseases. If this is the case the elucidation of the mechanisms will provide both a therapy and a generally preventive medicine for these diseases.

Of cause it is not yet possible to understand the fundamental mechanisms behind immunological self-nonself discrimination. However, following the holistic paradigm it is given that the intention has an essential influence on the immunological self-nonself discrimination, and thereby that the consciousness is deeply involved in the human health. A badly intention will lead to malfunction of the body's building plane and structures (8) and give the immune response, and capability to distinguish between self and nonself, bad conditions for optimal activity, while a good intention will involve a fit and sound body capable of survival. In such a body the conditions for the immunological self-nonself discrimination are optimal and capable of helping the body to stay fit. Such a body involves a kind of positive feed back mechanism, where a more fit body makes the conditions for the self-nonself better. This makes an even fitter body etc. On the other hand a bad intention involves a kind of negative feed back. Such persons that involve a negative feedback are very vulnerable to attack from bacteria, vira and other parasites, and thereby the non-genetic, somatic diseases often confront them. However, by the help of a therapist such persons can reverse a negative feed back to a positive feed back and by this activate the adult human metamorphosis (12,13) and strengthen the immunological self-nonself discrimination leading to fitness and survival.

Of cause such mechanisms are much more complicated then this. For example, for humans in a negative feed back cycle the etiology of the non-genetic, non-traumatic somatic diseases probably can be explained by too weak or too strong a self-nonself discrimination (see table 2) why the choices of the self-nonself discrimination could be the cause of somatic diseases. If this is the case, it will provide both a therapy and a generally preventive medicine against such diseases. Therefore it is most important to focus the future research on such problems and realize that a better understanding of the self-nonself discrimination could be the key to a better health. Such ideas are the general subject of this book.

Discussion

In our quest for a new understanding of the mysterious connection between life, matter and consciousness (6-13,15-17), we have now turned our interest towards the immune system. The structure held responsible for the structure and quality of human immune response is called the "immune system" but the problem is that it might not be a separate organ in the same way as the kidney or brain, but a much more integrated cell population completely depending on the state and informational conditions of the local tissue. Our motivation for the exploration of the immune system is our finding that quality of life, health and ability primarily is determined by our consciousness (18). This discovery has lead us to the interesting possibility that guided shifts in the state of consciousness and the development of self-insight with a more positive, responsible and constructive philosophy of life can be used to strengtehn the human immune response and help cure patients i.e. by reducing the number of recidivant infections. We have made a series of papers setting the strategy for alleviating healing many different physical, mental, existential, and sexual diseases and health problems (19-59). We also know that the discriminative ablitiy of the immuse system is closely connected to the patient recovering his experience of "sense of coherence" (SOC) (60-66). SOC is actually the experience that one conscious being is connected to the whole universe though our physical and mental existence (67); to bring the patient back to be an integrated part of the world seems to be the fundamental idea of all medicine, the tratition going all the way back to Hippocrates and his students (67).

Conclusions

The self-nonself discrimination is characterized by the ability of the organism to react toward the nonself-antigens and to avoid to react against self-antigens. A number of network theories appeared to describe biological systems. Professor Niels Jerne developed an immunological network model trying to explain the capability of a body to discriminate between organismic self and nonself through its immune system. We came to the conclusion that Jerne's network was not able to explain the immunological self-nonself discrimination. No model proposed in the past has been sufficient enough to explain such mechanisms satisfactory enough, but we think the holistic paradigm is able to give an appropriate explanation of the immunological self-nonself discrimination because it predicts an interaction between the psychical level and the intention.

Following the holistic paradigm the super orbital is governing everything in a biological body that demands an energetic source. This means that the immunological-systemic self-nonself discrimination is tuned by the consciousness (sourced by the super orbital). The immunological-systemic self-nonself discrimination is directed through the interaction between the intention and the systemic fractal levels, and the intention is sourced from the consciousness (read super orbital). Also a connection between the psychological self-nonself discrimination and the immunological-systemic self-nonself discrimination exists, that interacts in a united co-operation to fine tune the body to be fit and capable of survival. All this activity is governed from the super orbital of the human wholeness.

Following the holistic paradigm the interaction between the intention and the positional information of the systemic fractal levels are directly dependent on the individual human being's immediate psychically state. Thereby, the human self-nonself discrimination becomes an important part of the united system that controls the most important mechanisms leading to fitness and survival, namely human self-nonself discrimination. It is of an immensely importance for a human being to be able to fine tune his immunological self-nonself discrimination to be as optimal as possible, because only in this way he will be able to maintain the appropriate fitness necessary for survival, the main function of the evolution. The immunological-systemic self-nonself discrimination becomes a key function of human fitness and survival, but also a key function of the evolution. Therefore, the interaction between the immunological-systemic self-nonself discrimination and the adult human metamorphosis, in the human body, both have the mission of getting the individual human being capable of staying fit for fight and capable of survival, and to cooperate to insure the farther evolution of the human being.

At last, following the holistic paradigm it is a plausible hypothesis that choices of self-nonself discrimination throughout life are the cause of non-genetic, somatic diseases. If this is the case the elucidation of the mechanisms will provide both a therapy and a generally preventive medicine for such diseases.

References

[1] Davidson EH, Britten RJ. Regulation of gene expression: possible role of repetitive sequences. Science 1979;204(4397):1052-9.
[2] Marr D. A theory of cerebellar cortex. J Physiol 1969;202(2):437-70.
[3] Ballard R. General practice management of stabilised schizophrenics. Aust Fam Physician 1986;15(12):1616-21.
[4] Jones WP, Hoskins JC. Back propagation: A generalized delta learning rule. ByteMagazine 1987:155-62.
[5] Jerne NK. Towards a network theory of the immune system. Ann Immunol (Inst. Pasteur) 1974;125C:373–89.
[6] Ventegodt S, Hermansen TD, Kandel I, Merrick J. Human development XVII: Jerne's anti-idiotypic network theory cannot explain self-nonself discrimination. J Altern Med Res 2009;1(4):439-44.
[7] Ventegodt S, Hermansen TD, Flensborg-Madsen T, Rald E, Nielsen ML, Merrick J. Human development VII: A spiral fractal model of fine structure of physical energy could explain central aspects of biological information, biological organisation, and biological creativity. ScientificWorldJournal 2006;6:1434-40.
[8] Hermansen TD, Ventegodt S, Flensborg-Madsen T, Nielsen ML, Merrick J. Human development VIII: A theory of "deep" quantum chemistry and cell consciousness. Quantum chemistry controls genes and biochemistry to give cells and higher organisms consciousness and complex behavior. ScientificWorldJournal 2006;6:1441-53.
[9] Ventegodt S, Andersen NJ, Merrick J. Quality of life philosophy V. Seizing the meaning of life and becoming well again. ScientificWorldJournal 2003;3:1210-29.
[10] Ventegodt S, Hermansen TD, Flensborg-Madsen T, Nielsen ML, Merrick J. Human development IV: The living cell has information-directed self-organisation – and consciousness. ScientificWorldJournal 2006;6:1132-8.
[11] Ventegodt S, Hermansen TD, Flensborg-Madsen T, Nielsen ML, Merrick J. Human development VI: Supra-cellular morphogenesis - the origin of biological and cellular order. ScientificWorldJournal 2006;6:1424-33.
[12] Hermansen TD, Ventegodt S, Merrick J. Human development X: Explanation of the macro-evolution: top down evolution materialises consciousness. The origin of metamorphosis. ScientificWorldJournal 2006;6:1656-66.
[13] Ventegodt S, Hermansen TD, Clausen B, Nielsen ML, Merrick J. A theory for biological information, healing, and adult human metamorphosis for use in clinical holistic medicine XX: A theory for spontaneous healing: adult human metamorphosis. In press.
[14] Darwin C. The origin of species. New York: Gramercy, 1995.
[15] Hermansen TD, Ventegodt S, Rald E, Clausen B, Nielsen ML, Merrick J. Human development I: 20 fundamental problems of biology, medicine and neuro-psychology related to biological information. ScientificWorldJournal 2006;6:747-59.
[16] Ventegodt S, Hermansen TD, Nielsen ML, Clausen B, Merrick J. Human development II: We need an integrated theory for matter, life and consciousness to understand life and healing. ScientificWorldJournal 2006;6:760-6.
[17] Ventegodt S, Hermansen TD, Rald E, Flensborg-Madsen T, Nielsen ML, Clausen B, Merrick J. Human development III: Bridging brain-mind and body-mind. Introduction to "deep" (fractal, poly-ray) cosmology. ScientificWorldJournal 2006;6:767-76.
[18] Ventegodt S, Flensborg-Madsen T, Andersen NJ, Nielsen M, Mohammed M, Merrick J. Global quality of life (QOL), health and ability are primarily determined by our consciousness. Research findings from Denmark 1991-2004. Soc Indicator Res 2005;71:87-122.

[19] Ventegodt S, Merrick J. Clinical holistic medicine: Applied consciousness-based medicine. ScientificWorldJournal 2004;4:96-9.
[20] Ventegodt S, Morad M, Merrick J. Clinical holistic medicine: Classic art of healing or the therapeutic touch. ScientificWorldJournal 2004;4:134-47.
[21] Ventegodt S, Morad M, Merrick J. Clinical holistic medicine: The “new medicine”, the multi-paradigmatic physician and the medical record. ScientificWorldJournal 2004;4:273-85.
[22] Ventegodt S, Morad M, Merrick J. Clinical holistic medicine: Holistic pelvic examination and holistic treatment of infertility. ScientificWorldJournal 2004;4:148-58.
[23] Ventegodt S, Morad M, Hyam E, Merrick J. Clinical holistic medicine: Use and limitations of the biomedical paradigm ScientificWorldJournal 2004;4:295-306.
[24] Ventegodt S, Morad M, Kandel I, Merrick J. Clinical holistic medicine: Social problems disguised as illness. ScientificWorldJournal 2004;4:286-94.
[25] Ventegodt S, Morad M, Andersen NJ, Merrick J. Clinical holistic medicine Tools for a medical science based on consciousness. ScientificWorldJournal 2004;4:347-61.
[26] Ventegodt S, Morad M, Merrick J. Clinical holistic medicine: Prevention through healthy lifestyle and quality of life. Oral Health Prev Dent 2004;1:239-45.
[27] Ventegodt S, Morad M, Hyam E, Merrick J. Clinical holistic medicine: When biomedicine is inadequate. ScientificWorldJournal 2004;4:333-46.
[28] Ventegodt S, Morad M, Merrick J. Clinical holistic medicine: Holistic treatment of children. ScientificWorldJournal 2004;4:581-8.
[29] Ventegodt S, Morad M, Merrick J. Clinical holistic medicine: Problems in sex and living together. ScientificWorldJournal 2004;4:562-70.
[30] Ventegodt S, Morad M, Hyam E, Merrick J. Clinical holistic medicine: Holistic sexology and treatment of vulvodynia through existential therapy and acceptance through touch. ScientificWorldJournal 2004;4:571-80.
[31] Ventegodt S, Flensborg-Madsen T, Andersen NJ, Morad M, Merrick J. Clinical holistic medicine: A Pilot on HIV and Quality of Life and a Suggested treatment of HIV and AIDS. ScientificWorldJournal 2004;4:264-72.
[32] Ventegodt S, Morad M, Merrick J. Clinical holistic medicine: Induction of spontaneous remission of cancer by recovery of the human character and the purpose of life (the Life Mission). ScientificWorldJournal 2004;4:362-77.
[33] Ventegodt S, Morad M, Kandel I, Merrick J. Clinical holistic medicine: Treatment of physical health problems without a known cause, exemplified by hypertension and tinnitus. ScientificWorldJournal 2004;4:716-24.
[34] Ventegodt S, Morad M, Merrick J. Clinical holistic medicine: Developing from asthma, allergy and eczema. ScientificWorldJournal 2004;4:936-42.
[35] Ventegodt S, Morad M, Press J, Merrick J, Shek DTL Clinical holistic medicine: Holistic adolescent medicine. ScientificWorldJournal2004;4: 551-61.
[36] Ventegodt S, Solheim E, Saunte ME, Morad M, Kandel I, Merrick J. Clinical holistic medicine: Metastatic cancer. ScientificWorldJournal 2004;4:913-35.
[37] Ventegodt S, Morad M, Kandel I, Merrick J. Clinical holistic medicine: a psychological theory of dependency to improve quality of life. ScientificWorldJournal 2004;4:638-48.
[38] Ventegodt S, Merrick J. Clinical holistic medicine: Chronic infections and autoimmune diseases. ScientificWorldJournal 2005;5:155-64.
[39] Ventegodt S, Kandel I, Neikrug S, Merrick J. Clinical holistic medicine: Holistic treatment of rape and incest traumas. ScientificWorldJournal 2005;5:288-97.
[40] Ventegodt S, Merrick J. Clinical holistic medicine: Chronic pain in the locomotor system. ScientificWorldJournal 2005;5:165-72.
[41] Ventegodt S, Merrick J. Clinical holistic medicine: Chronic pain in internal organs. ScientificWorldJournal 2005;5:205-10.
[42] Ventegodt S, Kandel I, Neikrug S, Merrick J. Clinical holistic medicine: The existential crisis – life crisis, stress and burnout. ScientificWorldJournal 2005;5:300-12.

[43] Ventegodt S, Gringols G, Merrick J. Clinical holistic medicine: Holistic rehabilitation. ScientificWorldJournal 2005;5:280-7.
[44] Ventegodt S, Andersen NJ, Neikrug S, Kandel I, Merrick J. Clinical holistic medicine: Mental disorders in a holistic perspective. ScientificWorldJournal 2005;5:313-23.
[45] Ventegodt S, Andersen NJ, Neikrug S, Kandel I, Merrick J. Clinical holistic medicine: Holistic treatment of mental disorders. ScientificWorldJournal 2005;5:427-45.
[46] Ventegodt S, Merrick J. Clinical holistic medicine: The patient with multiple diseases. ScientificWorldJournal 2005;5:324-39.
[47] Ventegodt S, Clausen B, Nielsen ML, Merrick J. Clinical holistic medicine: Advanced tools for holistic medicine. ScientificWorldJournal 2006;6:2048-65.
[48] Ventegodt S, Clausen B, Merrick J. Clinical holistic medicine: The case story of Anna: I. Long term effect of child sexual abuse and incest with a treatment approach. ScientificWorldJournal 2006;6: 1965-76.
[49] Ventegodt S, Clausen B, Merrick J. Clinical holistic medicine: the case story of Anna. II. Patient diary as a tool in treatment. ScientificWorldJournal 2006;6:2006-34.
[50] Ventegodt S, Clausen B, Merrick J. Clinical holistic medicine: The case story of Anna. III. Rehabilitation of philosophy of life during holistic existential therapy for childhood sexual abuse. ScientificWorldJournal 2006;6:2080-91.
[51] Ventegodt S, Merrick J. Suicide from a holistic point of view. ScientificWorldJournal 2005;5:759-66.
[52] Ventegodt S, Clausen B, Omar HA, Merrick J. Clinical holistic medicine: Holistic sexology and acupressure through the vagina (Hippocratic pelvic massage). ScientificWorldJournal 2006;6:2066-79.
[53] Ventegodt S, Clausen B, Merrick J. Clinical holistic medicine: Pilot study on the effect of vaginal acupressure (Hippocratic pelvic massage). ScientificWorldJournal 2006;6:2066-79.
[54] Ventegodt S, Thegler S, Andreasen T, Struve F, Enevoldsen L, et al. Clinical holistic medicine: Psychodynamic short-time therapy complemented with bodywork. A clinical follow-up study of 109 patients. ScientificWorldJournal 2006;6:2220-38.
[55] Ventegodt S, Thegler S, Andreasen T, Struve F, Enevoldsen L, et al. Clinical holistic medicine (mindful, short-term psychodynamic psychotherapy complemented with bodywork) in the treatment of experienced impaired sexual functioning. ScientificWorldJournal 2007;7: 324-9.
[56] Ventegodt S, Thegler S, Andreasen T, Struve F, Enevoldsen L, et al. Clinical holistic medicine (mindful, short-term psychodynamic psychotherapy complemented with bodywork) improves quality of life, health, and ability by induction of Antonovsky-salutogenesis. ScientificWorldJournal 2007;7:317-23.
[57] Ventegodt S, Thegler S, Andreasen T, Struve F, Enevoldsen L, et al. Clinical holistic medicine (mindful, short-term psychodynamic psychotherapy complemented with bodywork) in the treatment of experienced physical illness and chronic pain. ScientificWorldJournal 2007;7:310-6.
[58] Ventegodt S, Thegler S, Andreasen T, Struve F, Enevoldsen L, et al. Clinical holistic medicine (mindful, short-term psychodynamic psychotherapy complemented with bodywork) in the treatment of experienced mental illness. ScientificWorldJournal 2007;7:306-9.
[59] Ventegodt S, Thegler S, Andreasen T, Struve F, Enevoldsen L, et al. Self-reported low self-esteem. Intervention and follow-up in a clinical setting. ScientificWorldJournal 2007;7:299-305.
[60] Flensborg-Madsen T, Ventegodt S, Merrick J. Sense of coherence and physical health. A Review of previous findings. ScientificWorldJournal 2005;5:665-73.
[61] Flensborg-Madsen T, Ventegodt S, Merrick J. Why is Antonovsky's sense of coherence not correlated to physical health? Analysing Antonovsky's 29-item sense of coherence scale (SOCS). ScientificWorldJournal 2005;5:767-76
[62] Flensborg-Madsen T, Ventegodt S, Merrick J. Sense of coherence and health. The construction of an amendment to Antonovsky's sense of coherence scale (SOC II ScientificWorldJournal 2006;6:2133-9.
[63] Flensborg-Madsen T, Ventegodt S, Merrick J. Sense of coherence and physical health. A cross-sectional study using a new SOC scale (SOC II). ScientificWorldJournal 2006;6:2200-11.
[64] Flensborg-Madsen T, Ventegodt S, Merrick J. Sense of coherence and physical health. Testing Antonovsky's theory. ScientificWorldJournal 2006;6:2212-9.

[65] Flensborg-Madsen T, Ventegodt S, Merrick J. Sense of coherence and health. The emotional sense of coherence (SOC-E) was found to be the best-known predictor of physical health. ScientificWorldJournal 2006;6: 2147-57.

[66] Ventegodt S, Flensborg-Madsen T, Andersen NJ, Merrick J. Life mission theory VII: Theory of existential (Antonovsky) coherence: a theory of quality of life, health and ability for use in holistic medicine. ScientificWorldJournal 2005;5:377-89.

[67] Jones WHS. Hippocrates. Vol. I–IV. London: William Heinemann, 1923-

Section VII:
"Energetic crisis" and "healing crisis"

Chapter XIX

The etiology of somatic diseases

The etiology of all non-genetic, non-traumatic, somatic diseases can be explained as coming from disturbances in the organism's biological informational system. On the cellular level the cells must know what to do and where to be, and lack of this information will cause either growth-disturbances or immunological disturbances (an either too strong or a too weak immunological self-nonself discrimination). We suggest that failure in the information system causes all somatic diseases that arises from lifestyle factors, and not from genetic or traumatic (physical environmental) factors. We recommend a therapy for these diseases based on rehabilitation of the function of the biological informational systems; such an intervention will also be a generally preventive medicine for humanity. In this paper we discuss the general cellular self-nonself discrimination and the specific immunological self-nonself discrimination. The understanding of cellular disorder as the origin of diseases leads to the natural question: why is the biological information disturbed? We suggest that emotions and repressed feelings in the human unconscious are also held by the biological informational system ("the body-mind") deminishing its ability to provide the cells of the organism with the information vital for their functioning. This allows philosophically for "the sence of coherence" – spiritual and emotinal health - as a fundamental basis for physical health.

Introduction

In this chapter we give our explanation of the non-genetic, non-traumatic, somatic disease etiology. These diseases stem from internal disorders in the organism and its information system. Disorders that make it difficult for the organism to take care of its normal functions in an orderly way. It is the case for most diseases that their etiologies (the first reasons) are unknown, but principally, the original reasons of disorder come from three sources: 1) Genetic disorders as DNA mutations and chromosomal defects. These kinds of disorders appear when a given genotype gives a bad phenotype, 2) Traumatic action from a physical influence, which could be an injury from outside coming factors as getting hit by a car or an infection by parasites or 3) Information disorders caused by disturbing influences between organs and/or organ systems. Conditioned learning is a very likely source for these information disturbances (1,2).

Lots of disease-groups belongs to the first group. A deletion of a gene on human chromosome 13 gives a specific kind of cancer (retinoblastoma). The presence of a mutation on both gene copies – one on each chromosome – each cell core contains a set of chromosomes from each of the parents – gives the autosomal recessive disease, Xeroderma pigmentosum (3). Many diseases fall into group 2, as asbestos in the lungs, agents that causes cancer (carcinogens) and radiation (3).

What is really interesting is that genetic and traumatic causes, apparently, only makes up small amounts of all cases of cancer and all other diseases there are identified several other causes (see for example (4) and American Cancer Society: http://www.cancer.org/downloads/STT/CancerFacts&Figures2002TM.pdf).

A major problem is that the etiology of diseases is not known. An example of this is that patients with a specific "genetic" disease have a specific "mutation" (a specific HLA-allele) more frequently then the population as a whole (5). The real connection between disease and its cause seems to be extremely complex and we think that a realistic interpretation would be that a "mutation" in the gene produces a defect product. But this seams to be wrong in many cases (5). In that case, the etiology and the connection to the HLA are unknown and cannot be connected to a genetic disorder.

Because the first two causes are consistent with a mechanical-reasonable interpretation of reality, they have, for a long time, been acknowledged as the reason for a lot of diseases. But the third cause has first recently been evident. In this context, experiments with rodents shows that the immune system can be conditioned in a Paulovish way (6,7). In the most famous experiments by Ader (6,7), he used a solution with a taste of saccharin – for the rat an unknown taste – as conditional stimulus and the effect of immuno-modulizing matters as unconditional stimulus. This made it possible for Ader to show that the humoral as the cellular immune response, had a corresponding modulation to the employed agent, when the rats tasted immunological neutral solution again. Concerning the immune suppression, the mortality of the rats was proportional with the amount of taste-liquid. These coordination experiments have not been able to determine the importance of the third cause. However, we assume that it is this cause that determines the huge variability, susceptible for the infection, concerning a row of diseases including the auto-immune and neoplasm.

A priori, we assume, that a disease has all three etiological elements – all three causes. But still, a persistent search for disruptions in the DNA, concerning both the big mental diseases (8), as the big somatic diseases (3) has not yet resulted in basic knowledge. Therefore, it is reasonable to suppose that these diseases alone can be understood from a mechanical reality interpretation. The documented occurrence of the placebo effect, and the general meaning of the psychological factor in almost all diseases (9), indicates a huge importance of the third etiological cause – a part of the emotional reality interpretation. Here we will discuss the etiology of the somatic diseases seen in the perspective of our holistic biological view.

The somatic diseases

Based on the brain function and the immunological self-nonself discrimination, we choose to divide the somatic diseases in the following five main groups:

- Diseases that directly are caused by a wrong neuro-regulation through hormones or the nervous system, among other things based in defective coordination between the intense communication between hypothalamus and the higher integrative areas in the brain. For instance, it is possible that these diseases are caused by a learning disturbance through the limbic system
- A large group of diseases that today is considered as autoimmune diseases (10)
- Neoplasia (dysplasia and dysregualtated cell growth), cancer, degenerative tissues disorder from general cellular disorder
- Infectious diseases
- Hyper-reactive diseases (i.e. allergi, excema, asthma)

A large group of diseases are today considered as autoimmune diseases (10). These are together with neoplasia, infectious diseases, and hyper-reactive diseases, the most non-genetic and non-traumatic diseases. For second to fifth main group, we set the following diagram (see table 1).

Table 1. AB. Results of imbalance in the self-non-self discrimination

Example A

Stimulus origin	Too low discrimination	Too high discrimination
Outer stimulus	Infection	Allergy
Inner stimulus	Cancer (?)	Autoimmunity

Example B

Stimulus origin	Too low discrimination	Too high discrimination
Outer stimulus	Infection	Hyperreactivity
Inner stimulus	Neoplasi (?)	Autoimmunity

Here exemplified by cancer in (A) and more generalised as neoplasi in (B). This diagram can be used for most non-genetic and non-traumatic diseases.

There are several problems related to this diagram: The existing immunologically knowledge is not able to explain the mechanism behind the ability of the immune system to distinguish between inner and outer stimuli – the self-nonself discrimination. It has not yet been possible to describe the regulation concerning to high or law immune-activity. Secondly, the relationship between the immune system and cancer is problematic. On the surface of cancer cell membranes, cancer specific markers are extremely rare. This should expectedly prevent that the T- and B-immune cells have an effect. The postulation that natural killer cells (NK-cells) take care of destruction of the cancer cells is doubtful, because naked mice, have lots of NK-cells even if they led the cancer grow. Ten years ago, almost all immunologists thought that the immune-surveillance theory was correct. Following this, thousands of cancer cells spontaneously arise every day, and every day these again, are eliminated by the immune system. However, today most immunologists believe that such mechanism, through immune cells, does not happen at all (11).

As cancer is usually developed in connection with lengthy dysplasi, it seems to arise from local problems in the information that maintains the tissue (3). However, it seems as if cancer arises from error in the information system that is responsible for the cell differentiation (12).

Errors in the immune functions can also be ascribed to an error of the level of differentiation (5). We believe that this is caused by the cell activity behind self-nonself differentiation that presupposes a really directed priming, namely, a beginning-differentiation.

Therefore, it is very interesting, that all non-neuro-regulatory diseases can be directed backwards to the information laying behind the control of the organisms cell differentiation. This is the same information that first controls the morphogenesis, and then maintains tissue and the organism (13,14).

A model for somatic diseases etiology build on the morphological, psychical, and immonological self-nonself discrimination

The Danish immunologist Niels Kaj Jerne (1911-1994) (15) suggested a network theory concerning the regulation of the immune system. But neither Jerne's, or any other immune network system hypothesis, seems to have any evidence (16). This can be caused by the difficulties that appears, when testing these hypothesis experimentally (this has been admitted by Jerne in personal communication in 1989). But, whatever happens, it is the case that such a network-hypothesis is not capable of explaining the immunological self-nonself discrimination (17). This is valid, since the network has to be complete to give any sense. But, after all, the hypothesis of Jerne has served its function by showing that the immune regulation does not happen at the cellular, but at the systemic levels.

If developing systems are disturbed, they have a considerable capability to correct themselves, as we have seen during the description of the morphogenesis. Such systems have a kind of morphological ability to discriminate between self and non-self, as biological systems have systemic information in their complex dynamism. This information is, as earlier described, a source to the morphological information. Almost all biological systems have the fundamental, immunologic capability to distinguish between self and non-self, the ability to distinguish our own biological system, from a stranger one. This ability exists both in single and multi-cellular organisms such as fungi or sponges (5). Existing phylogenic data has shown that the immunological self-nonself discrimination may arise from the same source as the morphological, namely the systemic information. It seems reasonable to believe that the evolutionary ancient developed immunologic self-nonself discrimination is the reason for the later developed extensions of the immune system. With mammalians and humans it also seems that the immunological response of an antigene requires a disturbance of the biological system. Such disturbance can be mediated commercially, by an adjuvant. Using an adjuvant, an immune response can be obtained, against almost all antigenes, even if it belongs to the organism itself. This shows that the regulation of the immune system happens on the superior systemic levels, and not on the cellular level, through selection.

As described earlier, the systemic information results in the capability of understanding whether a levels organization in an individual organism is suitable or not. When the organization of the levels is in coordination with the intension of the organism, these organizations can be understood as an expanded version of self, connected to positive emotions. If the opposite is the case, and the organization opposes the intension, this is understood as nonself in expanded sense, and connected to corresponding negative emotions as a consequence of a split in friend and enemy, and good and bad conditions, physiological as social and global. In this way, the psychical self-nonself discrimination has been developed.

The morphological, immunological, and psychical self-nonself discrimination fundamentally has the same systemic source and therefore the different aspects of self cannot be considered to be independent of each other. Conditioned learning can be understood as forced psychical self-nonself choices, across the biological intensions. The consequence of this may be that conditioned learning is disturbing the immunological as well as the morphological self-nonself discrimination. This disturbance reply to the etiological factors for diseases in the diagram described shown above.

The theoretic consequence of coordinated learning for development of diseases. If a child wants to survive, it has to adapt itself to parental reality and intensions. The parents manifest themselves opposite to the child through will, speech and action (18). Often the parents are in two different reality interpretations, the emotional and the rational. As a woman, the mother usually prefers to be in the emotional, while the father prefers to be in the rational interpretation. If the parents are really limited by their own conditioned learning, there can be a huge gap between their interpretations. Thereby, it is hard for them to communicate in agreement, and the child may meet the parents on its own conditions to get an acceptable communication (or the child takes over). This process is the nature of the conditionings – the child learns to reduce itself in order to a narrow spectrum of human characteristics.

When the human exists in the emotional interpretation, it manifests its degenerated intensions belonging to this. Usually this is the social sphere, as love between man and woman, family, friends, and the relation to themselves. When the human exists in the rational interpretation the intensions are manifested in a similar way. Usually, these intensions are resignations in relation to the outer worlds arrangements as: "The working life is horrible. I am a failure because I can not earn any money and get a good place in live".

The problems of the outer world areis that it forces us to use our senses. When children grow up to be young people, they want to leave home. Then, hard pressed, they have to leave the emotional interpretation to be able to manage the new problems. Simply, they learn that the emotional interpretation cannot help them to solve the important, practical problems, and therefore, they are forced to repress this. This means that they become an adult and reasonable person. By denying their emotions, they deny a great part of the biological life. This means that a great part of the biologic reality become non-self. The attention is applied to the brain, and away from the body. Signals from the body are not understood, since the representation of the whole is absolutely busy with the brain. A human in this condition, is extremely sensitive to the pictures of the body contained by the brain. This is because these pictures will be represented in the whole by great power. Thereafter, they will influence the body with a considerable strength, through information transmitting interactions.

As described above, the reality model in the brain is constructed as a consequence of the body functions (compare the model of Piaget) (19). The concrete aspects of the human life that can be suppressed through conditioning, give rise to pictures limited to these aspects. The sick and distorted pictures caused by the parents influence, can be thought of as corresponding to different parts of our body functions.

However, theoretical, conditioning can result in specific disturbances of the information conditions of the body. When a human is tied to a limited rational interpretation, it may generally be assumed that this disturb the information of the body. The degenerated intensions and the confused pictures, on the other hand, exists in the emotional interpretation, but an adaptation of feelings and reason to a healthy model of the reality in the brain and the regeneration of the degenerated intension in its whole.

The energetic, healing crisis ("kidney crisis")

In deep and intensive therapy with hypnosis, psychodynamic therapy, clinical holistic medicine and holistic sexology (20-66) patients can enter a physical, healing crisis, closely connected to the Meridian of the Kidney in Chinese medicine. The crisis can be extremely unpleassant with the sensation that the body will die physically; this idea can be further enhanced by the strange fact that the kidneys sometimes also physically temporarily shuts down its production of urine during the crisis. The crisis is often followed by radical and fast healing of physical illnesses and psychosexual problems strongly indicating that the physical and sexual problems were acturally caused on an informational level, not a chemical level in the body. The energetic, healing crisis, and the spontaneous healing after this crisis is important to know, as it is tempting to call a doctore to give morphin and other relevant treatment to the patients that suddenly seems be be extremely ill physically. The healing crisis normally last from 4 hours to 2 days depending on the ammount and quality of human and emotional healing ressources provided to the patient; no adverse effects have been registerd afterwards, and the healing seems complete.

Discussion

The way of fight and resignation – the way of integration or disintegration, health and diseases. Through its whole life, the individual search to adapt its varying life situations, through an ongoing modification of the intension and reality model in the brain. This ongoing modification of the intention may, if it develops itself as an ongoing degeneration, have a greater importance for the development of disease, then the basic conditions. Therefore, it is of a great importance to clarify the patterns of intension degeneration through a human life.

Though this pattern also has a meaning in childhood, the parents maintain their children in their interpretation of the world. Normally, the pattern is expressed when the youngster is drawn from home with its content of ideas, visions, and life strategy. If the youngster has great aims, or, has been conditioned with care, he inevitably meets great opposition on his way through realization of his goal. Often this will result in a great deal of defeat. If, on the other hand, the youngster is focused on fighting to get the things to succeed, he has difficulties to give up his ideas. However, he will then take up the fight again and turn in this way a lot of defeat to victory. A human that is focused on fighting, will usually be able to realize his goal. The learning process presupposes a subjection to the reality interpretation and the intensions through ongoing revision. The intension reacts, and the reality model is expanded to better being able to imply the reality, and the emotions. This human, through time, will integrate a great part of his coordinated human wholeness, but this strategy however, demands courage, because of the knowledge that this fight can result in defeat. Such a human often achieve to unfold life, but may live with the consciousness of defeat, which on the other hand is a fundamental biological factor, that is impossible to avoid.

The opposite pattern, the way of resignation, can lead to defeat and even death. However, this is almost never realised by a human forced to choose this way. This pattern is realized, through des-integrations cycles. Each time the individual suffers defeat these cycles are repeated in a narrowing spiral. If this youngster, that has not yet learned to take the circumstances of life as a challenge, decides to live as its parents, it has the change not to suffer great defeats in life. In this way, this person will be able to live throughout life, like his parents. But, when this person will encounter resistance and defeat, he will give up and compromise on goals and ideals, that are reduced depending upon the circumstances. With the lack of respect for the seriousness of the life, this person, at this stage, begins a new game, with a smaller goal. Now, heit has gone through a des-integration cycle, u. Usually, the first of a number. This person can be really well equipped by qualities, from its first step alone. But, when giving up, each time suffering defeats, this person will reduce itself. Instead of fighting to make reality better, this person allows himself to be coordinated by the reality, again and again. After a row of defeats and compromises, the intension of this human is extremely degenerated. Therefore, this human has limited itself to almost nothing. This means, that the unfolding of its life, has stopped. This human described above is therefore expected to get seriously ill.

Psychologically a trauma is leading to a reduction in the ambition for self-realisation and every time an intention that is important for the organism is closed down, that aspect of the organism is repressed from its life and consciousness. One way of phrasing this is to say that its status is converted from "self" to "nonself". When a person continously is using this pattern of resignation through life, gradually more and more of his biological reality is transformed into nonself. Over time it is possible that he gets somatically ill, as he is denying his life, body and its energies like sexuality. Choices taken after this destructive pattern has the effect of reducing his life. Therefore, when a person chooses not to fight, he really chooses the way of the victim. The biological intensions can be reduced to almost nothing. One stops running, working, walking, move… One gives up on partner, friends, to love people. All this goes on, until the person does not like life anymore. One gets despondent, tired, frail, weaken, weaker and ill. The modern human often lives a rather miserable life. He often dies after a longer period of disease and weakness, instead of dying satisfied after a fulfilled life.

Hypnosis as a way to gain control of the situation: The physical self-nonself discrimination affects the immunological self-nonself discrimination directly. Hypnosis is a phenomenon that can be understood from the principles of Figure 1.

When a human, totally adapts himself to another human's intension, he temporarily looses his own initiative and therefore if he is in a situation of hypnosis has to follow the commands from the hypnotist. It is the resistance from the hypnotized, and the hypnotisers own consciousness, that puts these limits and the reason why this situation is suitable for testing the frames of human possibilities. When hypnotized, the command not to react immunologically on an intra-dermal injection of tuberculin, effectsaffects the usual blush and swelling fails to appear (20). This shows that the immunological reaction has to do with the patient's own psychological control, in spite of the fact that the patient often does not believe this. We define this as the psychological self-nonself discrimination directly interfereinterferes with the immunological self-nonself discrimination.

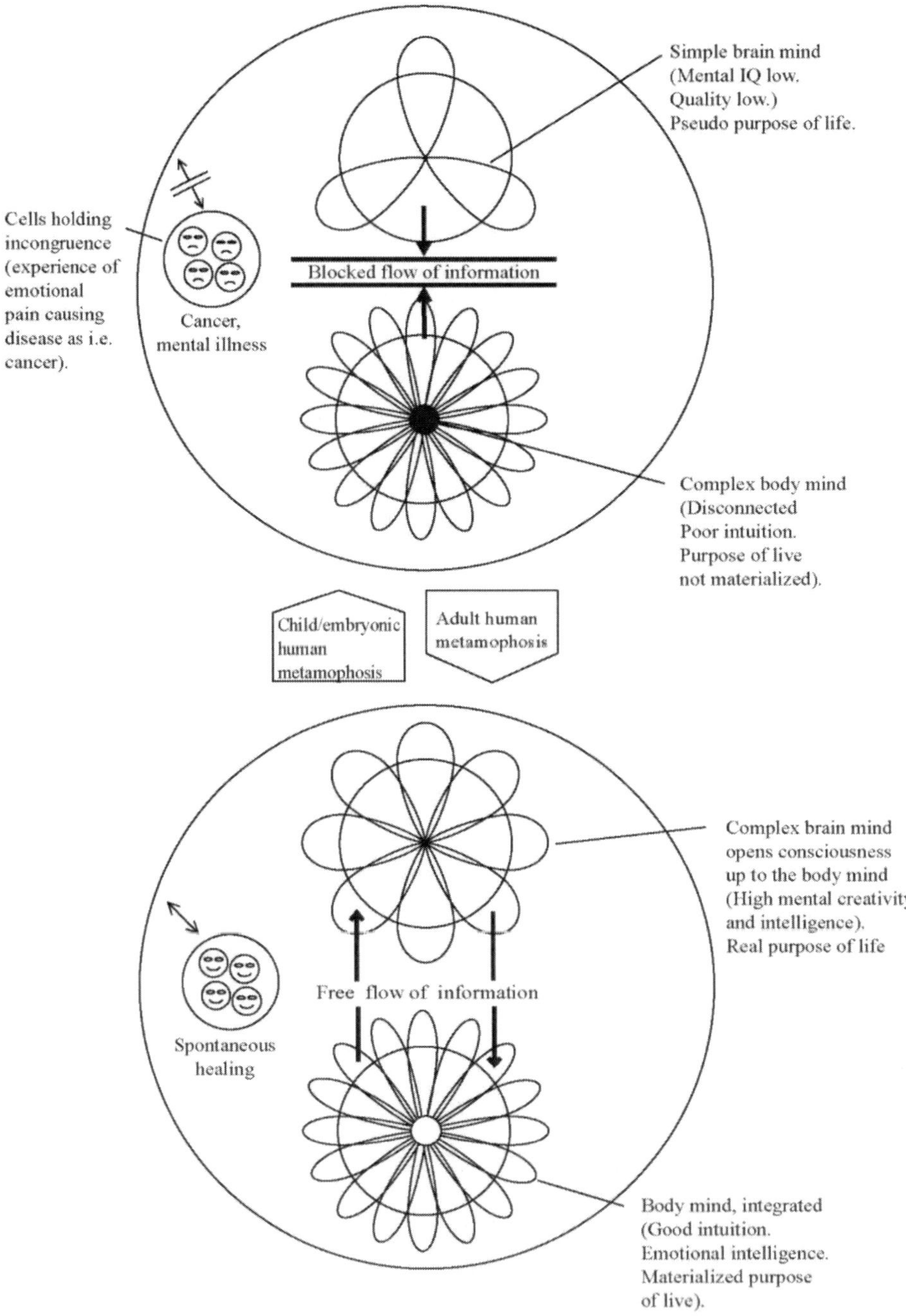

Figure 1. This figure illustrates what happens when two people are in confrontation.

From hypnosis, similar instances are known concerning the mind's influence on the immune system. It has been possible to hamper an allergic skin reaction using an inter-dermal egg test of an 18 years old girl (20) and other experiments that could hamper the allergic reactions associated with asthma and hay fever and treat a 12 years old boy allergic to cats (20). A number of dermatological diseases with an immunological origin have been treated by hypnosis, usually with a great amount of success. Hypnosis could cure half of 27 patients with chronically urticaria (nettle rash) (20). Other suffering has also been cured, as for example congenital ichthyosiform erhytrodermia hernia, which is a disease that causes a darkening of the skin. The etiology of this disease is unknown, and only hypnosis has been able to cure it (20). In addition, the inflammatory reaction following a burning can be controlled by the mind, since it can be hampered through hypnosis (20). Through hypnosis it has been shown that a patient with warts spread out over the body could be freed from these on the one side, if this was the demand from the hypnotiser (20). This seems to show, that the regulation of the immune system acts through a specific mechanism that can be controlled by the mind, and not only unspecifically, by means of hormones.

The representation of the patient's body in his brain, evidently, interacts with the body (21), so that an improved representation involves an improved condition. This means that a distorted representation – a bad picture – of the body in the brain may give disturbances of the body information. When a degenerated intension is characterized through distorted, sick pictures of the reality, it is likely that a wrong intension and reality interpretation could be the reason of disease.

What do we do when we get seriously ill? When we as patients are no longer able to solve the problems ourselves, we consult the physician that use the modern medicine and surgery. The education of a physician has its background in the reason-oriented, mechanical interpretation of life. This implies that the physician offers to help his patients with a "mechanical" surgical intervention: a physical or chemical solution. It is only rarely, that the physician sees himself in a true preventive role and when it comes to preventive advice, as more exercise and less alcohol, less fat, and less tobacco, it is seldom followed.

Physical diseases can be prevented in a far more effective and fundamental way, i.e. by means of regeneration of man's purpose of life through growth of self-knowledge. Regeneration of purpose through self-knowledge not only helps man to avoid psychiatric and somatic illness. In all areas of life man will gain constructive intensions, which are the best basis for his well being. Not only he himself and his relatives will benefit from this, but also the world at large, becoming populated by more whole and able people feeling responsibility.

This being said, we need a more thorough understanding of the factors leading to degeneration of purpose and the tools for the regeneration of it, before the direct use of the mental factor will be applied as a standard routine in daily medical praxis in a truly scientific way.

The physician shall prevent – of course – but until now it has not been understood how it should be done. When people are carrying on an intensely unhealthy way of life, this is simply a symptom that their life is not functioning. It is the duty of the physician to make people function, to teach them the good, unfolded, happy life. No doubt that exactly this above anything else produces healthy people. Therefore it is our sincere hope that the theoretical, biological basis for a preventive medicine will be a first step towards a happier mankind and a better world.

The possible etiology of most somatic diseases as a repression of the organisms fundamental intentions leading to emotional problems kept in the human unconscious and severe disturbances in the biological informational system leads to the conclusion that in the end most somatic diseases must be cured through a shift in the patient's consciousness. The suggestions that somatic diseases often are manifestations of impurities in the consciousness is in accordance with some recent research (22) and opens up for psychodynamic (23-25) and scientific holistic therapy as the rational cure also for physical (and mental) diseases, psychosexual developmental disturbances, and chronic and psychoform pain (26-66). It might also explain the most interesting connection found in so many studies now between the sense of coherence and somatic disease (67-73).

Conclusions

Through life an individual adapts to his various life situations by ongoing modification of his intension and the reality model in his brain. If he lets go of too many central intentions his life will become dull and he will lack energy. In the end the repressed and denied aspects of his life will cause so much disturbance in the biological information in his body that he can become sick, mentally or somatically. Therefore it is of a great importance for the doctor to clarify and investigate his patients to reveal this sad pattern of degeneration of intensions, if they are present. Aspects of life is repressed every time an intention, defined as a need for the person by his biology, is repressed and that side of the person labeled as" non-self". When an individual is living by this destructive pattern of resignation, gradually all his biological reality are transformed to nonself. Over time it is possible that he gets somatically ill. This means that when a person chooses not to fight for his values and dreams, he really chooses the way of the victim. This role-play goes on and on until he does not like his life anymore. He gets despondent and ill, and often dies after a longer period of disease and weakness, instead of dying satisfied after a fulfilled life.

Regeneration of the purpose of life (or the life mission) through self-insight not only helps man to avoid mental and somatic illness, but in all areas of life man will also gain constructive intensions as basis for his well-being. Not only himself and his relatives will benefit from this, but also the world at large. It is our hope that the understanding of somatic disease will lead to cures for many of the diseases that continue to plaque mankind, and to a preventive medicine that will be an important step forward.

References

[1] Beck A. Cognitive therapy and the emotional disorders. New York: Int Univ Press, 1976.
[2] Beck A. Cognitive therapy of depression. London: Guildford, 1979.
[3] Anderson JR. Muir's textbook of pathology. London: Edward Arnold, 1985.
[4] McPhee SJ, Lingappa VR, Ganong WF. Pathophysiology of disease. New York: McGraw Hill Med, 2002.
[5] Klein J. Natural history of the major histocompability complex. New York: John Wiley, 1986.
[6] Ader R. Animal models in the study of brain, behaviour and bodily disease. Res Publ Assoc Res Nerv Ment Dis 1981;59:11-26.
[7] Ader R, Cohen N. CNS-immune system interactions: Conditioning. BBS 1985;8(3):379-95.
[8] Meltzer HY. Psychofarmacology. The third generation of progress. New York: Raven, 1987.
[9] Darko DF. A brief tour of psychoneuroimmunology. Ann Allergy 1986;57(4):233-8.
[10] Roitt I, Brostoff J, Male D. Immunology. London: Gower, 1985.
[11] Otter WD. Immune surveillance and natural resistence: an evaluation. Cancer Immunol Immunother 1986;21(2):85-92.
[12] Gutkind JS. Signaling networks and cell cycle control : The molecular basis of cancer and other diseases. Totowa, NJ: Humana, 2000.

[13] Ventegodt S, Hermansen TD, Flensborg-Madsen T, Nielsen ML, Merrick J. Human development IV: The living cell has information-directed self-organisation – and consciousness. ScientificWorldJournal 2006;6:1132-8.
[14] Ventegodt S, Hermansen TD, Flensborg-Madsen T, Nielsen ML, Merrick J. Human development VI: Supra-cellular morphogenesis - the origin of biological and cellular order. ScientificWorldJournal 2006;6:1424-33.
[15] Jerne N. Towards a network theory of the immune system. Ann. Immunol. (Paris) 1974;125C(1-2):373-89.
[16] Ventegodt S, Hermansen TD, Rald E, Nielsen ML, Clausen B, Merrick J. Human development XVII: Jernes anti-idiotype network theory cannot explain self-nonself discrimination. ScientificWorldJournal 2006;6:1434-40.
[17] Paul WE. Fundamental immunology. New York: Raven, 1984.
[18] Ventegodt S, Hermansen TD, Flensborg-Madsen T, Rald E, Nielsen ML, et al. Human development IX: A model for the wholeness of man, his consciousness, and the collective consciousness. ScientificWorldJournal 2006;6:1454-9.
[19] Piaget J. The place of the sciences of man in the system of sciences. New York: Harper Torchbooks, 1974.
[20] Goldberg B. Hypnosis and the immune respons. Int J Psychosomatics 1985; 32(3):34-6.
[21] Ventegodt S, Andersen NJ, Merrick J. The life mission theory V. A theory of the anti-self and explaining the evil side of man. ScientificWorldJournal 2003;3:1302-13.
[22] Ventegodt S, Flensborg-Madsen T, Andersen NJ, Nielsen M, Mohammed M, Merrick J. Global quality of life (QOL), health and ability are primarily determined by our consciousness. Research findings from Denmark 1991-2004. Soc Indicator Res 2005;71:87-122.
[23] Leichsenring F, Rabung S, Leibing E. The efficacy of short-term psychodynamic psychotherapy in specific psychiatric disorders: a meta-analysis. Arch Gen Psychiatry 2004;61(12):1208-16.
[24] Leichsenring F. Are psychodynamic and psychoanalytic therapies effective? A review of empirical data. Int J Psychoanal 2005;86(Pt 3):841-68.
[25] Leichsenring F, Leibing E. Psychodynamic psychotherapy: a systematic review of techniques, indications and empirical evidence. Psychol Psychother 2007;80(Pt 2):217-28.
[26] Ventegodt S, Merrick J. Clinical holistic medicine: Applied consciousness-based medicine. ScientificWorldJournal 2004;4:96-9.
[27] Ventegodt S, Morad M, Merrick J. Clinical holistic medicine: Classic art of healing or the therapeutic touch. ScientificWorldJournal 2004;4:134-47.
[28] Ventegodt S, Morad M, Merrick J. Clinical holistic medicine: The "new medicine", the multi-paradigmatic physician and the medical record. ScientificWorldJournal 2004;4:273-85.
[29] Ventegodt S, Morad M, Merrick J. Clinical holistic medicine: Holistic pelvic examination and holistic treatment of infertility. ScientificWorldJournal 2004;4:148-58.
[30] Ventegodt S, Morad M, Hyam E, Merrick J. Clinical holistic medicine: Use and limitations of the biomedical paradigm ScientificWorldJournal 2004;4:295-306.
[31] Ventegodt S, Morad M, Kandel I, Merrick J. Clinical holistic medicine: Social problems disguised as illness. ScientificWorldJournal 2004;4:286-94.
[32] Ventegodt S, Morad M, Andersen NJ, Merrick J. Clinical holistic medicine Tools for a medical science based on consciousness. ScientificWorldJournal 2004;4:347-61.
[33] Ventegodt S, Morad M, Merrick J. Clinical holistic medicine: Prevention through healthy lifestyle and quality of life. Oral Health Prev Dent 2004;1:239-45.
[34] Ventegodt S, Morad M, Hyam E, Merrick J. Clinical holistic medicine: When biomedicine is inadequate. ScientificWorldJournal 2004;4:333-46.
[35] Ventegodt S, Morad M, Merrick J. Clinical holistic medicine: Holistic treatment of children. ScientificWorldJournal 2004;4:581-8.
[36] Ventegodt S, Morad M, Merrick J. Clinical holistic medicine: Problems in sex and living together. ScientificWorldJournal 2004;4:562-70.

[37] Ventegodt S, Morad M, Hyam E, Merrick J. Clinical holistic medicine: Holistic sexology and treatment of vulvodynia through existential therapy and acceptance through touch. ScientificWorldJournal 2004;4:571-80.
[38] Ventegodt S, Flensborg-Madsen T, Andersen NJ, Morad M, Merrick J. Clinical holistic medicine: A Pilot on HIV and quality of life and a suggested treatment of HIV and AIDS. ScientificWorldJournal 2004;4:264-72.
[39] Ventegodt S, Morad M, Merrick J. Clinical holistic medicine: Induction of spontaneous remission of cancer by recovery of the human character and the purpose of life (the Life Mission). ScientificWorldJournal 2004;4:362-77.
[40] Ventegodt S, Morad M, Kandel I, Merrick J. Clinical holistic medicine: Treatment of physical health problems without a known cause, exemplified by hypertension and tinnitus. ScientificWorldJournal 2004;4:716-24.
[41] Ventegodt S, Morad M, Merrick J. Clinical holistic medicine: Developing from asthma, allergy and eczema. ScientificWorldJournal 2004;4:936-42.
[42] Ventegodt S, Morad M, Press J, Merrick J, Shek DTL. Clinical holistic medicine: Holistic adolescent medicine. ScientificWorldJournal 2004;4:551-61.
[43] Ventegodt S, Solheim E, Saunte ME, Morad M, Kandel I, Merrick J. Clinical holistic medicine: Metastatic cancer. ScientificWorldJournal 2004;4:913-35.
[44] Ventegodt S, Morad M, Kandel I, Merrick J. Clinical holistic medicine: a psychological theory of dependency to improve quality of life. ScientificWorldJournal 2004;4:638-48.
[45] Ventegodt S, Merrick J. Clinical holistic medicine: Chronic infections and autoimmune diseases. ScientificWorldJournal 2005;5:155-64.
[46] Ventegodt S, Kandel I, Neikrug S, Merrick J. Clinical holistic medicine: Holistic treatment of rape and incest traumas. ScientificWorldJournal 2005;5:288-97.
[47] Ventegodt S, Morad M, Merrick J. Clinical holistic medicine: Chronic pain in the locomotor system. ScientificWorldJournal 2005;5:165-72.
[48] Ventegodt S, Merrick J. Clinical holistic medicine: Chronic pain in internal organs. ScientificWorldJournal 2005;5:205-10.
[49] Ventegodt S Kandel I, Neikrug S, Merrick J. Clinical holistic medicine: The existential crisis – life crisis, stress and burnout. ScientificWorldJournal 2005;5:300-12.
[50] Ventegodt S, Gringols G, Merrick J. Clinical holistic medicine: Holistic rehabilitation. ScientificWorldJournal 2005;5:280-7.
[51] Ventegodt S, Andersen NJ, Neikrug S, Kandel I, Merrick J. Clinical holistic medicine: Mental disorders in a holistic perspective. ScientificWorldJournal 2005;5:313-23.
[52] Ventegodt S, Andersen NJ, Neikrug S, Kandel I, Merrick J. Clinical holistic medicine: Holistic treatment of mental disorders. ScientificWorldJournal 2005;5:427-45.
[53] Ventegodt S, Merrick J. Clinical holistic medicine: The patient with multiple diseases. ScientificWorldJournal 2005;5:324-39.
[54] Ventegodt S, Clausen B, Nielsen ML, Merrick J. Clinical holistic medicine: Advanced tools for holistic medicine. ScientificWorldJournal 2006;6:2048-65.
[55] Ventegodt S, Clausen B, Merrick J. Clinical holistic medicine: The case story of Anna: I. Long term effect of child sexual abuse and incest with a treatment approach. ScientificWorldJournal 2006;6:1965-76.
[56] Ventegodt S, Clausem B, Merrick J. Clinical holistic medicine: the case story of Anna. II. Patient diary as a tool in treatment. ScientificWorldJournal 2006;6:2006-34.
[57] Ventegodt S, Clausen B, Merrick J. Clinical holistic medicine: The case story of Anna. III. Rehabilitation of philosophy of life during holistic existential therapy for childhood sexual abuse. ScientificWorldJournal 2006;6:2080-91.
[58] Ventegodt S, Merrick J. Suicide from a holistic point of view. ScientificWorldJournal 2005;5:759-66.
[59] Ventegodt S, Clausen B, Omar HA, Merrick J. Clinical holistic medicine: Holistic sexology and acupressure through the vagina (Hippocratic pelvic massage). ScientificWorldJournal 2006;6:2066-79.
[60] Ventegodt S, Clausen B, Merrick J. Clinical holistic medicine: Pilot study on the effect of vaginal acupressure (Hippocratic pelvic massage). ScientificWorldJournal 2006;6:2100-16.

[61] Ventegodt S, Thegler S., Andreasen T, Struve F, Enevoldsen L, et al. Clinical holistic medicine: Psychodynamic short-time therapy complemented with bodywork. A clinical follow-up study of 109 patients. ScientificWorldJournal 2006;6:2220-38.
[62] Ventegodt S, Thegler S, Andreasen T, Struve F, Enevoldsen L, et al. Clinical holistic medicine (mindful, short-term psychodynamic psychotherapy complemented with bodywork) in the treatment of experienced impaired sexual functioning. ScientificWorldJournal 2007;7:324-9.
[63] Ventegodt S, Thegler S, Andreasen T, Struve F, Enevoldsen L, et al. Clinical holistic medicine (mindful, short-term psychodynamic psychotherapy complemented with bodywork) improves quality of life, health, and ability by induction of Antonovsky-salutogenesis. ScientificWorldJournal 2007;7:317-23.
[64] Ventegodt S, Thegler S, Andreasen T, Struve F, Enevoldsen L, et al. Clinical holistic medicine (mindful, short-term psychodynamic psychotherapy complemented with bodywork) in the treatment of experienced physical illness and chronic pain. ScientificWorldJournal 2007;7:310-6.
[65] Ventegodt S, Thegler S, Andreasen T, Struve F, Enevoldsen L, et al. Clinical holistic medicine (mindful, short-term psychodynamic psychotherapy complemented with bodywork) in the treatment of experienced mental illness. ScientificWorldJournal 2007;7:306-9.
[66] Ventegodt S, Thegler S, Andreasen T, Struve F, Enevoldsen L, et al. Self-reported low self-esteem. Intervention and follow-up in a clinical setting. ScientificWorldJournal 2007;7:299-305.
[67] Flensborg-Madsen T, Ventegodt S, Merrick J. Sense of coherence and physical health. A Review of previous findings. ScientificWorldJournal 2005;5:665-73.
[68] Flensborg-Madsen T, Ventegodt S, Merrick J. Why is Antonovsky's sense of coherence not correlated to physical health? Analysing Antonovsky's 29-item sense of coherence scale (SOCS). ScientificWorldJournal 2005;5:767-76.
[69] Flensborg-Madsen T, Ventegodt S, Merrick J. Sense of coherence and health. The construction of an amendment to Antonovsky's sense of coherence scale (SOC II). ScientificWorldJournal 2006;6:2133-9.
[70] Flensborg-Madsen T, Ventegodt S, Merrick J. Sense of coherence and physical health. A cross-sectional study using a new SOC scale (SOC II). ScientificWorldJournal 2006;6:2200-11.
[71] Flensborg-Madsen T, Ventegodt S, Merrick J. Sense of coherence and physical health. Testing Antonovsky's theory. ScientificWorldJournal 2006;6:2212-9.
[72] Flensborg-Madsen T, Ventegodt S, Merrick J. Sense of coherence and health. The emotional sense of coherence (SOC-E) was found to be the best-known predictor of physical health. ScientificWorldJournal 2006;6:2147-57.
[73] Ventegodt S, Flensborg-Madsen T, Andersen NJ, Merrick J. Life Mission Theory VII: Theory of existential (Antonovsky) coherence: a theory of quality of life, health and ability for use in holistic medicine. ScientificWorldJournal 2005;5:377-89.

Chapter XX

A theory for accelerated, spontaneous existential healing (salutogenesis): Human adult metamorphosis

We have with a number of patients in intensive holistic therapy observed a process of accelerated, spontaneous healing of body, mind, sexuality and existence. The healing process seems to upgrade the purpose of life (life mission) of the patient, recover physical, mental, and spiritual character, deepen consciousness and self-insight and even sometimes "reshape" the patient's body. It is characterised by an introvert, visionary state of intense transformation that parallels what happens in the butterfly's pupae; therefore we have called the process "human adult metamorphosis". We believe that many interventions depending on personal development of talents, spontaneous healing, personal transformation and other forms of personal growth are all modifications of the basic process of human adult metamorphosis. Adaptation patterns from early childhood and even embryonic life that helped us to survive has given us as adults many irrational, neurotic or psychotic patterns of perception and behaviours, that seem to come from adaptation though "downgrade" metamorphosis. Even bodily problems like cosmetic problems with too small or too large breast or labia seems to be a product of the person not being in the optimal state and seem to be corrected, when the person enters metamorphosis and develops a natural deep understanding of life including sexuality. We presents a model, discuss its genetics, the therapeutic tools that induces it, the relation between complexity of cosmology and spontaneous metamorphosis, spontaneous healing of cancer after metamorphosis, induced recovery from schizophrenia with metamorphosis, body reshaping after metamorphosis, and shamanism and drug induced human adult metamorphosis.

Introduction

When the butterfly larva reach a certain point, it transforms into the pupae. All its organs melt down and cells becomes seemingly undifferentiated, able to make the completely new set of organs that composite the butterfly. The cells of the pupae seemingly have unlimited access to the genetic and biologic information about the adult individual. The amazing process of metamorphosis normally takes about 10 days. Known from biology not only butterflies have metamorphosis, but a large number of insects, vertebrates like frogs, and even unicellular organisms have it. Even humans go through metamorphosis early in life a certain number of stages during ontogenesis. In the womb a long series of morphic transformations remodels the growing embryo, so the whole embryogenesis might in fact be seen as one large metamorphic process. Therefore it is hardly shocking to find massive genetic equipment for metamorphosis in the human being, very similar to the genetics of the butterfly and the frog.

What might be much more shocking is that we appear to preserve the ability to enter the biological process of metamorphosis, even as adults. We have observed a spectrum of personal growth going from human development that often take decades, over healing that takes years, human transformation that takes months, to extremely accelerated personal growth and healing, where the patient in a few weeks goes through from a very poor state of being though a radical shift in personality, mind-set and even physical appearance, to a more conscious and much more functional state of being. A radical improvement in quality of life, self-assessed health, ability and talent seems to rise from a recovered existential coherence. The patient often change and reframe their purpose of life and release negative attitudes related to early life traumas, giving the patient a much more positive set of attitudes towards self, life at large and others. Philosophy of life is often drastically deepened, the patient starting to appreciate spiritual aspect of life. The cosmology is often growing more complex with the number of concepts in active use going many times up (1).

The healing connected to metamorphosis is often dramatic. Existential responsibility is often drastically improved, turning the mentally disturbed patient into a much more sound and capable person. If the patient receives a serious somatic disease like cancer, metamorphosis is often a sign of raised ability to survive and heal. If the patient had sexual and gender issues, these are often solved; if the problems were connected to specific organs like the breasts, the genitals, or the shape of the pelvis or another body part, quite surprisingly the involved organs are often also reshaping, in the same time as the patient's inner experience of the area is drastically improved. The healing process of human adult metamorphosis seems to attempt to correct any human flaw, being existential, mental, physical or sexual.

Our ability to induce the phenomenon of human adult metamorphosis is limited for several reasons. First we do not understand the process well; it is still a deep mystery under intense investigation in our laboratory. But it seems that human adult metamorphosis is under subconscious control of the patient and only when he or she feels completely safe and in good hands, then the possibility for transformation will be there. Another important aspect is that it almost always seems to generate from traumas early in life making the opening into the process an extremely painful process.

The process itself is not painful, but the patient is normally extremely introvert and unable to work or even sometimes speak or think for days or even a week or two. Therapeutic techniques that send the patient into intense old emotional pains seem to provoke it (2) and also philosophical training evolving the patients consciousness in the direction of deep cosmology, i.e. cosmology of seven or more rays (7-numbered bifurcations in the structural three of the cosmology, see (1)). We first saw the phenomenon when we combined safe space, advanced tool for holistic existential therapy, and cosmology with ray-numbers of 7 and higher.

We first saw human adult metamorphosis several years ago (3) with schizophrenic patients, who participated in our intense summer courses on quality of life, but they had kept their diagnosis secret in order not to be excluded from the course.

In the beginning we believed it to be provoked psychosis, which it can easily be mistaken for, but the intense healing effect of the episode of a few days of duration gave way to our suspicion that something else, something positive and even natural was going on. Years later we came to understand this episode as a first example of a new and extremely intense and rapid pattern of spontaneous healing, which we now call human adult metamorphosis.

Genetics of human adult metamorphosis

The humane genome project has revealed that human beings carry massive amount of master and metamorphic genes; the same genes that have been proven to be activated just before and during metamorphosis in many animal lines. In principle there seem to be no hindrance for human beings to go though a process of metamorphosis involving also the shape of the body, the patterns of behaviour etc., like the animals often does. Understanding the connection between the genes and the cellular and supra-cellular events in the organism has not been easy.

Ontogenesis itself is still quite poorly understood; it makes some sense that there is a cascade of events allowing the more complex structures to build on the less complex, very much as we believe evolution to have happened. The sequential recruiting of biological information making one pattern at the time available to the developing embryo gives us a safe feeling of a natural and well ordered informational process.

Understanding morphogenesis where the whole ontogenesis is repeated in a direct and explosive manner with no time to let one pattern build on another and letting the individual transform directly into something completely else is much more difficult to comprehend. How can this be done? How can the individual in a completely a-historical way suddenly address its full genetic potential and realise it from one day to another, transforming from one animal into a completely different? This is almost magic. The process of metamorphosis indicates a leap in evolution, symbolised by the insect suddenly getting wings. How did this happen evolutionary? Well, metamorphosis might be the model here: It happened because some insects began to shift, half though their life, into winged forms. They started to metamorphose!

The truth is that we never came to any understanding neither concerning the most basic principle of the creative side of evolution, nor even concerning the most basic principles of morphogenesis, in spite of 100 years of intensive scientific studies. In a way metamorphosis shows us how wrong we have been to take biological phenomena to be simple, linear events. The organism's system providing biological information to the cells seem to work in a global way, allowing all cells to shift at the same time. The informational system presents the whole blueprint for the whole individual, and maintain it that way; if the organism changes this blueprint, the shape, structure and behaviour of the whole organism follows. The shift of the blueprint is mysterious. That it happens demonstrates to us how the blueprint holds the biological order of the organism at any time. And when it changes, the whole organism changes with it. The problem is now, what initiates this shift in blueprint?

In lower organisms it might be something quite mechanical, like need for food or water, or age that provokes metamorphosis. But in humans the process seems to be under our subconscious control; we speculate that this change could be a function of a shift in the individual's basic intent, a shift in its purpose of life[4]. This opens up for the interesting possibility that patients, that become conscious of their deep existential structures, can enter metamorphosis at will.

Therapeutic tools that induces metamorphosis

In holistic existential therapy and holistic sexology, where both the development of feelings and the philosophy of the patient are addressed, the speed of personal development is very dependent on the skill of the therapist, the motivation and spirit of the patient. If the therapist is familiar with the steps and dynamical process of metamorphosis, and if the patient is highly motivated for this kind of therapy and see everything in life as learning events, the patient is likely to enter metamorphosis, when the supply of resources are sufficient.

If the patient has been arrested in psychosexual development since childhood, the process can be extremely dramatic, sending the patient into a transformativ state that is very similar to the butterfly's pupae: a silent, visionary and complete introvert state of being. Interestingly, this state has been described in the Native American ceremony called "the vision quest" (5).

Complexity of cosmology and human adult metamorphosis

The complexity of mind needs to be of the type we call "deep cosmology" (1), that allows brain-mind to merge with body-mind. Deep cosmology opens up for existential introspection. It is an interpretation of the world that is build on a structured web of concepts, organised around a core of key concepts, that is the abstract essence of the other concepts in use (6). The "depth" is a popular term refereeing to a high-bifurcation number in the conceptual tree of the cosmology that has fractal structure (7). The more complex the cosmology of the patient is, the more rapid the process of personal development (1).

Accelerated development and human adult metamorphosis

It will often be necessary to hospitalise the patient, if the traumas stems from early violent and sexual abuse, which often carry extremely negative emotional charge. The need for care and safety continues round the clock for days or weeks with these patients, while they re-live old traumas.

Interestingly the patients often relive and reverse metamorphosis in the childhood that made surviving an impossible childhood situation possible. Often these patients in such a metamorphosis get 100 or even 200 hours of treatment in the process of intense healing. This would normally take several years in standard therapy even with weekly sessions and this is actually what explains these patient's rapid development, radical changes, and dramatic healing.

Spontaneous healing of cancer and human adult metamorphosis

Most surprisingly spontaneous healing of cancer and other severe diseases seem to need human adult metamorphosis to happen. The rehabilitation of existential coherence (6) and human character (1) can be extremely fast, if it starts with the reorganisation of consciousness that comes from an upgrading of the purpose of life.

The downgrading from early childhood traumas seems to be a product of metamorphosis in childhood, where the child adapts with masterly creativity and no other intent than survival to fit into its family. The flattening of cosmology (1) with respect to complexity and bifurcation number (ray-number) hardens the child emotionally in order to be able to withstand even the roughest of handlings. But the feelings that force the child to go through metamorphosis are often extremely negative and painful, where the repressed feelings seem to be absorbed by the tissues of the body (8-10).

Cancer and other severe diseases like autoimmune diseases (11) tend to develop, where the tissues hold on to these emotions (12) and when the feelings are processed out of the body in holistic existential therapy the tumours seems to disappear by spontaneous apoptosis, that can only be explained as a consequence of the informational conditions of the cells being rebalanced.

Only with a metamorphosis into a deeper and more complex cosmology that mirrors the original metamorphosis out of the existential pain of the patient's childhood, can set the patient free at the deepest existential level, and this is what it takes for the patient to integrate the emotional pain and heal existence, as well as mind and body. As it took a metamorphosis to get into the trouble, only a re-metamorphosis (a healing metamorphosis) will get the patient out of the trouble again.

Case Story

36-year-old woman with breast cancer and many metastases to the liver, not responding to chemotherapy

This patient came to our research clinic several years ago and she is now a student at the Nordic School of Holistic Medicine. According to statistics and explained by her treating oncologist, she had 5.2 month left to live. She had a five years old daughter, so she desperately looked for a way to survive and entered our protocol for holistic treatment of cancer patients (12,13). She went though a mental revolution renewing her philosophy of life, and she worked with and integrated many strongly negative feelings towards her mother and father. She is well today (three years later) and her oncologists at the hospital now believe she will survive her cancer.

Induced recovery from schizophrenia and human adult metamorphosis

We know from many studies, often called "the recovery literature" that between 20% and 30% of the patients, who received the diagnosis of schizophrenia in their youth, do not have the disease in midlife. So they have spontaneously recovered. We have been able to induce healing in patients in the clinic by simply supporting the patients in the confrontation and integration of severe often very early life-traumas. The problem with the therapy has been the strangeness of it; the patients have almost always experienced severe violent or sexual abuse in the childhood, but the largest traumas have often appeared to be even before birth. It really is surprising that you can be almost completely arrested in your psychosexual development and 20 or 30 years later you can reclaim your life, transform and grow up to step forward – as yourself. The person you were meant to be. The core of this upgrade of the personality seems to be a sudden "remembrance" of what the meaning of life really is all about. The eternal now going all the way back to conception is suddenly here again and life re-established.

The most problematic side effect of the human adult metamorphosis with patients that have been violently or sexually abused is the reappearing of the traumas.

Case story

24-year old female university student, virgin, borderline, severe existential and sexual problems

A young female university student with a borderline personality, feeling like "not fitting into the world", unable to love herself and others, with no interest in sex at all. Completely unable to get close to boys, could not remember anything causing her severe problems. After a course in philosophy of life in which she engaged eagerly, she suddenly the day after the course entered into the state of metamorphosis.

We took her to the Research Clinic for Holistic Medicine and cared for her, providing a safe and loving space for her. For 10 days she was completely introverted in a visionary state of being what looked very much like an acute outbreak of schizophrenia. She was hallucinatory and had thousands of visualisations of all aspects of life, which became more and more evil and bloody as she sank deeper and deeper into the process of personal transformation. She said that when she closed her eyes her "eyeballs turned so she could look inside" and her consciousness obviously merged with the deep layer of existence Carl Gustav Jung (1875-1961) used to call "the collective consciousness".

We could only understand her visionary state as a state of intense healing and it was obvious to us that she went deep inside to find what she needed to heal and this way integrate "her shadow". Then everything came back to her: she recalled being raped by her 10-year older brother at age four.

After drawing hundreds of drawings in art therapy and having more that 20 hours of integrative timeline therapy, and about 20 sessions of individual philosophical training where she got accustomed to the logics and feeling of complex (high-ray) cosmology (1). She reappeared as almost normal. Most of her severe mental and existential problems had been solved.

If this woman had been treated for schizophrenia with sedative antipsychotic drugs her process of human adult metamorphosis would have been arrested and she would have learned to think of herself as a schizophrenic. She would have been stigmatised, marginalized, and maybe even institutionalised. Because we took care of her and let nature do its job, she transformed and healed spontaneously.

Body reshaping and human adult metamorphosis

After holistic existential therapy many patients have noticed a very different experience of their body. They have become more present in it, they have felt better in it, being more accepting, relaxing in the body, and feeling it stronger and more powerful. We know of cause that massage can do this, so that working with deep acceptance of body and sexuality combined with holistic bodywork can do this is not very surprising.

What has been a surprise though are the many patients reporting about experiences of physical change in their body. They often describe that their breast grow, their hips and whole body are reshaped, and often they also tell that they have observed great anatomical changes in their genitals, like a sudden reduction in the size of labia minores.

As the bodywork, as often done with the patient undressed, it has sometimes been possible for the therapist to agree to these changes, but as they have been so unexpected we are now only starting to scientifically document the body reshaping.

We expect most of the negative experiences of own body to be a product of downgrade metamorphosis, which can be alleviated with upgrade metamorphosis. Now we only need to get the process of metamorphosis under control so well that we can induce it also on cosmetic indication to replace plastic cosmetic surgery, which has so many unwanted side effects.

It seems that we actually are able to induce it with the most powerful techniques of holistic existential therapy, if the patient has a strong philosophical interest and if we have a set-up, where we take care of the patient for some weeks afterwards.

The prize of this treatment is believed to be only a fraction of the prise of cosmetic surgery, and instead of often-severe negative side effects, we expect large positive effects on the patient's quality of life, health and ability.

Sexuality in human adult metamorphosis

We have observed that many healing processes are associated with strong sexual feelings, both positive and negative. Often the patient is presenting him- or herself in a victimised position, strongly influenced by masochistic energies; during the proces of healing the patients mind is often filled with sexual symbolism, and often masochism is turned into sadism.

After the successful integration of both the masochistic and sadistic sexual energies and intentions, a natural ability of make love in a harmonic, caring and mutually satisfactioning way is emerging, often very much to the patients surprise, the surprise being larger, of the patients sex-life was never healthy, succesfulled and really satisfying.

The three steps masochism, sadism, and joyfull loving sexuality are seems to be parallel to Freud's, Jung's and Reich's idea of the infantile autosexual, immature sexual and genitally matur sexual patient. The psychosexual healing of the patient seems to be a key feature of human adult metamorphosis.

Drug induced metamorphosis and shamanism

The concept of human adult metamorphosis has made it possible for us to re-interpret many of the rituals of the pre-modern cultures. Especially we have been able to understand shamanism and the drug-induced transformational processes of peyote and Ayahuasca medicine of the Native Americans. The shaman induces transient or permanent metamorphosis using a combination of rituals and items symbolising the depth of the cosmology and hallucinogenic drugs to facilitate the visionary state related to the merging of the individual consciousness into the collective consciousness. In this way they intentionally create the transformational states of existence that we have observed as the patients enter spontaneously during their metamorphosis. The shamanistic key concepts and the shamanistic cosmology seems well adapted for doing this, and the cosmologies of the Native Americans seem to be sufficiently complex to induce human adult metamorphosis (often 8-rays) (1).

Discussion

For millenniums it seems that the pre-modern cultures have known how to take patients into accelerated personal development and existential healing. It now seems possible to accelerate the process of holistic healing by using the advanced tools from clinical holistic medicine. With the present understanding it might even be wise to start experimenting with the use of the hallucinogenic drugs for finding a scientific way to induce the intensive healing state of consciousness, we have called human adult metamorphosis. Such experiments should be focused on the chronically ill patients with shortened life expectancy, because of progressing illness or those who needs more support to enter the process of human adult metamorphosis than the normal tools for holistic medicine is offering.

Again it is important to remember that there are no effective drugs without side effects. The risk should always be balanced with the possible gain, deep and intense existential and emotional pain is always a follower of human adult metamorphosis, and sometimes the patients will not have the internal and external resources for integrating the pain.

The result is that the physician, who tries to cure a patient in spite of the obvious lack of resources, is causing the patient unnecessary sufferings, which is highly unethical. On the other hand we see both schizophrenic patients and cancer patient spontaneously healing so everybody seems to have extraordinary hidden resources that can be mobilised in holistic existential therapy. It is interesting that many young people are still using ecstasy, LSD and other drugs with the power of inducing transformation and metamorphosis – presumably many more than in the 1960s - and the reason for using them seems often to be healing more that recreation. The need in society for people to heal their existence is enormous as one of four of the citizens has severe existential problems at least in Denmark (13). As these drugs seems to have very few side effects and adverse effects compared to alcohol and tobacco, and as they will be used by young people for many years to come, and since it is most likely that they are benefiting the users helping then to heal existentially, it might be an idea to legalise the hallucinogenic drugs for medical use, so that the use of them can be followed by sound advice and supervision by their physician.

Let us underline that the research we are conducting at the Quality of Life Research Center and the Research Clinic for Holistic Medicine are not using any drugs. We want to be able to control the process so well that we can induce it by will, in cooperation with our patients. We want to induce healing just by intervening on human consciousness as research has shown that consciousness is determining health, quality of life and ability in general (14).

The many cures that we have developed for clinical holistic medicine might benefit dramatically from this new discovery (11-13,15-52), which basically is nothing but helping the patient back into life and into the sense of coherence (53-59). What we have witnessed is what Aaron Antonovsky (1923-1994) called salutogenesis and the ability to induce accelerated personal growth and existential healing of this kind is undoubtedly what made Hippocrates' medical system famous (60).

Conclusions

A deeper understand of the biological universe have lead us to a new understanding of human development. We have with the mastery of the advanced tools of clinical holistic medicine recently seen a highly accelerated new pattern of spontaneous healing, that we call human adult metamorphosis. It happens when the cosmology (1) of the patient is developed towards more complexity. We have come to learn about metamorphosis, because patients whose psychosexual development had been arrested since early childhood (often due to severe failure or violent and sexual abuse) spontaneously entered metamorphosis after training in philosophy of life. We have been able to understand human adult metamorphosis as the consequence of an inner reunion of the patient's brain-mind and body-mind. In the process of metamorphosis a new and deeper purpose of life is often found, and consciousness is completely reconstructed using lots of information from the deep layer of existence, what Jung called the collective consciousness. The informational system of the organism is rebalanced in this process and the cells and the tissues of the body seem to benefit from that. Often we see a healing of the mental and physical diseases and even a complete reshaping of the body is sometimes followed by the patient's claim of his or her new life.

We believe that personal development, development of talents, spontaneous healing, personal transformation and many more forms of personal growth is all modifications of the basic process of human adult metamorphosis. Survival patterns from early childhood and even embryonic life giving us irrational, neurotic or psychotic patterns of perception and behaviours also seem to come from adaptation though "downgrade" metamorphosis. Even most bodily problems like cosmetic problems with too small or too large breast seems to be a product of the person not being in the optimal state and can be corrected when the person enters metamorphosis and develops her natural deep understanding of life – what we call a cosmology sufficiently complex to embrace the body-mind and feelings.

So human adult metamorphosis seem to be the efficient medicine most patients need to heal. It can be induced with hallucinogenic drugs as the Native Americans have been doing for many centuries; we see the use of such drugs with the young as their attempt to get support to their healing and suggest that these drugs should be legalised for medical use. We also suggest that such treatment could help some cancer patients to survive.

We must strongly warn against giving sedative drugs to patients during the "learning crisis" of therapy; the process of human adult metamorphosis might get severely disturbed or even totally arrested by using such drugs. "Evil visualisations" and severe existential pain is a natural part of the process of healing and should be alleviated with attentive and loving care, not with drugs that will slow down the process of metamorphosis. We must warn that standard psychiatric antipsychotic treatment of a patient in developmental crisis might even cause severe damage to the patient.

References

[1] Ventegodt S, Kromann M, Andersen NJ, Merrick J. The life mission theory VI: A theory for the human character. ScientificWorldJournal 2004;4:859-80.
[2] Ventegodt S, Morad M, Andersen NJ, Merrick J. Clinical holistic medicine Tools for a medical science based on consciousness. ScientificWorldJournal 2004;4:347-61.
[3] Ventegodt S. Bevidsthedsmedicin set gennem lægejournalen. Copenhagen: Forskningscenterets Forlaget, 2003. [Danish]
[4] Ventegodt S. The life mission theory: A theory for a consciousness-based medicine. Int J Adolesc Med Health 2003;15(1):89-91.
[5] Foster S. Book of vision quest. New York: Fireside Books, 1989.
[6] Ventegodt S, Flensborg-Madsen T, Andersen NJ, Merrick J. Life Mission Theory VII: Theory of existential (Antonovsky) coherence: a theory of quality of life, health and ability for use in holistic medicine. ScientificWorldJournal 2005;5:377-89.
[7] Peitgen HO, Richter PH. The beauty of fractals. New York: Springer, 1988.
[8] Ventegodt S, Andersen NJ, Merrick J. Holistic Medicine IV: Principles of the holistic process of healing in a group setting. ScientificWorldJournal 2003;3:1294-1301.
[9] Ventegodt S, Merrick J, Andersen NJ, Bendix T. A combination of gestalt therapy, Rosen Body Work and Cranio Sacral therapy did not help in chronic whiplash-associated disorder (WAD) – results of a randomised clinical trial. ScientificWorldJournal 2004;4:1055-68.
[10] Ventegodt S, Morad M, Hyam E, Merrick J. Clinical holistic medicine: Holistic sexology and treatment of vulvodynia through existential therapy and acceptance through touch. ScientificWorldJournal 2004;4:571-80.
[11] Ventegodt S, Merrick J. Clinical holistic medicine: Chronic infections and autoimmune diseases. ScientificWorldJournal 2005;5:155-64.
[12] Ventegodt S, Solheim E, Saunte ME, Morad M, Kandel I, Merrick J. Clinical holistic medicine: Metastatic cancer. ScientificWorldJournal 2004;4:913-35.
[13] Ventegodt S. Quality of life in Denmark. Copenhagen: Forskningscenterets Forlag, 1995. [Danish]
[14] Ventegodt S, Flensborg-Madsen T, Andersen NJ, Nielsen M, Morad M, Merrick J. Global quality of life (QOL), health and ability are primarily determined by our consciousness. Research findings from Denmark 1991-2004. Soc Indicator Res 2005;71:87-122.
[15] Ventegodt S, Merrick J. Clinical holistic medicine: Applied consciousness-based medicine. ScientificWorldJournal 2004;4:96-9.
[16] Ventegodt S, Morad M, Merrick J. Clinical holistic medicine: Classic art of healing or the therapeutic touch. ScientificWorldJournal 2004;4:134-47.
[17] Ventegodt S, Morad M, Merrick J. Clinical holistic medicine: The "new medicine", the multi-paradigmatic physician and the medical record. ScientificWorldJournal 2004;4:273-85.
[18] Ventegodt S, Morad M, Merrick J. Clinical holistic medicine: Holistic pelvic examination and holistic treatment of infertility. ScientificWorldJournal 2004;4:148-58.
[19] Ventegodt S, Morad M, Hyam E, Merrick J. Clinical holistic medicine: Use and limitations of the biomedical paradigm ScientificWorldJournal 2004;4:295-306.

[20] Ventegodt S, Morad M, Kandel I, Merrick J. Clinical holistic medicine: Social problems disguised as illness. ScientificWorldJournal 2004;4:286-94.
[21] Ventegodt S, Morad M, Andersen NJ, Merrick J. Clinical holistic medicine Tools for a medical science based on consciousness. ScientificWorldJournal 2004;4:347-61.
[22] Ventegodt S, Morad M, Merrick J. Clinical holistic medicine: Prevention through healthy lifestyle and quality of life. Oral Health Prev Dent 2004;1:239-45.
[23] Ventegodt S, Morad M, Hyam E, Merrick J. Clinical holistic medicine: When biomedicine is inadequate. ScientificWorldJournal 2004;4:333-46.
[24] Ventegodt S, Morad M, Merrick J. Clinical holistic medicine: Holistic treatment of children. ScientificWorldJournal 2004;4:581-8.
[25] Ventegodt S, Morad M, Merrick J. Clinical holistic medicine: Problems in sex and living together. ScientificWorldJournal 2004;4:562-70.
[26] Ventegodt S, Flensborg-Madsen T, Andersen NJ, Morad M, Merrick J. Clinical holistic medicine: A pilot study on HIV and quality of life and a suggested treatment of HIV and AIDS. ScientificWorldJournal 2004;4:264-72.
[27] Ventegodt S, Morad M, Merrick J. Clinical holistic medicine: Induction of spontaneous remission of cancer by recovery of the human character and the purpose of life (the Life Mission). ScientificWorldJournal 2004;4:362-77.
[28] Ventegodt S, Morad M, Kandel I, Merrick J. Clinical holistic medicine: Treatment of physical health problems without a known cause, exemplified by hypertension and tinnitus. ScientificWorldJournal 2004;4:716-24.
[29] Ventegodt S, Morad M, Merrick J. Clinical holistic medicine: Developing from asthma, allergy and eczema. ScientificWorldJournal 2004;4:936-42.
[30] Ventegodt S, Morad M, Press J, Merrick J, Shek DTL. Clinical holistic medicine: Holistic adolescent medicine. ScientificWorldJournal 2004;4:551-61.
[31] Ventegodt S, Morad M, Kandel I, Merrick J. Clinical holistic medicine: a psychological theory of dependency to improve quality of life. ScientificWorldJournal 2004;4:638-48.
[32] Ventegodt S, Kandel I, Neikrug S, Merrick J. Clinical holistic medicine: Holistic treatment of rape and incest traumas. ScientificWorldJournal 2005;5:288-97.
[33] Ventegodt S, Merrick J. Clinical holistic medicine: Chronic pain in the locomotor system. ScientificWorldJournal 2005;5:165-72.
[34] Ventegodt S, Merrick J. Clinical holistic medicine: Chronic pain in internal organs. ScientificWorldJournal 2005;5:205-10.
[35] Ventegodt S, Kandel I, Neikrug S, Merrick J. Clinical holistic medicine: The existential crisis – life crisis, stress and burnout. ScientificWorldJournal 2005;5:300-12.
[36] Ventegodt S, Gringols G, Merrick J. Clinical holistic medicine: Holistic rehabilitation. ScientificWorldJournal 2005;5:280-7.
[37] Ventegodt S, Andersen NJ, Neikrug S, Kandel I, Merrick J. Clinical holistic medicine: Mental disorders in a holistic perspective. ScientificWorldJournal 2005;5:313-23.
[38] Ventegodt S, Andersen NJ, Neikrug S, Kandel I, Merrick J. Clinical holistic medicine: Holistic treatment of mental disorders. ScientificWorldJournal 2005;5:427-45
[39] Ventegodt S, Merrick J. Clinical holistic medicine: The patient with multiple diseases. ScientificWorldJournal 2005;5:324-39
[40] Ventegodt S, Clausen B, Nielsen ML, Merrick J. Clinical holistic medicine: Advanced tools for holistic medicine. ScientificWorldJournal 2006;6:2048-65.
[41] Ventegodt S, Clausen B, Merrick J. Clinical holistic medicine: The case story of Anna: I. Long term effect of child sexual abuse and incest with a treatment approach. ScientificWorldJournal 2006;6: 1965-76.
[42] Ventegodt S, Clausen B, Merrick J. Clinical holistic medicine: the case story of Anna. II. Patient diary as a tool in treatment. ScientificWorldJournal 2006;6:2006-34.
[43] Ventegodt S, Clausen B, Merrick J. Clinical holistic medicine: The case story of Anna. III. Rehabilitation of philosophy of life during holistic existential therapy for childhood sexual abuse. ScientificWorldJournal 2006;6:2080-91.

[44] Ventegodt S, Merrick J. Suicide from a holistic point of view.ScientificWorldJournal 2005;5:759-66.
[45] Ventegodt S, Clausen B, Omar HA, Merrick J. Clinical holistic medicine: Holistic sexology and acupressure through the vagina (Hippocratic pelvic massage). ScientificWorldJournal 2006;6: 2066-79.
[46] Ventegodt S, Clausen B, Merrick J. Clinical holistic medicine: Pilot study on the effect of vaginal acupressure (Hippocratic pelvic massage). ScientificWorldJournal 2006;6:2066-79.
[47] Ventegodt S, Thegler S, Andreasen T, Struve F, Enevoldsen L, et al. Clinical holistic medicine: Psychodynamic short-time therapy complemented with bodywork. A clinical follow-up study of 109 patients. ScientificWorldJournal 2006;6:2220-38.
[48] Ventegodt S, Thegler S, Andreasen T, Struve F, Enevoldsen L, et al. Clinical holistic medicine (mindful, short-term psychodynamic psychotherapy complemented with bodywork) in the treatment of experienced impaired sexual functioning. ScientificWorldJournal 2007;7:324-9.
[49] Ventegodt S, Thegler S, Andreasen T, Struve F, Enevoldsen L, et al. Clinical holistic medicine (mindful, short-term psychodynamic psychotherapy complemented with bodywork) improves quality of life, health, and ability by induction of Antonovsky-salutogenesis. ScientificWorldJournal 2007;7:317-23.
[50] Ventegodt S, Thegler S, Andreasen T, Struve F, Enevoldsen L, et al. Clinical holistic medicine (mindful, short-term psychodynamic psychotherapy complemented with bodywork) in the treatment of experienced physical illness and chronic pain. ScientificWorldJournal 2007;7:310-6.
[51] Ventegodt S, Thegler S, Andreasen T, Struve F, Enevoldsen L, et al. Clinical holistic medicine (mindful, short-term psychodynamic psychotherapy complemented with bodywork) in the treatment of experienced mental illness. ScientificWorldJournal 2007;7:306-9.
[52] Ventegodt S, Thegler S, Andreasen T, Struve F, Enevoldsen L, et al. Self-reported low self-esteem. Intervention and follow-up in a clinical setting. ScientificWorldJournal 2007;7:299-305.
[53] Flensborg-Madsen T, Ventegodt S, Merrick J. Sense of coherence and physical health. A Review of previous findings. ScientificWorldJournal 2005;5:665-73.
[54] Flensborg-Madsen T, Ventegodt S, Merrick J. Why is Antonovsky's sense of coherence not correlated to physical health? Analysing Antonovsky's 29-item sense of coherence scale (SOCS). ScientificWorldJournal 2005;5:767-76.
[55] Flensborg-Madsen T, Ventegodt S, Merrick J. Sense of coherence and health. The construction of an amendment to Antonovsky's sense of coherence scale (SOC II ScientificWorldJournal 2006;6:2133-9.
[56] Flensborg-Madsen T, Ventegodt S, Merrick J. Sense of coherence and physical health. A cross-sectional study using a new SOC scale (SOC II). ScientificWorldJournal 2006;6:2200-11.
[57] Flensborg-Madsen T, Ventegodt S, Merrick J. Sense of coherence and physical health. Testing Antonovsky's theory. ScientificWorldJournal 2006;6:2212-9.
[58] Flensborg-Madsen T, Ventegodt S, Merrick J. Sense of coherence and health. The emotional sense of coherence (SOC-E) was found to be the best-known predictor of physical health. ScientificWorldJournal 2006,6. 2147-57.
[59] Ventegodt S, Flensborg-Madsen T, Andersen NJ, Merrick J. Life Mission Theory VII: Theory of existential (Antonovsky) coherence: a theory of quality of life, health and ability for use in holistic medicine. ScientificWorldJournal 2005;5:377-89.
[60] Jones WHS. Hippocrates, Vol. I–IV. London: William Heinemann, 1923-1931.

Section VIII: Understanding evolution

Chaptr XXI

A theory of top-down macro-evolution and metamorphosis

Charles Darwin became famous for suggesting the mechanism of natural selection in his book "The origin of species". As DNA and DNA-mutation were discovered everybody could suddenly understand evolution. But remarkable evolutionary "jumps" like how the fly got its wings are still not accounted for, as DNA and biological form are not directly associated in a simple way - if at all. Another puzzling question address the evolution of non-local consciousness: What is the evolutionary advantage of it? We suggest that consciousness gives organisms freedom to change their form independently of the DNA, thus separating microevolution from macroevolution. We suggest that the development of consciousness gave life the dramatic possibility of metamorphosis through intent from a global, organic level. Darwin's evolutionary theory is "bottom up"; our "evolutionary hypothesis of metamorphosis it "top down". The evolution of the fly's wings is traditionally thought to have happened through natural selection. We do not deny natural selection, but suggest a parallel complementary pattern of "metamorphous top down evolution". Such evolutionary events involve decisions at a level of global information of the organism that controls master genes (possibly e.g. homeo-box and E-box genes). The top down evolution allows the species to assume new forms and organs with the same DNA. Top down evolution may be a main factor in the development of the planet's biodiversity. The fly-wing theory is not a substitute for Darwin's theory of natural selection, but a mechanism for speeding up evolution and allowing evolutionary jumps. Therefore, we think the fly-wing theory could be a very powerful tool for evolutionary scientists to explain the mechanisms of evolution in a general perspective, and also to explain individual mechanisms leading to the creation of varieties between individuals.

Introduction

Nature has through time succeeded in developing an almost infinite number of species, but it has been an inexplicable enigma how such diversity could embellish the surface of the earth. However, in 1907 Charles Darwin published his book, "The origin of species", in which he provided an explanation. He thought that diversity is created through a process, "natural selection", where an individual representing a trait most be fit for survival and able to out-compete other individuals not having such advantage. Darwin also operated with a special kind of natural selection he called "sexual selection". He discovered that the "conscious preference" is a main factor for the natural selection (sexual selection) to operate, because in most cases the females through evolution use their "consciousness" to select a mate. Darwin wrote: "if man can in a short time give elegant carriage and beauty to his bantams, according to his standard of beauty, I can see no good reason to doubt that female birds, by selecting, during thousands of generations, the most melodious or beautiful males, according to their standard of beauty, might produce a marked effect". His model has gained supporters all over the world from scientists to ordinary people. Natural selection is to a certain degree capable of explaining the huge diversity of species on the earth. Nevertheless, we believe that Darwin's theory of natural selection cannot explain such diversity by itself. Therefore, we propose a new hypothesis; the "Metamorphous top down" evolution, involving the development of advantageous traits that makes individuals fit for survival out-compete the others, and thereby participate in evolution of diversity. The idea is that these two kinds of evolutionary events interact with each other giving the scientists a very powerful tool to explain the great diversity of species on the planet earth. Most interestingly, metamorphosis is an ancient metaphor for personal development; many old books on human growth and transformation were up through history called "Books of metamorphosis".

The "fly-wing theory", or "theory of metamorphous top-down evolution"

As Charles Darwin described in his book, "The origin of species", he observed how species at the Galapagos Islands, when isolated, developed characteristic traits peculiar for the Galapagos population. On this, he developed his evolutionary "Bottom up" theory where he described the evolution as a function of "natural selection", involving development of diversity through evolution, in a specific order: Varieties → species → genera → families → higher taxa etc (1). On the other hand, our "Top down" hypothesis, proposes that the evolution of diversity may go in the opposite direction: Higher taxa → phyla → classes → order → species etc. The new hypothesis described in this paper is called "The fly-wing theory", because we think the fly illustrates how wings are developed through metamorphosis (2) in a "Top down" model of development, where the phylogenesis mirrors the ontogenesis (see below). Or, when e.g. the taxa is developed through Darwin's evolutionary "natural selection", we think an evolutionary shift happens that initiates a "Top down" evolution through metamorphosis. In this way, the individual organisms within the taxa, through metamorphosis, can differentiate to manifest the peculiarities that fit best to the needs within a specific habitat or microhabitat making up the daily resort of the concerned individual. Such event we call a "Metamorphous Top down" evolution. Through this, the evolutionary events that implement an evolutionary superior "Top down" evolution, as e.g. classes → order → species etc., may manifest the peculiar traits necessary to distinguish a species or varieties within a species. For example, the order Diptera contains gnats, mosquito and fly. These all have wings. Following "The fly-wing theory" these wings are developed from the division

Holometabola through metamorphosis: Holometabola – metamorphosis (development of wings) → Diptera [gnats, mosquito and fly]. E.g. in the fly such "Metamorphous Top down" development begins in the caterpillar that through metamorphosis develops to an adult fly with wings (2). The fly is chosen as a model for this hypothesis because the fruit fly is a model organism and thereby is one of the most documented existing species (2). However, this theory accounts for all organisms and involves the "Metamorphous Top down" evolutionary event.

"The fly-wing theory" does not exclude Darwin's evolutionary "Bottom up" theory, but combines a developmental history where the evolution goes in both directions through evolutionary interaction (see Figure 1). We think that the ability of a species to adjust to its specific needs by developing peculiar traits during "Metamorphous Top down" evolution, may have evolved during natural selection. This means that a specific genetic background, developed through natural selection may exist in all organisms. Such genetic qualities are proposed in a new hypothesis, "The Butterfly Theory" (manuscript submitted) forecasting the presumed possibility of existence of such genes in all organisms from bacteria to humans (3-5). Master genes like e.g. the homeobox genes and E-box genes are essential for metamorphosis in the fruit fly *Drosophila melanogaster* and conserved through the animal evolution (3-5). A great amount of homeobox master genes are involved in metamorphosis of plants, sponges, insects and animals (5-7). The homeobox gene family is also involved in human embryogenesis (8). All animals that have been examined including humans have at least one *Hox* gene (a homeobox gene) cluster. These genes show strong homology to the corresponding genes in *Drosophila*. In early developmental stages the Human Embryo is hard to distinguish from other mammalian embryos and it is segmented similar to the larva of a fruit fly (2). Such developmental identity indicates that metamorphous processes like those of a fly also could exist in humans (manuscript submitted). Furthermore, the master regulatory E-box genes controlling the circadian feed back loops, and thereby a master controller of cell proliferation, is identified through evolution from cyanobacteria to humans (5). Such genes are very important for the metamorphous events, and make the possibility of "Metamorphous Top down" evolution very probable. Such documentation supports the existence of the "Metamorphous Top dawn" evolution as an evolutionary event happening in all organisms from bacteria to humans (manuscript submitted).

When the fly develops its wings through metamorphosis (2), or e.g. a butterfly develops its adult form in a pupae or a tadpole develops to a toad, both through metamorphosis (9,10), or an adult human goes through metamorphosis to gain its correct adult traits (manuscript submitted), the evolution following "The fly-wing theory" shifts from a Darwinistic "Bottom up" evolution forced by "natural selection", to a "Metamorphous Top down" evolution directed by an inner drive in an organism to gain its most "desired" traits. Such inner drive may direct the master genes mentioned over and in (manuscript submitted), and other main regulatory genes, to master the "Metamorphous Top down" development. An ontogenetic development like this mirrors the phylogenetic development, because it "reverses" the "natural genetic selection" predicted by Darwin, so the single organism within the frame of its species uses its genetic background to express the most desired form needed by the species, or the most desired form within a micro habitat to make varieties within a species leading to the evolution of new species.

Evolutionary jumps could have happened through evolution of all organisms, even bacteria. Bacteria have very complex behavior and morphology, and why not "global level choices", and even consciousness, in its simplest form. Until now the shift between "natural selection" and "Metamorphous top down" evolution is mentioned as driven by "the most desired traits" and "very primitive and instinctive feelings". A plausible explanation for this is that the need for changes caused by environmental factors triggers a biochemical reaction that initiates the genetic control (we could call this a kind of primitive consciousness). This means that the shift between "natural selection" and "Metamorphous top down" development in reality could be directed as an interaction between the natural selection and the environmental needs continuously arising from the demands of survival in a specific habitat or microhabitat. Such a shift can be illustrated by the following formula: varieties -- natural selection → species – natural selection → genera – natural selection → families – natural selection → higher taxa – environmental influence → phyla -- environmental influence → classes -- environmental influence → order -- environmental influence → species -- environmental influence → varieties. Environmental factors of course, can influence at all stages of development, so the formula can be adjusted to the conditions at work in a specific situation and also look like e.g.: varieties -- natural selection → species -- environmental influence → varieties. The environmental powers that direct the evolutionary shift can also be explained by a simple formula: environmental influences → (quantum) biochemical reactions (intent, desire and primitive feelings) → genetic control → metamorphosis (see Figure 1).

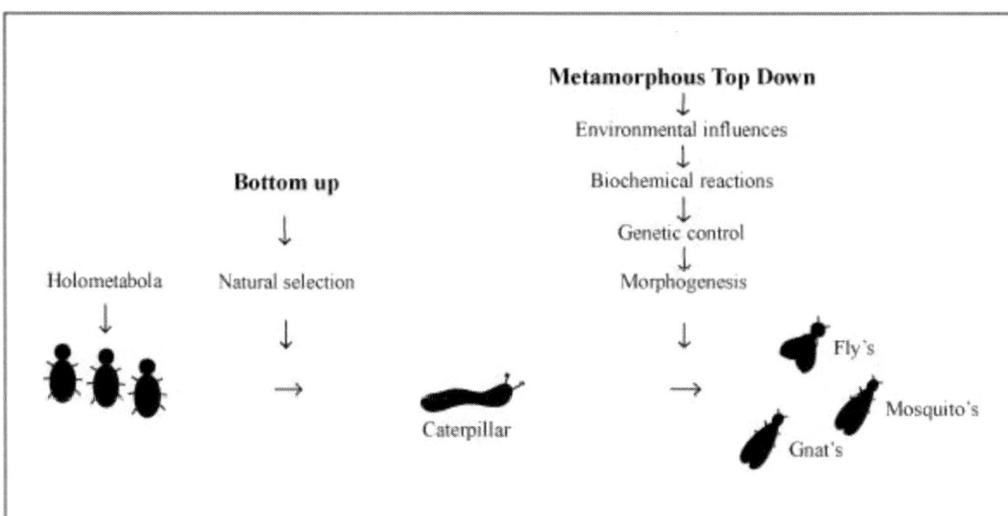

Figure 1. The figure illustrates the interaction between naturally selection and "Metamorphous Top down" development. Through "natural selection" individuals in a taxon, class order etc. (here illustrated by holometabola), can develop varieties in the population. Through "Metamorphous Top down" evolution a varying member of a population can develop its wings.

Discussion

In our long journey of exploring the scientific basis of biology and medicine we have come to analyse evolution. The fundamental problem seems to be what is happening in the living organism on an informational level. We know that proteins have interesting self-organizing features; we also know that a glass of even the smartest molecules will not organize themselves into a living being. The living being has consciousness and global order, and it seems that these are strangely connected. Only by a deeper insight in the nature of consciousness itself can we fully understand biological order, evolution and ontogenesis.

Such functions are really not so difficult to understand, because both "natural selection" and the "Metamorphous "Top down" evolution, in reality, have are both of the same nature. They can be described by the same formula, and thereby the shift only describes a reverse of the same function. This can be understood as: when natural selection rules, the evolution goes from: a single variety → lots of varieties → lots of species → lots of genera → lots of families → lots of taxa, and when metamorphous selection rules, the evolution goes from: a single taxon → lots of phyla → lots of classes → lots of orders → lots of species → lots of varieties. This shows that nature is directed by very simple rules that decide the variegation between all living organisms. This is a hypothesis that we believe may help us understand some of evolutionary phenomena that are hard to explain with natural selection alone. We hope that this will start a dialogue and more research on the subject.

References

[1] Darwin C. The origin of species. New York: Gramercy, 1979.

[2] Alberts B, Roberts K, Lewis J, Raff M, Walter P, Johnson A. Molecular biology of the cell, 4th ed. New York: Garland, 2002.

[3] Walter J, Gehring WJ. Master control genes in development and evolution: The homeobox story. New Haven, CT: Yale Univ, 1998.

[4] Stauber M, Jäckle H, Schmidt-Ott U. The anterior determinant bicoid of Drosophila is a derived Hox class 3 gene. Proc Natl Acad Sci USA 1999;96(7):3786-9.

[5] Hardin PE. Transcription regulation within the circadian clock: the E-box and beyond. J Biol Rhythms 2004;19:348-60.

[6] Seimiya M, Naito M, Watanabe Y, Kurosawa Y. Homeobox genes in the freshwater sponge Ephydatia fluviatilis. Prog Mol Subcell Biol 1998;19:133-55.

[7] Ito T. Transcriptional network during flower development: from homeotic genes to genes controlling morphogenesis in Arabidopsis. Tanpakushitsu Kakusan Koso 2005;50:228-38.

[8] Darras VM, Hume R, Visser TJ. Regulation of thyroid hormone metabolism during fetal development. TJ Mol Cell Endocrinol 1999;151:37-47.

[9] North G, French V. Insect wings. Patterns upon patterns. Curr Biol 1994;4:611-4.

[10] Paul BD, Buchholz DR, Fu L, Shi YB. Tissue and gene specific recruitment of steroid receptor coactivator-3 by thyroid hormone receptor during development. J Biol Chem 2005;280:27165-172.

Chapter XXII

Induction of Antonovsky salutogenesis can result in spontaneous healing and recovery

The spontaneous healing and recovery (salutogenesis) of patients have been observed by many physicians and medical researchers. Often the patients spontaneous healing from metastatic cancer, schizophrenia and other "incurable" diseases was induced by loving care combined with the patient's deep, existential self-exploration. In this chapter we present our final contribution to a psychobiological theory we hope can explain the power of salutogenesis: the (sub)conscious control of master genes (and protein transcription) related to metamorphosis.

We document the evolutionary connection between genes and proteins involved in insect and amphibian metamorphosis and human embryonic development (ontogenesis), and hypothesize that these may be responsible also for spontaneous self-healing in humans. Some of these genes and proteins imply a genetic master control that is similar in most organisms. We believe that a link between such genes and self-recovery exists in all organisms from bacteria to humans.

Aaron Antonovsky (1923-1994) discovered that salutogenesis could be induced by recovering the patient's sense of coherence (SOC) and we have seen several clinical examples of how SOC is followed by spontaneous recovery from physical and mental disease i.e. spontaneous remission of cancer. Philosophy of life, personality, character, purpose of life, self-esteem, quality of life, self-evaluated mental and physical health and ability to express and use personal talents seems to be radically improved, and soon after this we see the objective recovery.

One such case with spontaneous healing from cancer after clinical holistic therapy documented through PET-scan and CT-scan is presented. Salutogenesis might be controlled by the human genes for metamorphosis; if such genes are controlled by our (sub)consciousness this could explain the often-dramatic placebo effect used for cure in clinical holistic medicine and CAM.

Introduction

Spontaneous remission from cancer and other health conditions often happens, when quality of life is radically improved (1) and spontaneous recovery from schizophrenia can some time happen, when the patient for example meets a partner, finds love and improvement in quality of life (QOL) (2). Such phenomena are only a few examples of what is now called spontaneous healing (salutogenesis) and spontaneous remission and it has to this day defied scientific explanation. During the last decade our research team has studied quality of life and its relation to health and ability and recently we concluded an analysis of several large population studies conducted at the Quality of Life Research Center in Copenhagen, that consciousness - more specifically our positive or negative attitudes towards life, self and the world - seems to determine a person's level of health, QOL and ability (3). Using this understanding we have supported a number of patients in developing their consciousness and it seems that we have been able to induce spontaneous healing and even spontaneous recovery of cancer (4,5), mental illness (6,7), including schizophrenia (8).

Spontaneous self-recovery and human metamorphosis

In our research we have been using a combination of the most efficient therapeutic techniques known today from bodywork, psychotherapy, scientific CAM, and life-philosophical intervention strategies (9-31). The purpose has been to recover the patient's sense of coherence (SOC) according to the idea of Aaron Antonovsky (1923-1994), who found SOC crucial for inducing salutogenesis (32-37). Surprisingly some of our patients, using such techniques, have been able to go into such deep personal transformations that not only their minds, but also their bodies have started to change, reshape and heal (38).

The new matter of personal transformation we have observed is so radical that we have named it "adult human metamorphosis", because this is exactly what it looks like (see Figure 1) (39-41). The typical patient has often been arrested in psychosexual development, and presents him- or more often herself in an immature, irresponsible and often physically unappealing and even repelling way. Not only the body is sick or disproportioned in these patients, but also their feelings, mind, spirit, sexuality, and their ability to love themselves and others.

Many aspects of their human nature are still in a juvenile state in spite of their adult age, very much like a butterfly's larva that does not change shape, but stays in its early juvenile form, until the day it finally transforms into a pupa, and all the organs and characteristics of the larva melts down to allow the adult animal to appear. In a similar fashion the patient melts down in the therapeutic healing crisis we call "metamorphosis" (39-41) and the person let go of all mental and behavioural patterns and turns completely introvert. During the next days or weeks the patient enter what looks somewhat like a psychosis, with very little verbal capacity, disorganised thinking and very little ability; the patient is keeping control of reality so technically he/she is not psychotic in spite of the symbolic and often strange content of the consciousness (the "reality check" is preserved). Some patients are only a little disintegrated, while others are completely absorbed in the human transition from their old version of themselves to the new and improved one. In the middle of the treatment process a new purpose of life appears (42-46) and the patients normally experience extreme hope for the future and joy.

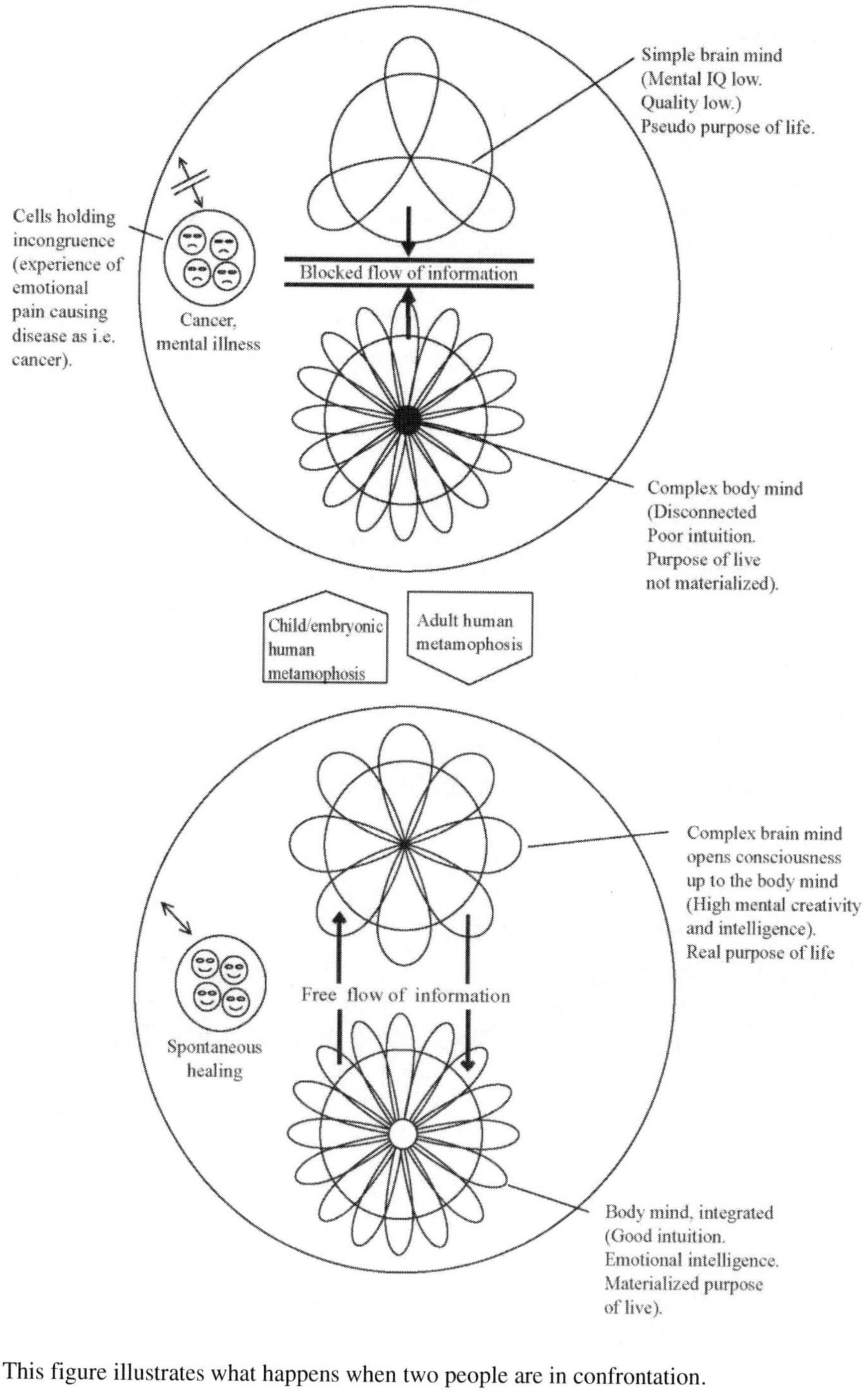

Figure 1. This figure illustrates what happens when two people are in confrontation.

After this important event the patien starts to reorganise his/her mind and a series of new thoughts and behavioural patterns seem to be recruited from a purely internal source, as in a dreamlike state. When the patient's consciousness and whole existence is finally reorganised and solid, he or she again turns outwards, re-appearing in the world as a complete new person. Philosophy of life, personality, character (45), purpose of life, self-esteem, quality of life (47), self-evaluated mental and physical health and ability to express and use personal talents (43) seem to be radically improved. What really happened seems to be the patient regaining coherence between life inside and the world outside - what we normally call existential coherence. In a process similar to a butterfly's metamorphosis from larva to adult individual, the patient goes from an arrested, immature state of being to an adult, mature state. In the most radical of processes, which seem to actually show the essence of the process of spontaneous healing, the process reveals itself as exactly that: adult human metamorphosis.

We believe that humans, like many other vertebrates, have kept the ability to go though metamorphosis. In fact we think that metamorphosis is an evolutionary event (48) and that human metamorphosis is controlled by our (sub)consciousness. We believe the process to be not only similar, but biologically identical to the metamorphosis we see in so many other animal species. A proof of that seems to be that not only mind and experience of life, but also often the physical body and its whole "energy" is being transformed. We have seen a number of times that female breasts, which stopped their growth early in puberty started to grow again, the labia minora of the female vulva have reshaped and grown smaller much to the patients surprise and satisfaction. We have seen female hips, thighs, and abdomens having reshaped into much more adult and sexually mature forms in this process. We have even seen a jaw reshape correcting very bad teeth during intensive healing sessions. Many therapists working with clinical holistic medicine have seen these shifts in body shape, and what we have witnessed is very surprising to us as we normally consider the body as a highly stable and conservative structure. Much more research with video documentation is needed to establish the biological limits of human body-reshaping during the process of salutogenesis.

So it seems that the pattern of adult human metamorphosis we have found might be a key to understanding the biology of spontaneous healing (salutogenesis). In this deepest and most radical of the processes the patient abandons completely the backbone of the mind in order to find the purpose of life (42-46). When we in early childhood are deeply traumatised i.e. by violence or sexual abuse we sometimes give up on our primary purpose of life and reorganise our existence and consciousness around a second less difficult and more manageable purpose of life (42). This transformation might even correspond to a different evolutionary strategy than Darwin's "bottom-up" theory of natural selection that could be called "metamorphous top down" evolution using consciousness as a vehicle for evolution (49).

A case story

The patient was a 45-year-old male Caucasian. He started in clinical holistic therapy (5) the very same day he was diagnosed with non-Hodgkin B-Lymphoma. His Ultrasound and CT-scan showed retroperitoneal tumors and his needle biopsy showed highly pathological cells (T06002 (marrow from crista,) M96003 (malignant lymphoma), M96103 (B-Lymphoma), P30990 (needle biopsy) P3A070 (enzyme histochemical examination) P3B000 (immune histochemical examination). He was in a very poor condition clinically, unable to work, unable to sleep due to constant fever, his self-assessed quality of life was low, his self-assessed mental health was low and he had severe problems with his partner including sexual problems. He was completely caught in the dark side and was tormented by the most evil obsessive fantasies about cutting a woman into pieces with a sharp knife. Every relation he has to other people seemed to be bad and he had big problems relating to self and to the surrounding world. He was most definitely low in his *sense of coherence*. The therapy focused on helping him back into contact with his self and other people. It took three month of intensive therapy to make this happen. He entered a huge personal crisis, where he mentally returned to childhood, cried constantly for a period of time as he re-experienced how he as a child was treated badly by both his father and mother. During the therapy he confronted his childhood reality and learned how he took all his parent's problems in (called "introjections" in the therapy) and since then lived from that crippled existential position, spoiling his life for so many years. This process is called "spontaneous regression" in clinical holistic medicine and it happens when a patient who intents to heal finally gets the resources necessary for the healing process. After the healing crisis (metamorphosis) he felt better, not feeling sick anymore and started quite suddenly to work again.

In the meantime the medical center established his exact diagnosis and had the chemotherapeutic treatment ready and he just needed a final check up to establish the tumor burden and exact health status before initiating the treatment. But the scan now was normal and the patient told us that the hospital doctor made a big cry out of surprise and could not believe that the patient's cancer had suddenly gone. The PET scan showed normal conditions of the metabolism and no area suspected for cancer. The conclusion from the diagnostic consultant reads: "Compared with the earlier CT scan (three month earlier) there is significant remission of earlier seen pathologically enlarged retro peritoneal lymphatic glands. No sign of changes due to malignancy. The CT-Scan shows complete remission of the retroperitoneal tumors found before".

The patient continued follow-up at the medical center, but the cancer did not reappear. Most interestingly the patient told the holistic therapist (SV) at that point that he now had gotten a completely new life – "as if he was part of a completely new movie". Now he was able to love and be close and intimate to other people. He was happy with his girlfriend – whom he hated before and wanted to separate from. He felt good about himself and at home in the world.

Self-healing as evolutionary basis for metamorphosis

The ability for self-healing (spontaneous salutogenesis) has presumably been necessary for survival all the way back to the bacteria; the eucariote cell has a well-known, quite extreme capacity for self-healing. During mitosis, if one of the microtubules that fixate the chromosomes is cut with a laser beam, the mitosis is completely arrested, and the cell enters a phase of spontaneous self-healing, identifying and repairing the destroyed molecule. First when the fibre is regenerated the mitosis will continue as reported by Henrik Flyvbjerg and co-workers (50). This ability to heal and recruit the biological information necessary for re-establishing the cell is also known in sponges, where the disorganisation of cells leads to a healing phase and the biological information of the form of the sponge is recruited. If the ability to self-heal presumably is a genetically programmed faculty originating from some of the most primitive organisms, the bacteria, it is easy to imagine the ability to recruit more complex packages of biological information allowing complete reorganisation and metamorphosis built on these early genes. Interestingly it seems that there are some genes that might be responsible for such functions even in the cyanobacteria (51). The transcriptional feedback loop's main-regulatory clock proteins of circadian rhythms are identified through evolution from cyanobacteria to humans. This implies a main-genetic control, similar in most organisms from bacteria to humans (51).

Genes and proteins possibly involved in human metamorphosis

We propose that genes involved in metamorphosis of animals (such as the butterfly caterpillar transformation to butterfly and the amphibian *Xenopus laevis* transformation from tadpole to toad) are also involved in human metamorphosis. This could be the master genes, such as the huge amounts of homeobox master genes known to participate in development and metamorphosis in insects and animals throughout the evolution (52), which are also involved in human development (53). The homeobox transcription factors are essential for the embryogenesis and thought to manage the regenerative cell differentiation in organisms (54).

The *Drosophila* homeobox genes are essential for metamorphosis in *Drosophila* and they are conserved through the animal evolution (55). We think that the homeobox genes and other master genes could be important candidates for mastering the development of human morphogenesis. For instance the *MyoD* family of genes include master regulatory E-box binding proteins able to convert a fibroblast to a myoblast (56). When the *MyoD1* is expressed in high concentrations in fibroblasts, these are able to subvert the gene control and convert to a muscle cell (57). MyoD transcription factors play a main role in human myoblast differentiation (56). In *Drosophila* a member of the *myoD* family, *Nautilus*, is involved in muscle differentiation of embryos (58). Proteins such as E-box and related master regulatory proteins are involved in the control of regulatory feed back loops of circadian rhythms through evolution, from cyanobacteria to humans (51). Haudin concluded: "Sequence comparisons reveal that oscillator components are not well conserved between animals, plants, fungi, and bacteria. Nevertheless, the mechanism underlying circadian oscillator functions has been conserved throughout phylogeny as one or more molecular feedback loops in gene expression". Other regulatory master genes as the *Drosophila eyeless* gene (a *Pax6* gene), that today make a model system in defining the multistage roles of the *Pax6* gene in eye and olfactory development throughout evolution (59) has a human pendant, *Aniridia* (60). We think that such master regulatory genes and master regulatory proteins could play an important role in human metamorphosis (48).

Also Amphibians have master regulatory genes and proteins evolutionarily related to human equivalents. The thyroid hormone receptors (TRs) functions as master control factors in amphibians and humans (61). The amphibian metamorphosis depends on thyroid hormones (THs) and TRs. The TRs are able to repress or activate gene activation when ligands bind

(62). This regulation is dependent on the presence of co-repressors and co-activators – as e.g. the steroid receptor co-activator family (SRCs). E.g. SRC3 is shown to be involved in a gene/tissue specific development interacting with TR during tadpole metamorphosis (63) – in the presence or absence of THs. The TRs are also expressed during human development (64).

In amphibians the TH family T1-4 induces gene expression during transformation and metamorphosis (63,65). In humans T1-3 are active, T1 in various tissues and T3 in liver and HeLa cells. The human thyroid hormones are present through endogenous development of the human fetus (64).

The SRC family SRC1/SRC2/SRC3 is traditionally co-activating hormones involved in development. In amphibians their genes function as adaptors to recruit other co-activators such as e.g. p300/CPB (64). The *SRC3* gene is expressed in all analysed tadpole organs that undergo extensive transformation through amphibian metamorphosis (63,65). In humans the *SRC1-3* genes function as a co-activator for nuclear receptor mediated transcriptional activation involved in gene regulation see (66,67).

Another family of mastering genes involved in human development and amphibian metamorphosis are the genes coding for the aryl hydrocarbon receptor nuclear translocators ARNT1 and ARNT2, transcription factors mediating the cellular response to various developmental signals in humans and amphibians (68,69). In humans, master genes as *T-bet*, *GATA-3* and *Foxp3* control the immune response. These genes are master regulators of the T cells (70,71). Such regulatory master genes could be activated during human metamorphosis through spontaneous self-healing. In this way the immune system could be triggered to attack aggressive cancer cells resulting in spontaneous healing. In the same way, master genes that control the brain development as the *Hox* genes, and others, could gain the recovery from schizophrenia. *Hox* genes are homeobox genes preserved through animal evolution and found in most animals controlling development of body and central nervous system (72-74).

The participation of such master genes and proteins in development and morphogenesis through the evolution makes the hypothesis of human metamorphosis very likely, because these could be linked to direction of subconscious spontaneous self-healing in human beings through regulation of all genes needed for spontaneous self-recovery in any cases such as hormone genes, hormone receptor genes, and all other genes needed for development and metamorphosis.

To gain control over master genes and proteins, and the process of cell division, it is necessary to be able to control the DNA and other cell structures important for cell segregation through the cell division in order to unwind and supercoil areas of the DNA for the accessibility of single stranded DNA and DNA replication. These processes are directed by the helicases that unwind stable duplex DNA (75) and the topoisomereses that supercoil the DNA to gain stable duplex DNA (76). Furthermore the main proteins for chromosome segregation SMC-containing cohesin, condensin, APC/C ubiquitin ligase, securin-separase complex, kinetochore microtubule destabilizers or regulators (77) are important for gaining control over the process of cell segregation, and thereby metamorphosis. These main proteins, and others, of cell control play a very important role in most aspects of nucleic acid metabolism and cell division (75-77). Thus, it is easy to recognize that metamorphosis is a very complicated process that includes a lot of coordination through regeneration and self-recovery all necessary to be able to gain control over the process of metamorphosis.

Of course all the regulatory master genes and proteins presented in this paper are only examples of genes and proteins possibly involved in human metamorphosis. Lots of other genes and proteins could likewise be candidates, but we have chosen only to give some examples in this paper, in order to illustrate the possibility of this hypothesis and to explain these phenomena. We believe that a human metamorphosis can take place corresponding to the metamorphosis of a tadpole to a toad, a metamorphosis involving spontaneous self-healing of human beings through (sub)conscious control of master genes.

Conclusions

We have observed a new pattern of intense spontaneous, existential healing (salutogenesis) in humans with so many similarities to the metamorphosis of animals, that we expect that we have observed adult human metamorphosis. The global transformation of body and mind is so radical that it seems to be able to explain even spontaneous remission of cancer and recovery in schizophrenia. We suggest that the phenomenon has a genetic basis in the many master genes and genes coding for proteins controlling cell segregation, all necessary for metamorphosis, preserved in the human genome. We have suggested some of the most likely master regulatory genes and proteins of human spontaneous self-healing and think there is a possible link between master genes and human metamorphosis as a plausible hypothesis. Many animals enter metamorphosis when it is beneficial for survival and self-healing obviously is. Human transformation might be very useful for survival if it allows the patient to draw on genetic potentials to recover from disease or allows the child to adapt to a rough childhood to come-back into living fully as an adult when the danger is over. Such processes are exactly what can be observed through the clinical sessions, where most of the patients entering physical changes (metamorphosis) have been either extremely sick or sexually or violently abused as children.

Could the spontaneous self-healing in humans, possibly related to animal metamorphosis, be an evolutionary trait that stems from the self-healing systems of unicellular organisms? We think so. We hope that future research should concentrate on finding a link between genes and metamorphosis causing spontaneous self-healing, in all organisms from bacteria to humans. This connection indicates that genetic control of all cases of self-recovery described in this paper and all other cases could be mastered during the action of master genes and genes necessary to gain cell control, controlled through the human (sub)consciousness.

References

[1] Spiegel D, Bloom JR, Kraemer HC, Gottheil E. Effect of psychosocial treatment on survival of patients with metastatic breast cancer. Lancet 1989;2:888-91.

[2] Jørgensen P. Recovery and insight in schizophrenia. Acta Psychiatr Scand 1995;92:436-40.

[3] Ventegodt S, Flensborg-Madsen T, Andersen NJ, Nielsen M, Morad M, Merrick J. Global quality of life (QOL), health and ability are primarily determined by our consciousness. Research findings from Denmark 1991–2004. Soc Indic Res 2005;71:87-122.

[4] Ventegodt S, Morad M, Hyam E, Merrick J. Clinical holistic medicine: induction of spontaneous remission of cancer by recovery of the human character and the purpose of life (the life mission). ScientificWorldJournal 2004;4:362-77.

[5] Ventegodt S, Solheim E, Saunte ME, Morad M, Kandel I, Merrick, J. Clinical holistic medicine: Metastatic cancer. ScientificWorldJournal 2004;4:913-35.
[6] Ventegodt S, Andersen NJ, Neikrug S, Kandel I, Merrick J. Clinical holistic medicine: holistic treatment of mental disorders. ScientificWorldJournal 2005;5:427-45.
[7] Ventegodt S, Andersen NJ, Neikrug S, Kandel I, Merrick J. Clinical holistic medicine: Mental disorders in a holistic perspective. ScientificWorldJournal 2005;5:313-23.
[8] Ventegodt S, Kandel I, Merrick J. Clinical holistic medicine (mindful short term psychodynamic psychotherapy complimented with bodywork) in the treatment of schizophrenia (ICD10-F20) and other psychotic mental diseases (ICD10-F2/DSM-IV-code 295). ScientificWorldJournal 2007;7:1987-2008.
[9] Ventegodt S, Merrick J. Clinical holistic medicine: Applied consciousness-based medicine. ScientificWorldJournal 2004;4:96-9.
[10] Ventegodt S, Morad M, Merrick J. Clinical holistic medicine: Classic art of healing or the therapeutic touch. ScientificWorldJournal 2004;4:134-47.
[11] Ventegodt S, Morad M, Merrick J. Clinical holistic medicine: The "new medicine", the multi-paradigmatic physician and the medical record. ScientificWorldJournal 2004;4:273-85.
[12] Ventegodt S, Morad M, Merrick J. Clinical holistic medicine: Holistic pelvic examination and holistic treatment of infertility. ScientificWorldJournal 2004;4:148-58.
[13] Ventegodt S, Morad M, Hyam E, Merrick J. Clinical holistic medicine: Use and limitations of the biomedical paradigm ScientificWorldJournal 2004;4:295-306.
[14] Ventegodt S, Morad M, Kandel I, Merrick J. Clinical holistic medicine: Social problems disguised as illness. ScientificWorldJournal 2004;4:286-94.
[15] Ventegodt S, Morad M, Andersen NJ, Merrick J. Clinical holistic medicine Tools for a medical science based on consciousness. ScientificWorldJournal 2004;4:347-61.
[16] Ventegodt S, Morad M, Merrick J. Clinical holistic medicine: Prevention through healthy lifestyle and quality of life. Oral Health Prev Dent 2004;1:239-45.
[17] Ventegodt S, Morad M, Hyam E, Merrick J. Clinical holistic medicine: When biomedicine is inadequate. ScientificWorldJournal 2004;4:333-46.
[18] Ventegodt S, Morad M, Merrick J. Clinical holistic medicine: Holistic treatment of children. ScientificWorldJournal 2004;4:581-8.
[19] Ventegodt S, Morad M, Merrick J. Clinical holistic medicine: Problems in sex and living together. ScientificWorldJournal 2004;4:562-70.
[20] Ventegodt S, Flensborg-Madsen T, Andersen NJ, Morad M, Merrick J. Clinical holistic medicine: A pilot on HIV and quality of life and a suggested treatment of HIV and AIDS. ScientificWorldJournal 2004;4:264-72.
[21] Ventegodt S, Morad M, Merrick J. Clinical holistic medicine: Induction of spontaneous remission of cancer by recovery of the human character and the purpose of life (the Life Mission). ScientificWorldJournal 2004;4:362-77.
[22] Ventegodt S, Morad M, Kandel I, Merrick J. Clinical holistic medicine: Treatment of physical health problems without a known cause, exemplified by hypertension and tinnitus. ScientificWorldJournal 2004;4:716-24.
[23] Ventegodt S, Morad M, Merrick J. Clinical holistic medicine: Developing from asthma, allergy and eczema. ScientificWorldJournal 2004;4:936-42.
[24] Ventegodt S, Morad M, Press J, Merrick J, Shek DTL. Clinical holistic medicine: Holistic adolescent medicine. ScientificWorldJournal 2004;4:551-61.
[25] Ventegodt S, Morad M, Kandel I, Merrick J. Clinical holistic medicine: a psychological theory of dependency to improve quality of life. ScientificWorldJournal 2004;4:638-48.
[26] Ventegodt S, Kandel I, Neikrug S, Merrick J. Clinical holistic medicine: Holistic treatment of rape and incest traumas. ScientificWorldJournal 2005;5:288-97.
[27] Ventegodt S, Merrick J. Clinical holistic medicine: Chronic pain in the locomotor system. ScientificWorldJournal 2005;5:165-72.
[28] Ventegodt S, Merrick J. Clinical holistic medicine: Chronic pain in internal organs. ScientificWorldJournal 2005;5:205-10.

[29] Ventegodt S, Kandel I, Neikrug S, Merrick J. Clinical holistic medicine: The existential crisis – life crisis, stress and burnout. ScientificWorldJournal 2005;5:300-12.
[30] Ventegodt S, Gringols G, Merrick J. Clinical holistic medicine: Holistic rehabilitation. ScientificWorldJournal 2005;5:280-7.
[31] Ventegodt S, Merrick J. Clinical holistic medicine: The patient with multiple diseases. ScientificWorldJournal 2005;5:324-39.
[32] Flensborg-Madsen T, Ventegodt S, Merrick J. Sense of coherence and physical health. A Review of previous findings. ScientificWorldJournal 2005;5:665-73.
[33] Flensborg-Madsen T, Ventegodt S, Merrick J. Why is Antonovsky's sense of coherence not correlated to physical health? Analysing Antonovsky's 29-item sense of coherence scale (SOCS). ScientificWorldJournal 2005;5:767-76.
[34] Flensborg-Madsen T, Ventegodt S, Merrick J. Sense of coherence and health. The construction of an amendment to Antonovsky's sense of coherence scale (SOC II). ScientificWorldJournal 2006;6:2133-9.
[35] Flensborg-Madsen T, Ventegodt S, Merrick J. Sense of coherence and physical health. A cross-sectional study using a new SOC scale (SOC II). ScientificWorldJournal 2006;6:2200-11.
[36] Flensborg-Madsen T, Ventegodt S, Merrick J. Sense of coherence and physical health. Testing Antonovsky's theory. ScientificWorldJournal 2006;6:2212-9.
[37] Flensborg-Madsen T, Ventegodt S, Merrick J. Sense of coherence and health. The emotional sense of coherence (SOC-E) was found to be the best-known predictor of physical health. ScientificWorldJournal 2006;6:2147-57.
[38] Ventegodt S, Andersen NJ, Merrick J. Holistic Medicine III: The holistic process theory of healing. ScientificWorldJournal 2003;3:1138-46.
[39] Ventegodt S, Clausen B, Merrick J. Clinical holistic medicine: The case story of Anna: I. Long term effect of child sexual abuse and incest with a treatment approach. ScientificWorldJournal 2006;6: 1965-76.
[40] Ventegodt S, Clausen B, Merrick J. Clinical holistic medicine: the case story of Anna. II. Patient diary as a tool in treatment. ScientificWorldJournal 2006;6:2006-34.
[41] Ventegodt S, Clausen B, Merrick J. Clinical holistic medicine: The case story of Anna. III. Rehabilitation of philosophy of life during holistic existential therapy for childhood sexual abuse. ScientificWorldJournal 2006;6:2080-91.
[42] Ventegodt S. The life mission theory: A theory for a consciousness-based medicine. Int J Adolesc Med Health 2003;15:89-91.
[43] Ventegodt S, Andersen NJ, Merrick J. The life mission theory III. Theory of talent. ScientificWorldJournal 2003;3:1286-93.
[44] Ventegodt S, Andersen NJ, Merrick J. The life mission theory V. Theory of the anti-self (the shadow) or the evil side of man. ScientificWorldJournal 2003;3:1302-13.
[45] Ventegodt S, Andersen NJ, Merrick J. The life mission theory VI: A theory for the human character. ScientificWorldJournal 2004;4:859-88.
[46] Ventegodt S, Flensborg-Madsen T, Andersen NJ, Merrick J. The life mission theory VII. Theory of existential (Antonovsky) coherence: a theory of quality of life, health, and ability for use in holistic medicine. ScientificWorldJournal 2005;5:377-89.
[47] Ventegodt S, Merrick J, Andersen NJ. Quality of life theory I. The IQOL theory: an integrative theory of the global quality of life concept. ScientificWorldJournal 2003;3:1030-40.
[48] Hermansen TD, Ventegodt S, Merrick J. Human Development X: Explanation of Macroevolution — Top-Down Evolution Materializes Consciousness. The Origin of Metamorphosis. ScientificWorldJournal 2006;6:1656–66.
[49] Hermansen TD, Ventegodt S, Kandel I, Merrick J. Human development XXI: How the fly got its wings. The "fly-wing theory" explains how top down evolution gave the fly its wings, and how the Self's intent materializes biological order and body-form during fylogenesis and ontogenesis (metamorphosis). Int J Child Health Hum Dev, in press.
[50] Ventegodt S. Livsfilosofi der helbreder. Copenhagen: Forskningscenterets Forlag, 1999. [Danish]
[51] Hardin PE. Transcription regulation within the circadian clock: the E-box and beyond. J Biol Rhythms 2004;19:348-60.

[52] Walter J, Gehring WG. Master control genes in development and evolution: The homeobox story. New Haven, CT: Yale Univ Press, 1998.
[53] Arcioni L, Simeone A, Guazzi S, Zappavigna V, Boncinelli E, Mavilio F. The upstream region of the human homeobox gene HOX3D is a target for regulation by retinoic acid and HOX homeoproteins. EMBO J 1992;11:265-77.
[54] Hogeveen KN, Sassone-Corsi P. Homeobox galore: when reproduction goes RHOX and roll. Cell 2005;120:287-8.
[55] Stauber M, Jäckle H, Schmidt-Ott U. The anterior determinant bicoid of Drosophila is a derived Hox class 3 gene. Dev Biol 1999;96:3786-9.
[56] Jane DT, Morvay LC, Koblinski J, Yan S, Saad FA, et al. Evidence that E-box promoter elements and MyoD transcription factors play a role in the induction of cathepsin B gene expression during human myoblast differentiation. Biol Chem 2002;383:1833-44.
[57] Alberts B, Roberts K, Lewis J, Raff M, Walter P, Johnson A. Molecular biology of the cell, 4th ed. New York: Garland, 2002.
[58] Michelson AM, Abmayr SM, Bate M, Arias AM, Maniatis T. Expression of a MyoD family member prefigures muscle pattern in Drosophila embryos. Genes Dev 1990;4:2086-97.
[59] Pan D, Rubin GM. Targeted expression of teashirt induces ectopic eyes in Drosophila. Proc Natl Acad Sci USA 1998;95:15508-12.
[60] Heyningen V, Williamson KA. PAX6 in sensory development. Hum Mol Genet 2002;11:1161-7.
[61] Wolffe AP, Shi Y-B. A hypothesis for the transcriptional control of amphibian metamorphosis bythe thyroid hormone receptor. Am Zoologist 1999;39:807-17.
[62] Paul BD, Shi Y-B. Distinct expression profiles of transcriptional coactivators for thyroid hormone receptors during Xenopus laevis metamorphosis. Cell Res 2003;13:459-64.
[63] Paul BD, Buchholz DR, Fu L, Shi Y-B. Tissue- and gene-specific recruitment of steroid receptor coactivator-3 by thyroid hormone receptor during development. J Biol Chem 2005;280:27165-72.
[64] Darras VM, Hume R, Visser TJ. Regulation of thyroid hormone metabolism during fetal development. Mol Cell Endocrinol 1999;151:37-47.
[65] Paul BP, Fu L, Buchholz DR, Shi Y-B. Coactivator recruitment is essential for liganded thyroid hormone receptor to initiate amphibian metamorphosis. Mol Cell Biol 2005;25: 5712-24.
[66] Moore JMR, Galicia SJ, McReynolds AC, Nguyen N-H, Scanlan TS, Guy RK. Quantitative Proteomics of the Thyroid Hormone Receptor-Coregulator Interactions. J Biol Chem 2004;279:27584-90.
[67] Alper O, Bowden ET. Novel insights into c-Src : Anti-cancer drugs. Curr Pharm Des 2005;11:1119-30.
[68] Rowatt AJ, DePowell JJ, Powell WH. ARNT gene multiplicity in amphibians: characterization of ARNT2 from the frog Xenopus laevis. J Exp Zoolog B Mol Dev Evol 2003;300:48-57.
[69] Stolze I, Berchner-Pfannschmidt U, Freitag P, Wotzlaw C, Rössler J, et al. Hypoxia-inducible erythropoietin gene expression in human neuroblastoma cells. Blood 2002;100: 2623-8.
[70] Ho IC, Vorhees P, Marin N, Oakley BK, Tsai SF, Orkin SH, Leiden JM. Human GATA-3: a lineage-restricted transcription factor that regulates the expression of the T cell receptor alpha gene. EMBO J 1991;10:1187-92.
[71] Neurath MF, Weigmann B, Finotto S, Glickman J, Nieuwenhuis E, et al. The transcription factor T-bet regulates mucosal T cell activation in experimental colitis and Crohn's disease. J Exp Med 2002;195:1129-43.
[72] Hogeveen KN, Sassone-Corsi P. Homeobox galore: when reproduction goes RHOX and roll. Cell 2005;120:287-8.
[73] Gaufo GO, Wu S, Capecchi MR. Contribution of Hox genes to the diversity of the hindbrain sensory system. Development 2004;131:1259-66.
[74] Capecchi MR. Hox genes and mammalian development. Cold Spring Harb Symp Quant Biol 1997;62:273-81.
[75] Tuteja N, Tuteja R. Prokaryotic and eukaryotic DNA helicases. Eur J Biochem 2004;271:1835-48.
[76] Zubay GL. Biochemistry, 4th ed. Columbus, OH: McGraw-Hill. 1998.
[77] Yanagida M. Basic mechanism of eukaryotic chromosome segregation. Philos Trans R Soc Lond B Biol Sci 2005;360:609-21.

Section IX: Acknowledgements

Chapter XXIII

About the authors

Søren Ventegodt, MD, MMedSci, EU-MSc is the director of the independent Quality of Life Research Center in Copenhagen, and Research Clinic for Holistic Medicine and Sexology, Copenhagen. He teaches holistic medicine, psychiatry and sexology at the Nordic School of Holistic Health and Denmark. Since 2006 he has been director for and lecturer at Inter-University College, International Campus, Denmark in collaboration with Inter-University Consortium for Integrative Health Promotion, Inter-University College Graz, Austria and the Austrian Ministry of Education, Science and Culture. He is a popular speaker throughout Scandinavia with more than 1000 presentations and for the last 10 years he has taught quality of life in the Danish Army. As holistic physician he has treated over 500 chronically ill patients and given more than 10.000 hours of holistic therapy. He has published numerous scientific or popular articles and a number of books on holistic medicine, quality of life and quality of working life. He is reviewer for more than 10 scientific journals. 130 of his scientific papers on quality of life and holistic medicine are listed on www.pubmed.gov making him one of the most productive scientists in these fields. His most important scientific contributions are the theory of holistic healing, the integrated quality of life (QOL) theory, the comprehensive SEQOL questionnaire, the very short QOL5 questionnaire, the life mission theory, and the ongoing Danish Quality of Life Research Survey, 1991-94 in connection with follow-up studies of the Copenhagen Perinatal Birth Cohort 1959-61 initiated at the University Hospital of Copenhagen by the late professor of pediatrics, Bengt Zachau-Christiansen, MD, PhD. E-Mail: ventegodt@livskvalitet.org. Website: www.livskvalitet.org

Tyge Dahl Hermansen, MSc in biology from the University of Copenhagen and former research assistant at the Quality of Life Research Center, Copenhagen, now at the Institute for Conservation Biology, Department of Biological Sciences, University of Wollongong, Wollongong, New South Wales in Australia where he is a PhD candidate. Research areas are ecology, molecular biology, physiology and life quality. Experimental experience in mangroves, fruit flies, yeast, bacteria, cnidarians and theoretical research in human life quality. Technical knowledge in the areas of ecology, animal physiology, molecular biology, biotechnology, biochemistry, cell biology and immunology. Publications in popular scientific as well as international journals. E-mail: tdhermansen@speedpost.net

Joav Merrick, MD, MMedSci, DMSc, is professor of pediatrics, child health and human development affiliated with Kentucky Children's Hospital, University of Kentucky, Lexington, United States and the Divisionepartment of Pediatrics, Hebrew University Hadassah Medical Centers, Mount Scopus Campus, Jerusalem, Israel, the medical director of the Division for Intellectual and Developmental DisabilitiesMental Retardation, Ministry of Social Affairs and Social Services, Jerusalem, the founder and director of the National Institute of Child Health and Human Development. Numerous publications in the field of pediatrics, child health and human development, rehabilitation, intellectual disability, disability, health, welfare, abuse, advocacy, quality of life and prevention. Received the Peter Sabroe Child Award for outstanding work on behalf of Danish Children in 1985 and the International LEGO-Prize ("The Children's Nobel Prize") for an extraordinary contribution towards improvement in child welfare and well-being in 1987. E-mail: jmerrick@zahav.net.il; Home-page: http://jmerrick50.googlepages.com/home

Chapter XXIV

About the quality of life research center in Copenhagen, Denmark

The Quality of Life Research Center in Copenhagen was established in 1989, when the physician Søren Ventegodt succeeded in getting a collaboration started with the Department of Social Medicine at the University of Copenhagen in response to the project "Quality of life and causes of disease". An interdisciplinary "Working group for the quality of life in Copenhagen" was established and when funds were raised in 1991 the University Hospital of Copenhagen (Rigshospitalet) opened its doors for the project.

The main task was a comprehensive follow-up of 9,006 pregnancies and the children delivered during 1959-61. This Copenhagen Perinatal Birth Cohort was established by the a gynecologist and a pediatrician, the late Aage Villumsen, MD, PhD and the late Bengt Zachau-Christiansen, MD, PhD, who had made intensive studies during pregnancy, early childhood and young adulthood. The cohort was during 1980-1989 directed by the pediatrician Joav Merrick, MD, DMSc, who established the Prospective Pediatric Research Unit at the University Hospital of Copenhagen and managed to update the cohort for further follow-up register research, until he moved to Israel. The focus was to study quality of life related to socio-economic status and health in order to compare with the data collected during pregnancy, deliverty and early childhood.

The project continued to grow and later in 1993, the work was organized into a statistics group, a software group that developed the computer programs for use in the data entry and a group responsible for analysis of the data.

Quality of life research center at the university medical center

The Quality of Life Center at the University Hospital generated grants, publicity with research and discussions among the professionals leading to the claim that quality of life was significant for health and disease. It is obvious that a single person cannot do much about his/her own disease, if it is caused by chemical defects in the body or outside chemical-physical influences. However, if a substantial part of diseases are caused by a low quality of life, we can all prevent a lot of disease and operate as our own physicians, if we make a personal effort and work to improve our quality of life. A series of investigations showed that this was indeed possible. This view of the role of personal responsibility for illness and health would naturally lead to a radical re-consideration of the role of the physician and also influence our society.

Independent quality of life research center

In 1994, The Quality of Life Research Center became an independent institution located in the center of the old Copenhagen. Today, the number of full-time employees have grown. The Research Center is still expanding and several companies and numerous institutions make use of the resources, such as lectures, courses, consulting or contract research. The companies, which have used the competence of the reseach center and its tools on quality of life and quality of working life, include IBM, Lego, several banks, a number of counties, municipalities, several ministries, The National Defense Center for Leadership and many other management training institutions, along with more than 300 public and private companies. It started in Denmark, but has expanded to involve the whole Scandinavian area.

The center's research on the quality of life have been through several phases from measurement of quality of life, from theory to practice over several projects on the quality of life in Denmark, which have been published and received extended public coverage and public impact in Denmark and Scandinavia. The data is now also an important part of Veenhoven's Database on Happiness at Rotterdam University in the Netherlands.

New research

Since The Quality-of-Life Research Center became independent a number of new research projects were launched. One was a project that aimed to prevent illness and social problems among the elderly in one of the municipalities by inspiring the elderly to improve their quality of life themselves. Another a project about quality of life after apoplectic attacks at one of the major hospitals in Copenhagen and the Danish Agency for Industry granted funds for a project about the quality of work life.

Quality of life of 10,000 danes

There is a general consensus that many of the diseases that plague the Western world (which are not the result of external factors such as starvation, micro-organisms, infection or genetic defects) are lifestyle related and as such, preventable through lifestyle changes. Thus increasing time and effort is spent on developing public health strategies to promote "healthy" lifestyles. However, it is not a simple task to identify and dispel the negative and unhealthy parts of our modern lifestyle even with numerous behavioural factors that can be readily highlighted harmful, like the use of alcohol, use of tobacco, the lack of regular exercise and a high fat, low fibre diet.

However there is more to Western culture and lifestyle than these factors and if we only focus on them we can risk overlooking others. We refer to other large parts of our life, for instance the way we think about and perceive life (our life attitudes, our perception of reality and our quality of life) and the degree of happiness we experience through the different dimensions of our existence. These factors or dimensions can now, to some degree, be isolated and examined. The medical sociologist Aaron Antonovsky (1923-1994) from the Faculty of Health Sciences at Ben Gurion University in Beer-Sheva, who developed the salutogenic model of health and illness, discussed the dimension, "sense of coherence", that is closely related to the dimension of "life meaning", as perhaps the deepest and most important dimension of quality of life. Typically, the clinician or researcher, when attempting to reveal a connection between health and a certain factor, sides with only one of the possible dimensions stated above. A simple, one-dimensional hypothesis is then postulated, like for instance that cholesterol is harmful to circulation. Cholesterol levels are then measured, manipulated and ensuing changes to circulatory function monitored. The subsequent result may show a significant, though small connection, which supports the initial hypothesis and in turn becomes the basis for implementing preventive measures, like a change of diet. The multi-factorial dimension is therefore often overlooked.

In order to investigate this multifactorial dimension a cross-sectional survey examining close to 10,000 Danes was undertaken in order to investigate the connection between lifestyle, quality of life and health status by way of a questionnaire based survey. The questionnaire was mailed in February 1993 to 2,460 persons aged between 18-88, randomly selected from the CPR (Danish Central Register) and 7,222 persons from the Copenhagen Perinatal Birth Cohort 1959-61.

A total of 1,501 persons between the ages 18-88 years and 4,626 persons between the ages 31-33 years returned the questionnaire (response rates 61.0% and 64,1% respectively). The results showed that health had a stronger correlation to quality of life ($r= 0.5$, $p<0.0001$), than it had to lifestyle ($r=0.2$, $p< 0.0001$).

It was concluded that preventable diseases could be more effectively handled through a concentrated effort to improve quality of life rather than through n approach that focus solely on the factors that are traditionally seen to reflect an unhealthy life style.

Collaborations across borders

The project has been developed during several phases. The first phase, 1980-1990, was about mapping the medical systems of the pre-modern cultures of the world, understanding their philosophies and practices and merging this knowledge with western biomedicine. A huge task seemingly successfully accomplished in the Quality of Life (QOL) theories, and the QOL philosophy, and the most recent theories of existence, explaining the human nature, and especially the hidden resources of man, their nature, their location in human existence and the way to approach them through human consciousness.

Søren Ventegodt visited several countries around the globe in the late 1980s and analysed about 10 pre-modern medical systems and a dozen of shamans, shangomas and spiritual leaders noticing most surprisingly similarities, allowing him together with about 20 colleagues at the QOL Study Group at the University of Copenhagen, to model the connection between QOL and health. This model was later further developed and represented in the integrative QOL theories and a number of publications. Based on this philosophical breakthrough the Quality of Life Research Center was established at the University hospital. Here a brood cooperation took place with many interested physicians and nurses from the hospital.

A QOL conference in 1993 with more than 100 scientific participants discussed the connection between QOL and the development of disease and its prevention. Four physicians collaborated on the QOL population survey 1993. For the next 10 years the difficult task of integrating bio-medicine and the traditional medicine went on and Søren Ventegodt again visited several centers and scientists at the Universities of New York, Berkeley, Stanford and other institutions. He also met people like David Spiegel, Dean Ornish, Louise Hay, Dalai Lama and many other leading persons in the field of holistic medicine and spirituality.

Around the year 2000 an international scientific network started to take form with an intense collaboration with the National Institute of Child Health and Human Development (NICHD) in Israel, which has now developed the concept of "Holistic Medicine". We believe that the trained physician today has three medical toolboxes: the manual medicine (traditional), the bio-medicine (with drugs and pharmacology) and the consciousness-based medicine (scientific, holistic medicine). What is extremely interesting is that most diseases can be alleviated with all three sets of medical tools, but only the bio-medical toolset is highly expensive. The physician, using his hands and his consciousness to improve the health of the patient by mobilising hidden resources in the patient can use his skills in any cultural setting, rich or poor.

Contact person

Director Søren Ventegodt, MD, MMedSci, MSc
Quality of Life Research Center, Frederiksberg Alle 13A, 2tv, DK-1820 Copenhagen V
Denmark
E-mail: ventegodt@livskvalitet.org; Website: www.livskvalitet.org

Support our research

The Quality-of-Life Research Center, Copenhagen, is a private research center, which for the last 20 years has taken international leadership in research on quality of life and holistic medicine. A large number of scientific publications many of which are listed in www.pubmed.gov, the world's most recognized medical database, has made the research center among the leading research institutes of holistic health worldwide. The research center has collected more than 6 million USD for its research making it the leading fundraiser for research in holistic medicine in Scandinavia. Through these funding we have managed to establish an international cooperation that currently involves researchers from all continents.

Research just cost a lot of money, and we need more research to be done. You can help us by giving your contribution to the Quality of Life Research Center (account with Danske Bank SWIFT: DABADKKK; IBAN: DK7330003113434252).

It is important for us that you understand that your contribution, even if it is just 10 USD, has great importance for us in this phase. Our goal is the establishment of the first major research hospital in holistic medicine.

The establishment of this research hospital is estimated to cost around 10 million dollars. It will be able to document the efficacy of scientific holistic medicine for a wide variety of patient groups, including patients with schizophrenia and cancer. A holistic-medical research hospital will also carry out comparative studies of the efficacy of biomedicine and holistic medicine for a number of diseases. It could thus generate much of the knowledge being sought for in this book.

Please visit the website www.livskvalitet.org and feel free to write to Søren Ventegodt MD, director of the research center (email: ventegodt@livskvalitet.org).

Endowments for the quality of life research center

We have been running for 20 years and the international research group affiliated with the research center will for the next many years continue the research. We will therefore also need continued support. We kindly encourage you to bequeath to the research center, owned by Søren Ventegodt MD (CPR 260361-0493).

If you want to do this we offer you free legal assistance for drafting of wills. Please write to Søren Ventegodt (email: ventegodt@livskvalitet.org) if you need more information on the research or if you have special requests for use of the resources you wish to bequeath, for example, that they should be earmarked for research into cancer or schizophrenia.

You are also welcome to call Soren Ventegodt at +45 33 14 11 13 to hear more about the research center's plans for future research and development.

Chapter XXV

About the National Institute of Child Health and Human development in Israel

The National Institute of Child Health and Human Development (NICHD) in Israel was established in 1998 as a virtual institute under the auspicies of the Medical Director, Ministry of Social Affairs and Social Services in order to function as the research arm for the Office of the Medical Director. In 1998 the National Council for Child Health and Pediatrics, Ministry of Health and in 1999 the Director General and Deputy Director General of the Ministry of Health endorsed the establishment of the NICHD and in 2011 the National Institute of Child Health and Human Development became affiliated with the Division of Pediatrics, Hadassah Hebrew University Medical Centers, Jerusalem, Israel.

Mission

The mission of a National Institute for Child Health and Human Development in Israel is to provide an academic focal point for the scholarly interdisciplinary study of child life, health, public health, welfare, disability, rehabilitation, intellectual disability and related aspects of human development. This mission includes research, teaching, clinical work, information and public service activities in the field of child health and human development.

Service and academic activities

Over the years many activities became focused in the south of Israel due to collaboration with various professionals at the Faculty of Health Sciences (FOHS) at the Ben Gurion University of the Negev (BGU). Since 2000 an affiliation with the Zusman Child Development Center at the Pediatric Division of Soroka University Medical Center has resulted in collaboration around the establishment of the Down Syndrome Clinic at that center. In 2002 a full course on "Disability" was established at the Recanati School for Allied Professions in the

Community, FOHS, BGU and in 2005 collaboration was started with the Primary Care Unit of the faculty and disability became part of the master of public health course on "Children and society". In the academic year 2005-2006 a one semester course on "Aging with disability" was started as part of the master of science program in gerontology in our collaboration with the Center for Multidisciplinary Research in Aging. From 2011 teaching second, fourth and six year medical students at Hadassah Hebrew University Hadassah Medical Centers, Jerusalem.

Research activities

The affiliated staff havestaff has over the years published work from projects and research activities in this national and international collaboration. In the year 2000 the International Journal of Adolescent Medicine and Health and in 2005 the International Journal on Disability and Human development of Freund Publishing House (London and Tel Aviv), in the year 2003 the TSW-Child Health and Human Development and in 2006 the TSW-Holistic Health and Medicine of the Scientific World Journal (New York and Kirkkonummi, Finland), all peer-reviewed international journals were affiliated with the National Institute of Child Health and Human Development. From 2008 also the International Journal of Child Health and Human Development (Nova Science, New York), the International Journal of Child and Adolescent Health (Nova Science) and the Journal of Pain Management (Nova Science) affiliated and from 2009 the International Public Health Journal (Nova Science) and Journal of Alternative Medicine Research (Nova Science).

National collaboration

Nationally the NICHD works in collaboration with the Faculty of Health Sciences, Ben Gurion University of the Negev; Department of Physical Therapy, Sackler School of Medicine, Tel Aviv University; Autism Center, Assaf HaRofeh Medical Center; National Rett and PKU Centers at Chaim Sheba Medical Center, Tel HaShomer; Department of Physiotherapy, Haifa University; Department of Education, Bar Ilan University, Ramat Gan, Faculty of Social Sciences and Health Sciences; College of Judea and Samaria in Ariel and in 2011 affiliation with Center for Pediatric Chronic Illness and Center for Down Syndrome, Department of Pediatrics, Hebrew University Hadassah Medical Center, Mount Scopus Campus, Jerusalem.

International collaboration

Internationally with the Department of Disability and Human Development, College of Applied Health Sciences, University of Illinois at Chicago; Strong Center for Developmental Disabilities, Golisano Children's Hospital at Strong, University of Rochester School of Medicine and Dentistry, New York; Centre on Intellectual Disabilities, University of Albany,

New York; Centre for Chronic Disease Prevention and Control, Health Canada, Ottawa; Chandler Medical Center and Children's Hospital, Kentucky Children's Hospital, Section of Adolescent Medicine, University of Kentucky, Lexington; Chronic Disease Prevention and Control Research Center, Baylor College of Medicine, Houston, Texas; Division of Neuroscience, Department of Psychiatry, Columbia University, New York; Institute for the Study of Disadvantage and Disability, Atlanta; Center for Autism and Related Disorders, Department Psychiatry, Children's Hospital Boston, Boston; Department of Paediatrics, Child Health and Adolescent Medicine, Children's Hospital at Westmead, Westmead, Australia; International Centre for the Study of Occupational and Mental Health, Düsseldorf, Germany; Centre for Advanced Studies in Nursing, Department of General Practice and Primary Care, University of Aberdeen, Aberdeen, United Kingdom; Quality of Life Research Center, Copenhagen, Denmark; Nordic School of Public Health, Gottenburg, Sweden, Scandinavian Institute of Quality of Working Life, Oslo, Norway; Centre for Quality of Life of the Hong Kong Institute of Asia-Pacific Studies and School of Social Work, Chinese University, Hong Kong.

Targets

Our focus is on research, international collaborations, clinical work, teaching and policy in health, disability and human development and to establish the NICHD as a permanent institute in Israel in order to conduct model research and together with the four university schools of public health/medicine in Israel establish a national master and doctoral program in disability and human development at the institute to secure the next generation of professionals working in this often non-prestigious/low-status field of work. For this project we need your support. We are looking for all kinds of support and eventually an endowment.

Support for our work

Contact iIn the Uunited Sstates

In the United States the Israel Foundation for Human Development was created in order to support the work of the National Institute of Child Health and Human Development in Israel. It is possible to send donations to the Israel Foundation for Human Development Inc, which is a recognized tax-exempt organization in the United States (charitable non-for-prifit organization with 501c (3) tax exempt number 56-230-6116). Checks for the Foundation in the United States can be send to Israel Foundation for Human Development Inc., President Arlene Feldman, 2 Lawrence Street, New Hyde Park, New York 11040. Phone: 516-352-3596. E-mail: AFeldman@FarrellFritz.com

Contact in Israel

Professor Joav Merrick, MD, MMedSci, DMSc
Medical Director, Division for Mental Retardation
Ministry of Social Affairs, PO Box 1260
IL-91012 Jerusalem, Israel
E-mail: jmerrick@zahav.net.il

Chapter XVI

About the book series "Health and human development"

Health and human development is a book series with publications from a multidisciplinary group of researchers, practitioners and clinicians for an international professional forum interested in the broad spectrum of health and human development.

- Merrick J, Omar HA, eds. Adolescent behavior research. International perspectives. New York: Nova Science, 2007.
- Kratky KW. Complementary medicine systems: Comparison and integration. New York: Nova Science, 2008.
- Schofield P, Merrick J, eds. Pain in children and youth. New York: Nova Science, 2009.
- Greydanus DE, Patel DR, Pratt HD, Calles Jr JL, eds. Behavioral pediatrics, 3 ed. New York: Nova Science, 2009.
- Ventegodt S, Merrick J, eds. Meaningful work: Research in quality of working life. New York: Nova Science, 2009.
- Omar HA, Greydanus DE, Patel DR, Merrick J, eds. Obesity and adolescence. A public health concern. New York: Nova Science, 2009.
- Lieberman A, Merrick J, eds. Poverty and children. A public health concern. New York: Nova Science, 2009.
- Goodbread J. Living on the edge. The mythical, spiritual and philosophical roots of social marginality. New York: Nova Science, 2009.
- Bennett DL, Towns S, Elliot E, Merrick J, eds. Challenges in adolescent health: An Australian perspective. New York: Nova Science, 2009.
- Schofield P, Merrick J, eds. Children and pain. New York: Nova Science, 2009.
- Sher L, Kandel I, Merrick J. Alcohol-related cognitive disorders: Research and clinical perspectives. New York: Nova Science, 2009.
- Anyanwu EC. Advances in environmental health effects of toxigenic mold and mycotoxins. New York: Nova Science, 2009.
- Bell E, Merrick J, eds. Rural child health. International aspects. New York: Nova Science, 2009.
- Dubowitz H, Merrick J, eds. International aspects of child abuse and neglect. New York: Nova Science, 2010.

- Shahtahmasebi S, Berridge D. Conceptualizing behavior: A practical guide to data analysis. New York: Nova Science, 2010.
- Wernik U. Chance action and therapy. The playful way of changing. New York: Nova Science, 2010.
- Omar HA, Greydanus DE, Patel DR, Merrick J, eds. Adolescence and chronic illness. A public health concern. New York: Nova Science, 2010.
- Patel DR, Greydanus DE, Omar HA, Merrick J, eds. Adolescence and sports. New York: Nova Science, 2010.
- Shek DTL, Ma HK, Merrick J, eds. Positive youth development: Evaluation and future directions in a Chinese context. New York: Nova Science, 2010.
- Shek DTL, Ma HK, Merrick J, eds. Positive youth development: Implementation of a youth program in a Chinese context. New York: Nova Science, 2010.
- Omar HA, Greydanus DE, Tsitsika AK, Patel DR, Merrick J, eds. Pediatric and adolescent sexuality and gynecology: Principles for the primary care clinician. New York: Nova Science, 2010.
- Chow E, Merrick J, eds. Advanced cancer. Pain and quality of life. New York: Nova Science, 2010.
- Latzer Y, Merrick, J, Stein D, eds. Understanding eating disorders. Integrating culture, psychology and biology. New York: Nova Science, 2010.
- Sahgal A, Chow E, Merrick J, eds. Bone and brain metastases: Advances in research and treatment. New York: Nova Science, 2010.
- Postolache TT, Merrick J, eds. Environment, mood disorders and suicide. New York: Nova Science, 2010.
- Maharajh HD, Merrick J, eds. Social and cultural psychiatry experience from the Caribbean Region. New York: Nova Science, 2010.
- Mirsky J. Narratives and meanings of migration. New York: Nova Science, 2010.
- Harvey PW. Self-management and the health care consumer. New York: Nova Science, 2011.
- Ventegodt S, Merrick J. Sexology from a holistic point of view. New York: Nova Science, 2011.
- Ventegodt S, Merrick J. Principles of holistic psychiatry: A textbook on holistic medicine for mental disorders. New York: Nova Science, 2011.
- Greydanus DE, Calles Jr JL, Patel DR, Nazeer A, Merrick J, eds. Clinical aspects of psychopharmacology in childhood and adolescence. New York: Nova Science, 2011.
- Bell E, Seidel BM, Merrick J, eds. Climate change and rural child health. New York: Nova Science, 2011.
- Bell E, Zimitat C, Merrick J, eds. Rural medical education: Practical strategies. New York: Nova Science, 2011.
- Latzer Y, Tzischinsky. The dance of sleeping and eating among adolescents: Normal and pathological perspectives. New York: Nova Science, 2011.
- Deshmukh VD. The astonishing brain and holistic consciousness: Neuroscience and Vedanta perspectives. New York: Nova Science, 2011.
- Bell E, Westert GP, Merrick J, eds. Translational research for primary healthcare. New York: Nova Science, 2011.
- Shek DTL, Sun RCF, Merrick J, eds. Drug abuse in Hong Kong: Development and evaluation of a prevention program. New York: Nova Science, 2011.
- Ventegodt S, Hermansen TD, Merrick J. Human Development: Biology from a holistic point of view. New York: Nova Science, 2011.

- Ventegodt S, Merrick J. Our search for meaning in life. New York: Nova Science, 2011.

Contact

Professor Joav Merrick, MD, MMedSci, DMSc
Medical Director, Medical Services
Division for Mental Retardation
Ministry of Social Affairs
PO Box 1260
IL-91012 Jerusalem, Israel
E-mail: jmerrick@zahav.net.il
Home-page: http://jmerrick50.googlepages.com/home

Index

F

G

H

I

J

K

L

M

N

O

P

Q

R

S

T

U

W